A Designer's Guide to Built-in Self-Test

FRONTIERS IN ELECTRONIC TESTING

Consulting Editor
Vishwani D. Agrawal

Books in the series:

Boundary-Scan Interconnect Diagnosis
J. de Sousa, P.Cheung
ISBN: 0-7923-7314-6
Essentials of Electronic Testing for Digital, Memory, and Mixed Signal VLSI Circuits
M.L. Bushnell, V.D. Agrawal
ISBN: 0-7923-7991-8
Analog and Mixed-Signal Boundary-Scan: A Guide to the IEEE 1149.4 Test Standard
A. Osseiran
ISBN: 0-7923-8686-8
Design for At-Speed Test, Diagnosis and Measurement
B. Nadeau-Dosti
ISBN: 0-79-8669-8
Delay Fault Testing for VLSI Circuits
A. Krsti_, K-T. Cheng
ISBN: 0-7923-8295-1
Research Perspectives and Case Studies in System Test and Diagnosis
J.W. Sheppard, W.R. Simpson
ISBN: 0-7923-8263-3
Formal Equivalence Checking and Design Debugging
S.-Y. Huang, K.-T. Cheng
ISBN: 0-7923-8184-X
On-Line Testing for VLSI
M. Nicolaidis, Y. Zorian
ISBN: 0-7923-8132-7
Defect Oriented Testing for CMOS Analog and Digital Circuits
M. Sachdev
ISBN: 0-7923-8083-5
Reasoning in Boolean Networks: Logic Synthesis and Verification Using Testing Techniques
W. Kunz, D. Stoffel
ISBN: 0-7923-9921-8
Introduction to I_{DDQ} Testing
S. Chakravarty, P.J. Thadikaran
ISBN: 0-7923-9945-5
Multi-Chip Module Test Strategies
Y. Zorian
ISBN: 0-7923-9920-X
Testing and Testable Design of High-Density Random-Access Memories
P. Mazumder, K. Chakraborty
ISBN: 0-7923-9782-7
From Contamination to Defects, Faults and Yield Loss
J.B. Khare, W. Maly
ISBN: 0-7923-9714-2
Efficient Branch and Bound Search with Applications to Computer-Aided Design
X.Chen, M.L. Bushnell
ISBN: 0-7923-9673-1
Testability Concepts for Digital ICs: The Macro Test Approach
F.P.M. Beenker, R.G. Bennetts, A.P. Thijssen
ISBN: 0-7923-9658-8
Economics of Electronic Design, Manufacture and Test
M. Abadir, A.P. Ambler
ISBN: 0-7923-9471-2
I_{DDQ} Testing of VLSI Circuits
R. Gulati, C. Hawkins
ISBN: 0-7923-9315-5

A Designer's Guide to Built-In Self-Test

Charles E. Stroud

University of North Carolina at Charlotte

Kluwer Academic Publishers

Boston / Dordrecht / London

Distributors for North, Central and South America:
Kluwer Academic Publishers
101 Philip Drive
Assinippi Park
Norwell, Massachusetts 02061 USA
Telephone (781) 871-6600
Fax (781) 681-9045
E-Mail: kluwer@wkap.com

Distributors for all other countries:
Kluwer Academic Publishers Group
Post Office Box 322
3300 AH Dordrecht, THE NETHERLANDS
Telephone 31 786 576 000
Fax 31 786 576 474
E-Mail: services@wkap.nl

Electronic Services < http://www.wkap.nl>

Library of Congress Cataloging-in-Publication Data

Stroud, Charles E.
A designer's guide to built-in self-test / Charles E. Stroud.
p.cm.--(Frontiers in electronic testing)
Includes bibliographical reference and index.
ISBN 1-4020-7050-0
1.Electronic apparatus and appliances--Testing. 2.Electronic apparatus appliances--Design and construction. 3.Automatic checkout equipment. I.Title.II. Series

Printed on acid-free paper.

Printed in the United States of America.

Table of Contents

Preface

In 1980, I was designing my first three chips at Bell Labs for the No. 5 Electronic Switching System (5ESS). These chips included a 3.5 μm NMOS standard cell device and two 4.5 μm bipolar gate arrays. The NMOS chip took me three months to design, and another three months to write test vectors that barely achieved 95% single stuck-at gate-level fault coverage. I decided that there had to be a better way to test these devices than the monotony of manually performing path sensitization day after day to determine what vectors I needed to detect a handful of faults. Even worse was running fault simulations only to see a miserably small increase in fault coverage.

That year, I attended the IEEE International Test Conference (ITC) for the first time and learned a great deal about design for testability. But the one thing I remember the most was a presentation by Richard Sedmak on Built-In Self-Test which received the best paper award for that year (actually the term BIST was not yet in use and Rich referred to it as "self-verification"). While listening to his talk about designing chips that could test themselves, I thought to myself, "This is crazy! It will never work! You can't tell if the chip is lying to you or not!" Then I thought, "Wait a minute - if I think that the chip can lie and say that it is good when it is really faulty, then I must believe that it can be designed to test itself. The real challenge has to be figuring out a way to make sure the chip is not lying." From that point on, I was fascinated with BIST.

In 1981, I was faced with designing a chip containing embedded RAM (a 2.5 μm CMOS standard cell device with 8K-bits of embedded RAM). I had heard horror stories from other designers about the difficulty of writing vectors to test embedded

RAMs. I decided I would attempt to design a self-test circuit for the RAM to avoid having to write the external test vectors. That is how I started working in BIST, being fortunate to have been in the right place at the right time in the very early days of BIST. I was also fortunate to be a bit naive, believing in BIST at a time when most people did not think it would work. I designed a self-test circuit for embedded RAMs which had a counter-based test pattern generator, a signature analysis-based output response analyzer, and a counter-based test controller in my fourth chip. At a testability review of the chip with Vishwani Agrawal and Ray Mercer from the Bell Labs at Murray Hill, NJ, I presented the self-test circuit for the RAM for the first time. That was also the first time I had met either Vishwani or Ray and I was overwhelmed by their encouragement and support. The BIST circuit worked as planned when the chip was fabricated and later was implemented in many more chips by other designers at Bell Labs. Vishwani was working on an automated scan design system and had just specified the scan support features in the automatic RAM generator for all standard cell-based VLSI devices at Bell Labs. He suggested that the BIST option should also be included in the RAM generator. After some refinements, the BIST circuitry was integrated into the RAM generator by Duane Aadsen at Bell Labs in Allentown, PA.

My next BIST target was general sequential logic. I had studied the Built-In Logic Block Observer (BILBO) with its problems of test scheduling and register self-adjacency and I began to experiment with BILBO during a cost reduction redesign of one of my high volume chips. I tried operating all BILBOs in the MISR mode of operation during a single test session and achieved good fault coverage. Finding primitive polynomials for the different size registers was a pain so I tried making one big BILBO for the whole chip. When I couldn't find a primitive polynomial for the big BILBO, I removed the characteristic polynomial and by accident connected the output of the last flip-flop to the input of the first, creating a circular chain of BIST flip-flops. The fault coverage was even higher than before and this BIST approach for general sequential logic became known as Circular BIST and widely used in Bell Labs. Incorporating both the logic BIST and RAM BIST approaches in the redesign effort, this 1985 chip became the first completely self-testing chip at Bell Labs. In 1987, I worked on the first mixed-signal BIST approach at Bell Labs. I didn't begin publishing papers on any of this work until 1986 when Ed McCluskey of Stanford University convinced me that other people would be interested.

I incorporated BIST circuitry of various types in 16 of the 21 chips I designed at Bell Labs. I experimented with each new chip, trying different BIST approaches for different types of circuits in different parts of the chips. I also helped other designers incorporate BIST into many of their chips. As a result of the more than 30 production chips I worked on that incorporated BIST, I learned a lot of valuable lessons with many of these lessons learned through the hard knocks of making mistakes or failing to anticipate certain issues that had not been considered before. These chips included gate array, standard cell, and full custom devices, with design rules from 3.5 μm to 0.9 μm, sizes ranging from about 10,000 to 400,000 transistors, and running at speeds from 2MHz to 200MHz. While these chips are not large or fast by today's standards, they were pushing the limits of technology in terms of area and performance throughout the 1980s and very early 1990s. In most of those chips, I made the BIST accessible during system-level operation and testing. As a result, I was able to gain valuable experience in all of the advantages (as well as many of the disadvantages) associated with BIST and system-level use of BIST. Since leaving Bell Labs, I have continued research and development of BIST for digital and mixed-signal VLSI as well as BIST for CPLDs and FPGAs. This work has been funded by AT&T Bell Laboratories, Lucent Technologies, Agere Systems, Cypress Semiconductor, NSF, DARPA, NSA, US Air Force, and the US Army.

One of the aspects of BIST that I enjoyed the most was that, while I designed the system function for the good of the project according to system specifications, I designed the BIST circuitry for myself according to my specifications. The primary reason for incorporating BIST in my early chip designs was to save myself a lot of time and effort in test vector development. That continued to be my goal in later chips as well, even when BIST capabilities were sought by system diagnostic software developers. While I enjoyed the creativity and satisfaction of designing the system function to meet the system requirements, I found the creativity and satisfaction of BIST design even more enjoyable since it was something I was designing for myself. The ability to use the BIST approach for the good of the project to reduce manufacturing and system-level testing time and cost was simply 'icing on the cake'.

This book is primarily intended for designers, test engineers, product engineers, system diagnosticians, and managers. Many engineers have heard about BIST but are not quite sure what it is, how it works, or if and how they should go about applying it to their system. Other engineers know some of the basics of BIST but are not familiar with all of the techniques and issues. Twenty years ago I would have loved to see a book entitled "BIST for the Grunts in the Trenches" that gave a simple, but fair treatment of BIST at all levels of testing. While there are some good books on testing and DFT that also discuss BIST, there has been only one book that focuses on BIST [32]. As a designer, I found it difficult to wade through the theory and math. As a result, this book is not primarily intended to be a classroom textbook with lots of theory, but rather a book to help the practicing engineer with examples and discussion of some of the key issues they face. For students reading this book, I hope that this book can offer some valuable insight into the issues and considerations that practicing engineers and managers must address. Background chapters are included that contain an overview of fault modeling, detection, and simulation (Chapter 2) and design for testability (Chapter 3) for those that have limited background in testing. However, these chapters also contain information that is essential to a complete understanding and efficient implementation of BIST. I also include footnotes throughout that contain tidbits, tips, and "war" stories that I have encountered in my years working with BIST.

Just as Vishwani Agrawal (Agere Systems, formerly Bell Labs) and Ray Mercer (Texas A&M Univ., formerly with Bell Labs) had a big influence on my BIST career, there were several others who encouraged my BIST work along the way. Since this book is largely a result of the influence of these good people, it is only proper to acknowledge them. They include: Harry Kalvonjian (Agere Systems, formerly Bell Labs), Ed McCluskey (Stanford Univ.), Tom Williams (Synopsys, formerly with IBM), Jacob Abraham (Univ. of Texas - Austin, formerly with Univ. of Illinois), Jacob Savir (New Jersey Institute of Technology, formerly with IBM), Vinod Agarwal (President/CEO of LogicVision, formerly with McGill Univ.) and last but certainly not least, Miron Abramovici (Agere Systems, formerly Bell Labs). Finally, most of us owe a great deal to those professors that took the interest and time to shape our lives for the better through our undergraduate and graduate studies. For me they include: Gene Bradley, Ray Distler, and Chun Ro (all are Professor Emeritus of Univ. of Kentucky), and my undergraduate and MSEE advisor at Univ. of Kentucky, Lee Todd (President of Univ. of Kentucky, formerly President/CEO of DataBeam).

I have been fortunate over the past 10 years to work with outstanding undergraduate and graduate students in BIST related research and development. Since some of the material in this book is a result of our time spent together, they include: Jamie Bailey, Ping Chen, Chandan Das, Ming Ding, Travis Ferry, Gretchen Gibson, Carter Hamilton, Ping He, Piyumani Karunaratna, Varma Konala, Matt Lashinsky, Eric Lee, Brandon Lewis, Wai Khuan Long, Kristi Maggard, Jeremy Nall, Bob Puckett, Andy Russ, Kripa Sankarananayanan, Brandon Skaggs, Thomas Slaughter, Shannon Spencer, Malissa Sullivan, Joe Tannehill, Andrew Taylor, Stan Tungate, Nick Vocke, Sajitha Wijesuriya, and Yingchang Yang.

I would like to thank the people that assisted in preparing this book: my wife Ramona; Matt Lashinsky, Chris Mays, and Jeremy Nall (students at UNC-Charlotte); Miron Abramovici (Bell Labs, Murray Hill, NJ), Loren Charnley (Managing Partner, Dyna-Zign, Inc., Charlotte, NC), and Rafic Makki (UNC-Charlotte). A special thanks to Carl Harris of Kluwer Academic Publishers for encouraging me to pursue this book. Finally, it was pleasantly ironic and appropriate to have Vishwani Agrawal as the consulting editor; my sincerest thanks to Vishwani for his guidance in this effort as well as for his being a role model for twenty years.

Charles E. Stroud, Professor
Dept. of Electrical & Computer Engineering
University of North Carolina at Charlotte

About the Author

Charles E. Stroud served in the U.S. Army, 101st Airborne Division near Phu Bai, Republic of Vietnam, 1971-1972, prior to his college education. He received B.S. and M.S. degrees in Electrical Engineering from the University of Kentucky in 1976 and 1977, respectively. He then joined Bell Labs in Naperville, IL, where he worked for 15 years. His primary areas of experience were in VLSI and printed circuit board (PCB) design, CAD tool development, and VLSI testing (primarily in BIST). He designed 21 production VLSI devices and 3 production PCBs for telephone switching systems and computers. He developed CAD tools for behavioral model synthesis of VLSI devices and programmable logic-based PCBs, automatic implementation of BIST approaches in VLSI devices, VLSI timing analysis, and fault simulation. He received the Bell Labs Distinguished Member of Technical Staff Award in 1987. While working at Bell Labs, he received his Ph.D. in Electrical Engineering and Computer Science from the University of Illinois at Chicago in 1991.

Stroud left Bell Labs for one year in 1979 to help start up a company that has since became known as DataBeam in Lexington, KY. He returned to Bell Labs in 1980. Stroud left Bell Labs again in 1993 to join the faculty in the Dept. of Electrical Engineering at the University of Kentucky where he worked for 7 years. His teaching and research focused in the areas of VLSI, FPGA, and CPLD design and testing. His research and development program was funded by the National Science Foundation, the Defense Advanced Research Projects Agency, the U.S. Air Force, the U.S. Army, AT&T Bell Laboratories, Lucent Technologies, and Cypress Semiconductor.

Stroud joined the faculty in the Dept. of Electrical and Computer Engineering at the University of North Carolina at Charlotte as a full Professor in 2000 where he teaches VLSI design and testing courses. His research continues to focus on VLSI and FPGA/CPLD design and testing with primary emphasis on BIST and BIST-based diagnosis. His research and development program is currently funded by the National Science Foundation, the Defense Advanced Research Projects Agency, the National Security Agency, the US Air Force, Agere Systems, and Lattice Semiconductor.

Stroud is a member of Eta Kappa Nu, Tau Beta Pi, and a Senior Member of IEEE. He has served on the Technical Program Committees for IEEE International Test Conference (4 years), IEEE International Application Specific Integrated Circuits Conference (3 years), ACM/IEEE International Workshop for Hardware-Software Co-Design (2 years), IEEE International On-Line Testing Workshop (5 years), and the ACM/SIGDA International Symposium on Field Programmable Gate Arrays (1 year). He served as the Design for Testability Editor of IEEE Design & Test of Computers for 4 years and as Associate Editor of IEEE Transactions on VLSI Systems for 2 years.

He has authored/co-authored over 100 journal and conference papers in the area of VLSI and FPGA/CPLD design and testing, primarily in the area of BIST. He received Best Technical Paper Award at the 1988 ACM/IEEE Design Automation Conference for a paper on Circular BIST. He also received Best Paper Award at the 2001 IEEE Automatic Test Conference (AUTOTESTCON). During his academic career, he has received 6 outstanding teaching awards including 5 department-level awards and one college-level award. He holds 10 U.S. patents (with 5 more pending) and numerous international patents for various BIST techniques for VLSI and FPGAs.

To Ramona

About the Cover

The microchip on the cover is a Time Slot Interchange for telephone switching systems consisting of approximately 100,000 transistors, designed by the author. The chip was designed in 1981-1982 and was the first chip at Bell Labs to incorporate Built-In Self-Test (BIST). The BIST circuitry is located in the lower portion of the chip as indicated in the floorplan below and is used to test the 8K-bit Random Access Memory (RAM)... (and I will put a few other tid-bits as well).

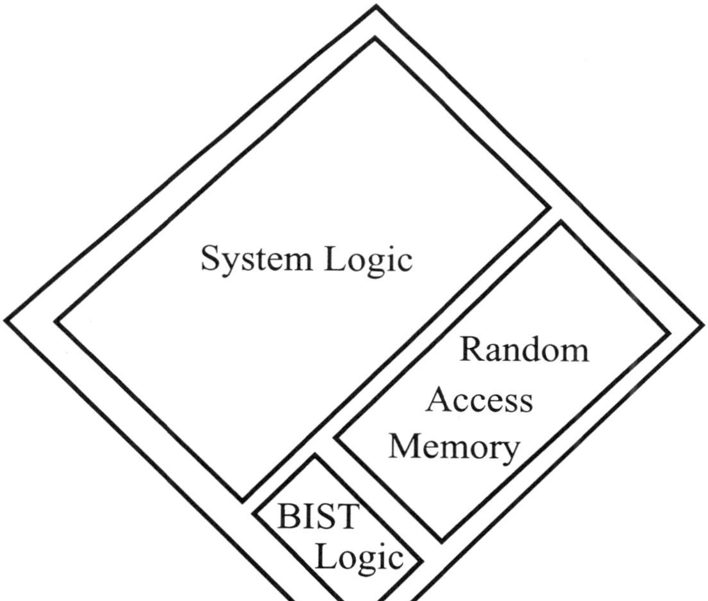

CHAPTER 1 An Overview of BIST

This chapter presents a basic introduction to Built-In Self-Test (BIST) including what it is and how it works, as well as the primary advantages and disadvantages of BIST. The intent is to establish the basic principles of BIST and how it fits into the overall testing process. Therefore, we begin with an overview of testing in order to understand some the important aspects of testing that are central to the motivation, implementation and application of BIST.

1.1 An Overview of Testing

There are three phases in the life-cycle of a product where testing is of critical importance. First, during the design phase, testing in the form of design verification seeks to detect and identify *design errors* in order to ensure that the manufactured product will correctly perform its intended function. Next, during the manufacturing phase, testing seeks to detect any manufacturing *defects* that would prevent a given part or product (integrated circuit, printed circuit board, etc.) from providing quality operation and performance in the final system for which it is destined. Finally, during system operation, testing seeks to detect any *faults* sustained during operation that would produce erroneous operation of that system. In some systems, testing also seeks to identify the faulty component for replacement in order to regain fault-free operation of the system.

The quality and quantity of testing during the life-cycle of a product is a function of properties of the product itself. For example, a low-cost, high-volume consumer product (such as a simple 4-function calculator) may never undergo testing beyond that of manufacturing tests such that when a fault occurs during operation, the product is discarded (thrown away) and a new one is purchased. On the other hand, a high-reliability, high-availability, complex system (such as a telephone switching system) will undergo rigorous and frequent testing to ensure that the system is operating properly. When faults are detected, diagnostic procedures are used for identification and replacement of faulty components. Most systems fall somewhere in between these two ends of the testing spectrum. As a result, designers, product engineers, test engineers, and managers must consider the possible product life-cycle for their components to ensure that the customer's requirements and perception of quality are met.

There are many testing issues that must be addressed during the design and development of a product, but the ultimate goal is to provide quality testing in a cost effective manner. This goal has become more difficult to achieve as Very Large Scale Integrated (VLSI) circuit and printed circuit board (PCB) component densities increase to the extent that many companies report that testing costs are sometimes as much as 55% of the total product cost [75]. This includes product development cost, where test development accounts for half the total development cost, as well as production costs where the testing cost accounts for half of the total manufacturing cost. As a result, one of the most fertile targets for reducing the cost of a product (in order to increase profits or to provide more competitive pricing) is the reduction of testing time and cost while continuing to maintain high test quality standards to minimize the number of defective parts shipped [50]. A product that is difficult to test will cost more to test and will be a much more difficult target for reduction of its associated testing cost. The best way to ensure that a product is testable (with reasonable testing costs) is to consider Design for Testability (DFT) from the very beginning, during the design phase of the product life-cycle.

1.1.1 Product Development and Design Verification

Design verification is an important testing aspect of the product development process illustrated in Figure 1.1. The intent is to verify that the design meets the system requirements and specifications.[1] Typically, a number of different simulation techniques are used including high-level simulation through a combination of behavioral

1. As most experienced designers have observed, we seldom see complete system requirements and specifications available at the beginning of the design cycle as the textbooks claim. Instead, we spend considerable time developing the requirements and specifications during the course of the design process.

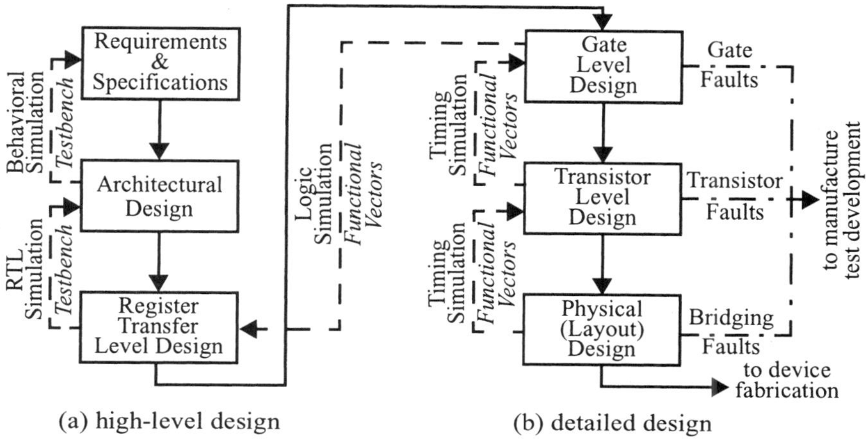

FIGURE 1.1 Product development and design verification.

modeling and testbenches. Testbenches are behavioral models that emulate the surrounding system environment to provide input stimuli to the design under test and process the output responses during simulation [37]. Register Transfer Level (RTL) models of the detailed design are then developed and verified with the same testbenches that were used for verification of the architectural design, in addition to testbenches that target design errors in the RTL description of the design [255]. With sufficient design verification at this point in the design process, the functional vectors can be captured in the RTL simulation and used for subsequent simulations of the more detailed levels of design including the synthesized gate-level design, transistor-level design, and the physical design (or layout). These latter levels of design abstraction (gate, transistor, and physical layout) provide the ability to perform additional design verification through logic, switch-level, and timing simulations. These three levels of design abstraction also provide the basis for fault models that will be used to evaluate the effectiveness of manufacturing tests [344].

Changes in system requirements or specifications late in the design cycle jeopardize the schedule as well as the quality of the design. Late changes to a design represent one of the two most significant risks to the overall project, the other risk being insufficient design verification. The quality of the design verification process is dependent on the ability of the testbenches, functional vectors, and the designers who analyze the simulated responses to detect design errors. Therefore, any inconsistency observed during the simulations at the various levels of design abstraction should be carefully studied to determine if there exists a potential design error that must be corrected before design verification continues.

1.1.2 Manufacturing Test

Once a design has been verified and is ready for manufacturing, testing at the various points throughout the manufacturing process seeks to find defects that may have resulted during manufacturing. For VLSI devices, the first testing encountered during the manufacturing process is wafer-level testing as illustrated in Figure 1.2. Each chip (or *die*) on the wafer is tested and when a faulty chip is encountered, the wafer-level test machine squirts a dot of ink onto the chip to indicate it is faulty [69]. The wafer is then cut apart and the devices that passed the wafer-level tests are packaged. The packaged devices are then retested; this is typically referred to as device-level testing or package testing. Wafer and device-level testing seek to detect any defects that could have been sustained during the fabrication or packaging processes. As a result, the test vectors used are typically the functional vectors developed during the design verification in conjunction with additional test vectors that target specific potential defect sites. These defect sites are extracted from the gate, transistor, and physical designs. The effectiveness of the complete set of test vectors used by the VLSI test

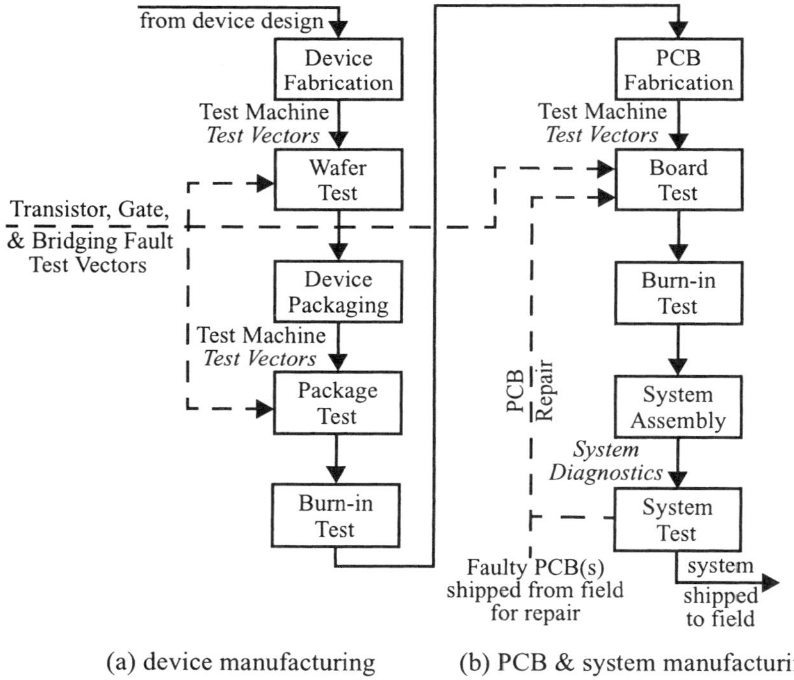

FIGURE 1.2 Manufacturing tests.

machines can be evaluated through *fault simulations* where the various fault models (gate, transistor, and physical faults) are emulated to determine if the set of test vectors used for manufacturing tests will indeed detect these possible defects. Functional vectors exercise the design in a system-like manner based on the specifications for which it was designed. Therefore, functional vectors are based on *specification-oriented testing* (SPOT). Test vectors, on the other hand, target specific fault sites or fault models and, as a result, may exercise the design in ways that it was not intended to be used during actual system operation. As a result, test vectors are often based on *defect-oriented testing* (DOT).

Another important step in the manufacturing testing process is *burn-in testing* [121], where the intent is to find those devices that have marginal defects that will lead to 'infant mortality' and low reliability [48]. In this type of testing, the VLSI device is put under stress conditions including temperature and voltage at the high end of the ratings for that device. However, to fully stress the part, the device must also be processing data at a high effective data rate. Because these burn-in tests may run for long periods of time, this type of testing is very expensive when test machines are required to apply external test vectors and monitor the output responses for failures.

For a VLSI manufacturer that sells packaged devices only, this point might incorrectly be considered the end of the product life-cycle such that only wafer and package levels of testing are considered during the product design and development process. However, the real life-cycle of a VLSI device is considered by many to begin at this point. Ultimately the VLSI device will become one of several, or many, devices on a PCB. The PCB, in turn, may be one of many PCBs in a system. Therefore, testing of the assembled PCB as well as the assembled unit or system is also necessary to ensure that the final system is working properly. For PCB testing (also referred to as board-level testing), the same set of test vectors used to test a VLSI device at the wafer and device-level is often used to test that VLSI device once it has been assembled on the PCB. In some cases, burn-in testing of PCBs is also performed at this point. It is difficult to use the set of test vectors for that device beyond board-level testing and PCB burn-in testing. In the absence of any special testing feature or capability, testing the assembled system (referred to as system-level testing) is typically performed in the normal system mode of operation which results in functional test patterns at the device and PCB input/output (I/O) pins [165].

Most complex PCBs and systems are too expensive to discard when defects are detected during the manufacturing tests. In this case, a diagnostic procedure (or some similar testing mechanism) is required to identify the faulty component for replacement in order to obtain a working PCB and/or system. In some cases, PCBs that have been determined to be faulty in the field (due to faults sustained during system operation) are shipped back to the manufacturing facility for repair. This repair process requires diagnostic testing to identify the faulty component as well as re-testing after the repair process is complete in order to determine if the PCB has been successfully repaired and is capable of proper operation in the system.

1.1.3 System Operation

When a complex system is first shipped to the field for installation, system functional tests (which often include system diagnostic tests) are performed prior to putting the system into service (see Figure 1.3). Once in service, many complex systems are routinely tested to ensure fault-free operation. For example, telephone switching systems are typically duplex systems with one of the duplicated units operating in the active mode (processing telephone calls), while the other unit is in the standby mode. Each day (usually early in the morning when voice and data traffic is low), the standby unit is taken out of service and tested. If the standby unit passes all tests, it is put back into service and made the active unit while the previously active unit is taken out of service for testing. If the previously active unit passes all tests, it is brought back into service as the standby unit. If a fault is detected when testing either of these units, diagnostic testing is performed to identify the faulty PCB. After the faulty PCB is replaced, the unit is re-tested to ensure that no additional faults are detected such that the repaired unit is now functional. Faulty PCBs are typically shipped back to the manufacturing facility for repair. As a result, the faulty PCB re-enters the manufacturing testing sequence as shown in Figure 1.2.

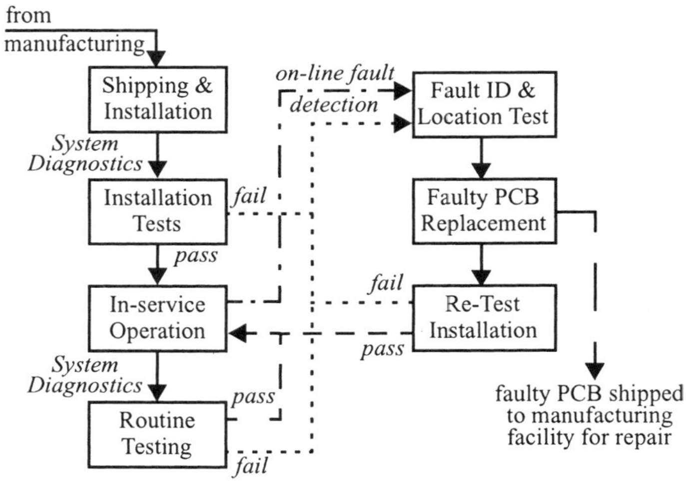

FIGURE 1.3 Testing during system operation.

Due to the high-reliability and high-availability of telephone systems, the daily testing described above cannot ensure that faults will not occur during peak traffic periods. Therefore, on-line testing, in the form of concurrent fault detection circuits (CFDCs) using error detection and/or error correction codes, is used to determine when a fault occurs during system operation [122]. These CFDCs typically use error detection codes such as parity, Cyclic Redundancy Check (CRC), Berger, Bose-Lin, and checksum codes or error correction codes such as Hamming codes. When errors are detected in the active unit by a CFDC, the standby unit is immediately switched to the active mode of operation and the unit in which the fault was detected is taken out of service for testing and diagnosis. In this way, fault tolerant operation is obtained in the telephone switching system with a typical interruption in service of only about 125 microseconds. While most systems do not target this level of reliability or availability, there are many that require routine testing and diagnosis. Of course, there are some disposable systems that do not undergo any form of testing in the field other than the consumer determining that the system is faulty, throwing it away, and buying a new one. As a designer, one may not always know the intended system application, but it is important that potential system applications be considered during the design phase to help ensure efficient and cost effective testing throughout the life-cycle of a given VLSI device or PCB.

1.1.4 The Testing Problem

Testing has become a more difficult problem over the past two to three decades. While the number of I/O pins for most VLSI devices has increased by an order of magnitude, the number of transistors contained in many VLSI devices has increased by four orders of magnitude [50]. Surface mounted components have become the norm, facilitating mounting components on both sides of PCBs as well as allowing micro vias and floating vias on the PCBs. This makes *in-circuit* testing (where a "bed-of-nails" test fixture provides access to the I/O pins of each VLSI device on the PCB) no longer feasible for many PCBs [34]. Embedded core functions have become common in VLSI devices with the embedded nature of these functions making them more difficult to test. All of these factors serve to reduce the accessibility to the VLSI devices on PCBs as well as the circuitry contained within the VLSI devices. This reduced accessibility also reduces the testability of the VLSI devices and PCBs unless specific steps (such as incorporating BIST) are taken to avoid testability problems.

Device sizes have become larger, creating a larger area to sustain manufacturing defects. Feature sizes (the sizes of transistors and wires) have become smaller, increasing the number and types of defects that can occur during the fabrication process. For example, thinner wire segments lead to more opens in those segments while wire segments that are closer together lead to more shorts between those segments.

Higher operating frequencies associated with the smaller feature sizes also add to the testing problem in that defects that reduce the frequency of operation of the circuit must be detected along with those that cause erroneous operation.

Higher levels of design abstraction, including VHDL and Verilog descriptions, along with automated synthesis tools have removed the designer from gate and transistor-level familiarity with their design. It is not surprising that test development costs have grown from about 25% to around 50% of the total product development costs in just over ten years and that manual test development for a VLSI device typically requires 1 to 2 technical head-count years. In addition, more expensive test machines are required to handle the larger number of I/O pins, higher operating frequencies, and the larger sets of test vectors typically associated with the more complex VLSI devices. As a result, test machines in excess of one million dollars are common place.

Predictions for the future indicate that the testing problem will only get worse by the year 2014 [248]. For example, test machines for leading edge VLSI devices will cost more than twenty million dollars. The cost associated with testing each transistor in a VLSI device will be greater than the cost to manufacture that transistor. Some of the successful testing techniques that have been used up to this point in time may not meet the testing needs in the near future. Finally, due to the growing complexity of VLSI devices and PCBs, the ability to provide some level of fault diagnosis (information regarding the location, and possibly the type, of the fault or defect) during manufacturing testing is needed to assist failure mode analysis (FMA) for yield enhancement and repair procedures. BIST is considered to be one of the primary solutions to these predicted and growing testing problems [248].

1.2 What Is BIST? How Does It Work?

The basic idea of BIST, in its most simple form, is to design a circuit so that the circuit can test itself and determine whether it is "good" or "bad" (fault-free or faulty, respectively). This typically requires that additional circuitry and functionality be incorporated into the design of the circuit to facilitate the self-testing feature. This additional functionality must be capable of generating test patterns as well as providing a mechanism to determine if the output responses of the circuit under test (CUT) to the test patterns correspond to that of a fault-free circuit. One of the first definitions of BIST (referred to as self-verification at the time) was given as:

> "... *the ability of logic to verify a failure-free status automatically, without the need for externally applied test stimuli (other than power and the clock), and*

without the need for the logic to be part of a running system." - Richard M. Sedmak, in [244]

While this earlier definition is fairly explicit, a recent definition is more general, in some aspects, in that BIST is:

"*... any of the methods of testing an integrated circuit (IC) that uses special circuits designed into the IC. This circuitry performs test functions on the IC, and signals whether the parts of the IC covered by the BIST circuits are working properly.*" - SEMATECH Official Dictionary, Rev 5.0 on-line version at www.sematech.org

1.2.1 Basic BIST Architecture

A representative architecture of the BIST circuitry as it might be incorporated into the CUT is illustrated in the block diagram of Figure 1.4. This BIST architecture includes two essential functions as well as two additional functions that are necessary to facilitate execution of the self-testing feature while in the system. The two essential functions include the test pattern generator (TPG) and output response analyzer (ORA). While the TPG produces a sequence of patterns for testing the CUT, the ORA compacts the output responses of the CUT into some type of *Pass/Fail* indication. The other two functions needed for system-level use of the BIST include the test controller (or BIST controller) and the input isolation circuitry. The significance and importance of these two functions will be discussed in greater detail in Chapter 6. Aside from the normal system I/O pins, the incorporation of BIST may also require additional I/O pins for activating the BIST sequence (the *BIST Start* control signal), reporting the results of the BIST (the *Pass/Fail* indication), and an optional indication

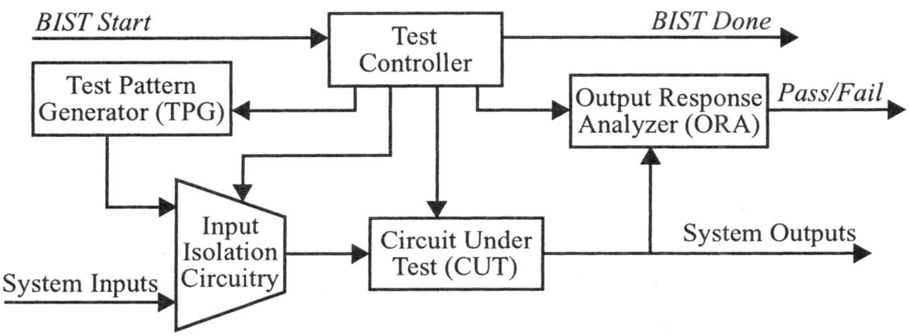

FIGURE 1.4 Basic BIST architecture.

(*BIST Done*) that the BIST sequence is complete and that the BIST results are valid and can be read to determine the fault-free/faulty status of the CUT.

1.2.2 A Simple BIST Design

As an example of a BIST implementation, assume that a set of test vectors has already been developed for the CUT and we want to use those test vectors to test the CUT during the BIST sequence. The input test patterns could be stored in a Read Only Memory (ROM) with a counter used to read the test patterns from the ROM for application to the CUT, as illustrated in Figure 1.5. This combination of counter and ROM would constitute the TPG function in the BIST circuitry. A similar structure could be used to construct the ORA function where the expected fault-free circuit output responses to the input test patterns (obtained from logic simulation) are also stored in a ROM. As the expected responses are read from the ORA ROM, they are compared to the actual output response of the CUT for each test vector being applied by the TPG. Any mismatch detected by the comparator is latched to indicate a failure has occurred during the BIST sequence. A single counter, clocked by the same clock signal as the input clock to the CUT, can be used to read the data from both ROMs and can also function as the test controller.

In this simple but detailed BIST design, the *BIST Start* signal is assumed to be active high. During normal system operation, the input isolation circuitry (a multiplexer in this example) selects the system inputs to be applied to the CUT which performs its intended system function. The inactive *BIST Start* signal resets the counter used to address the TPG and ORA ROMs via the active low *clear* input to the *N*-bit binary counter. The inactive *BIST Start* signal also resets the active high *BIST Done* and

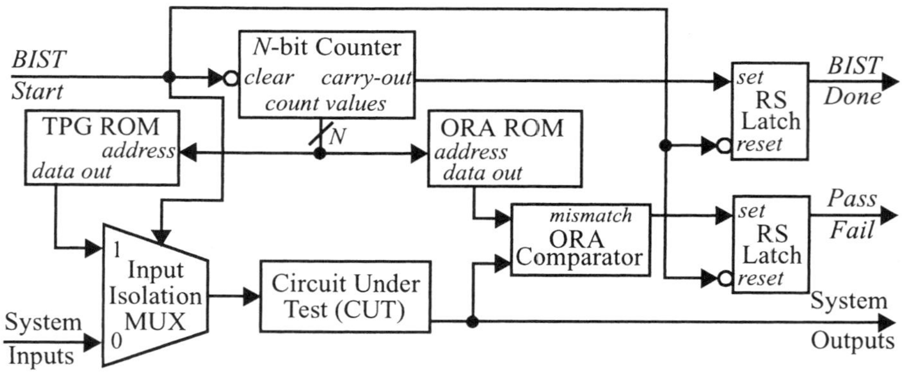

FIGURE 1.5 A simple BIST design.

Pass/Fail indicators via the active low *reset* inputs to the two synchronous Reset/Set (RS) latches. The BIST sequence is activated by setting *BIST Start* to a logic 1, at which time the counter begins to count, in turn, addressing and reading the input test patterns from the TPG ROM and the corresponding expected output responses from the ORA ROM. As the input test patterns read from the TPG ROM are applied to the CUT via the input isolation multiplexer, the output responses of the CUT are compared to the expected output responses being read from the ORA ROM. The ORA comparator will produce a logic 1 in the case of any mismatch between the expected and actual output responses, which will produce a logic 1 at the output of the *Pass/Fail* RS latch via the active high *set* input.

When the *N*-bit counter has finished counting, all of the test patterns stored in the TPG ROM have been applied to the CUT. All of the CUT's output responses, meanwhile, have been compared to the expected output responses stored in the ORA ROM. Therefore, when the counter rolls over, the active high *carry-out* from the counter will set the RS latch producing a logic 1 on *BIST Done* to indicate that the BIST sequence has completed. At this point, the *Pass/Fail* indication can be checked to determine the fault-free (*Pass/Fail* = 0) or faulty (*Pass/Fail* = 1) status of the CUT. Note that the BIST sequence will continue to repeat until the *BIST Start* input is set to a logic 0, which resets the *N*-bit counter along with the *BIST Done* and *Pass/Fail* RS latches. As a result, the BIST circuitry stops operating (and dissipating power) since the *N*-bit counter is being held in the all zeros state, and normal system operation resumes. It is important to note that the two RS latches must be synchronous (clocked by the system clock) in order to prevent transient glitches at the output of the comparator or the *carry-out* of the counter from inadvertently setting the latches. Otherwise, a glitch on the output of the comparator could cause a fault-free CUT to appear faulty. Conversely, a glitch on the *carry-out* output of the counter could cause a faulty CUT to appear fault-free if the *Pass/Fail* result is read immediately after the *BIST Done* goes active and the glitch occurs prior to the completion of the BIST sequence and prior to the detection of the fault by the comparator.

This type of BIST architecture is rarely used in practical applications since it requires conventional test vector development and considerable circuit area for the TPG and ORA ROMs as well as the counter. This is particularly true when the set of test vectors and expected output responses is large in terms of the number of bits and number of vectors. The approach has other drawbacks as well. A minor design change to the CUT could require regeneration of the test vectors and expected output responses which, in turn, would require re-programming the ROMs in the best case and resizing the ROMs and counter in the worst case. As a result, the minor design change to the CUT could require a major change in the BIST implementation. In addition, the comparator will not be completely tested during the BIST sequence such that additional testing would be required to insure that the comparator is fault-free (as will be discussed in Chapter 5). However, this example does illustrate the overall idea of a BIST architecture as well as the system-level operation of the approach.

1.3 Advantages and Disadvantages of BIST

Given that BIST enables a circuit (VLSI device or PCB) to test itself, one of the main advantages of BIST is that it can easily be used at all levels of testing. This *vertical testability*, as it is often referred to, means that the same testing approach (BIST in this case) can be used at wafer and device-level testing, for all manufacturing testing, and even for system-level testing in the field. When BIST is applied to a VLSI device, detection of a fault by the BIST circuitry will not only indicate that a fault exists in the system, but will also identify that VLSI device as faulty. Therefore, BIST offers an inherent diagnostic resolution to the CUT with which it is associated. Vertical testability and high diagnostic resolution are two of the major advantages of BIST.

Recall that in the simple BIST design example, the *N*-bit counter used to read the TPG and ORA ROMs was clocked by the same clock as the CUT. As a result, all testing by the BIST circuitry is performed at the system clock frequency. This is typically referred to as *at-speed* testing and is important since it facilitates the detection of faults that lead to excessive delay (referred to as delay faults) in an otherwise working CUT. During manufacturing testing, the clock frequency can be varied to determine the maximum speed at which the CUT will perform correctly, such that the BIST can be used at manufacturing testing for speed sorting the devices or to determine the timing margins of the fabricated devices under various voltage and temperature conditions. Since a new test pattern is applied during each clock cycle while the output response of the CUT is checked by the ORA, testing occurs at its maximum rate which helps to minimize testing time for that device; this in turn helps to minimize testing cost.

Testing cost can sometimes be significantly reduced when BIST is incorporated into a VLSI device or PCB by eliminating the need for expensive external test machines [50]. The cost of these test machines is typically driven by the number of test vectors that must be stored, the speed at which the input test patterns must be applied and the output responses monitored, as well as the number of I/O pins that must be serviced. With BIST, the test patterns are produced internally by the TPG and the output responses are checked internally by the ORA. Therefore, the only I/O pins that must be serviced by external sources are power, ground, clock, and a small number of pins for initiating BIST and retrieving the *Pass/Fail* results at the end of the BIST sequence. This opens the door for low cost test machines that service only a few I/O pins with very little test vector memory for each of those I/O pins. Another important benefit of BIST is for burn-in testing, where the VLSI device or PCB is put under stress conditions, including high voltage at high and low temperatures, to expose defects that will lead to infant mortality [48]. To fully stress the part, the device or PCB must also process data at a high effective data rate for long periods of time. With the self-contained TPG and ORA functions provided by BIST, burn-in testing can be performed more economically than having test machines applying external test patterns and monitoring the output responses for failures since only power and clock signals must be applied to the device or PCB once the BIST sequence is initiated.

Most BIST approaches also eliminate the time and expense associated with test vector development as a result of the embedded TPG and ORA circuits. This savings in test development can carry over to all levels of testing. In many applications, the savings in test development time is greater than the additional design time required for the BIST circuitry. This can translate to an overall reduction in time-to-market in some cases.

BIST does have associated costs that must be taken into account, primarily due to the additional circuitry that must be incorporated into the device. First, there is the penalty of the BIST circuitry increasing the area of the overall device; this is typically referred to as the *area overhead* associated with the BIST approach. Larger chip area results in fewer chips per wafer and, in turn, higher cost per chip. Larger chip area also results in more area susceptible to manufacturing defects which lowers the yield and further increases the cost of the chip [341]. Even in the case of pad-limited VLSI devices where the area of the device is unaffected by the incorporation of BIST, the active area inside the device is increased and more susceptible to defects; hence the yield is still affected by the area overhead. The larger area typically results in performance penalties due to longer signal routing paths as the system function is spread out to make room for the BIST circuitry. This performance penalty is usually minor compared to the insertion of BIST circuitry into the critical timing paths of the CUT (for example, the input isolation circuitry in our simple BIST design example in Section 1.2). In some applications, performance penalties can be of more critical concern than area overhead. Similarly, increased power dissipation due to high data activity during the BIST sequence can present a problem in some system applications such as those with power consumption or temperature restrictions.

Perhaps the most significant disadvantage of BIST comes in the design stage. In addition to designing and verifying proper operation of the intended system function, we are now faced with the task of designing and verifying proper operation of the BIST system. In other words, we must design two systems, a BIST system on top of the intended system, and both systems must work in order for the project to succeed. This can increase the total design time and effort for the device or product. In this sense, BIST does add risk to a project that must be taken into consideration.

The advantages and disadvantages discussed thus far are summarized in Table 1.1. Despite the disadvantages, most BIST applications case studies have demonstrated that the benefits of BIST almost always make up for the costs of BIST (including the additional design effort, increase in chip area, and potential risk to the project). This is particularly true when the BIST circuitry is designed to be accessible and operational in the system for off-line system-level testing. For example, consider the number of times a device may be tested over its life-cycle. The device may be tested only about ten times during manufacturing testing; this includes wafer-level test, package test, burn-in test, device-level test at the board-level, as well as unit and system-level tests before the product is shipped to the field. A significant savings in testing cost must be associated with the BIST approach during manufacturing testing in order to overcome the cost of the area overhead and additional design effort required for BIST. However, once the product is in the field and operational, the device may undergo testing as often and as regularly as once or twice a day. In such cases, manufacturing testing accounts for only a small percentage of the total number of times the chip is tested during its life-cycle. As a result, system-level use of BIST provides the best arena for overcoming the costs of BIST and reaping the profits from investing in BIST. That is another purpose of this book - to show not only how to implement BIST but also how to maximize the effectiveness of BIST while minimizing the penalties and risks.[1] Therefore, we will return to these advantages and disadvantages of BIST, as well as discuss new ones, in the subsequent chapters.

TABLE 1.1 Summary of advantages and disadvantages of BIST.

Advantages	**Disadvantages**
vertical testability (wafer to system)	area overhead
high diagnostic resolution	performance penalties
at-speed testing	additional design time & effort
reduced need for external test equipment	additional risk to project
reduced test development time & effort	
more economical burn-in testing	
reduced manufacturing test time & cost	
reduced time-to-market	

1. It is interesting to note that the responsibility for the disadvantages of BIST falls primarily on the shoulders of the designers, while the benefits offered by BIST are reaped by practically all other people associated with the project including managers, test engineers, product engineers, diagnosticians, and even the marketing/sales staff.

CHAPTER 2

Fault Models, Detection, and Simulation

This chapter presents a basic overview of testing and testability in terms of fault modeling, fault simulation, and fault detection. The intent is to provide the necessary background needed for a thorough understanding of the implementation and evaluation of various BIST approaches.

2.1 Testing and Fault Simulation

The mechanics of testing, as illustrated in Figure 2.1, are similar at all levels of testing, including design verification. A set of input stimuli is applied to a circuit and the output response of that circuit is compared to the known good output response, or expected response, to determine if the circuit is "good" or "faulty". During the design

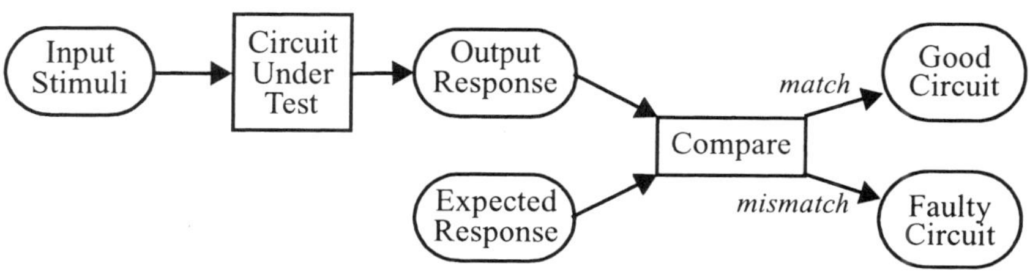

FIGURE 2.1 Basic testing flow.

verification process, the input stimuli are the design verification vectors intended to determine whether the design functions properly, according to requirements and specifications. The design verification vectors (often produced by testbenches) are applied to the hardware description language (HDL) abstraction of the circuit in the simulation environment to produce the output responses of the design to the input stimuli. The comparison of the output response to the expected response is usually done by the designer (perhaps via the testbench) with the expected response being a mental collection of the designer's interpretation of the requirements and specifications. Alternative approaches include computer-aided design (CAD) based techniques such as formal verification or automatic test pattern generation (ATPG) based methods [63]. A "faulty" circuit in this case would indicate that there is either a design error or a problem with the design verification vectors that prevented manipulation of the circuit in the way the designer had originally intended. A "good" circuit, on the other hand, does not mean that the design is free of errors. This requires sufficient design verification be performed to obtain a high level of confidence that the design is error-free (or at least that there is a relatively low probability of any remaining design error). Obtaining the "warm, fuzzy feeling" that sufficient design verification has been obtained is a difficult step for the designer. While the ultimate determination is whether or not the design works in the system, fault simulation can provide a rough quantitative measure of the level of design verification much earlier in the design process. Fault simulation also provides valuable information on portions of the design that need further design verification (as we shall discuss shortly).

The design verification vectors are often used as functional vectors during manufacturing testing. The test machine, also referred to as the automatic test equipment (ATE), applies the test vectors to the fabricated circuit and compares the output responses to the expected responses obtained from the design verification simulation environment for the fault-free (and hopefully, design error-free) circuit. A "faulty" circuit is now considered to be a circuit with manufacturing defects. In the case of a VLSI device, the chip may be discarded or it may be investigated by failure mode analysis (FMA) for yield enhancement [50]. In the case of a PCB, FMA may be performed for yield enhancement or the PCB may undergo further testing for fault location and repair. A "good" circuit is assumed to be defect-free, but this assumption is only as good as the quality of the tests being applied to the manufactured design. Once again, fault simulation provides a quantitative measure of the quality of the set of tests.

Fault simulation also follows the basic testing flow illustrated in Figure 2.1 with one exception. In addition to the input test vectors, a list of faults that are to be emulated in the CUT is also applied to the fault simulation environment [4]. The fault simulator emulates each fault (or set of faults) in the list and applies the test vectors. Each output response is then compared to the associated expected response obtained from a fault-free circuit simulation. If a mismatch is obtained for any test vector, the faulty circuit has produced an erroneous response compared to the fault-free circuit and the fault (or set of faults) is considered to be *detected*. As soon as a fault is detected, most fault simulators will stop simulating that fault (this is often referred to as *fault dropping*) and restart simulation at the beginning of the set of test vectors with the next fault in the list being emulated. If there are no mismatches in the output response for the complete set of test vectors, then the fault is *undetected*. At this point, the next fault (or set of faults) is emulated in the CUT and the simulation restarts at the beginning of the set of test vectors. Once the fault simulation is complete (all faults have been emulated), the fault coverage can be determined for the set of test vectors. The fault coverage, *FC*, is a quantitative measure of the effectiveness of the set of test vectors in detecting faults, and in its most basic form is given by:

$$FC = \frac{D}{T} \tag{2.1}$$

where D is the number of detected faults and T is the total number of faults in the fault list. Obviously 100% fault coverage is desired but in practice, fault coverage in excess of 95% is considered by some companies to be acceptable for manufacturing testing. For design verification, the fault coverage can not only give the designer a rough quantitative measure of how well the design has been exercised, but the undetected fault list can provide valuable information on those subcircuits that have not been exercised as thoroughly as other subcircuits. For example, the fault naming convention in most fault simulators maintains the hierarchical naming convention of the design. Therefore, the designer can quickly scan the undetected fault list to determine the subcircuit(s) with the most undetected faults and target those subcircuit(s) for additional design verification. While this is a valuable tool for the designer in obtaining feedback on the quality of design verification, it should be remembered that fault coverage only gives the designer a measure of how well the circuit has been exercised and not a guarantee the circuit is free of design errors.

Logic and timing simulation during the design verification process can take considerable time for large circuits. Fault simulation time will take even longer since each fault or set of faults must be simulated. Fault simulation is not complete until all faults in the fault list have been emulated and simulated with the set of test vectors. Therefore, fault simulation time is not only a function of the size of the circuit and the number of test vectors to be simulated, but also a function of the number of faults to be simulated and the type of fault simulator used. As a result, fault simulation time can be a major concern during the design and test development processes. There are a number of techniques used to improve fault simulation time. For example, like most test machines that stop as soon as a faulty device is detected (often referred to as *trip-on-first-failure*) and move on to the next chip to be tested, most fault simulators employ fault dropping to stop simulating a fault once it is detected and continue to simulate only the undetected faults. The type of fault simulator used also makes a big difference in the fault simulation time [4]. For example, serial fault simulators are slow because they emulate only one fault in the circuit at a time. Parallel fault simulators significantly reduce the fault simulation time by taking advantage of the multiple-bit computer words and the fact that simulation of a digital circuit is binary (0 and 1 logic values). As a result, each bit in a computer word used for the fault simulation can represent a different faulty circuit. In a 32-bit machine, a parallel fault simulator can emulate either 31 or 32 faults in parallel; 31 faults in simulators that emulate the fault-free circuit in one of the bit positions and 32 faults in simulators that compare the faulty circuit response with the stored results of a previous fault-free circuit simulation [288]. There are other types of fault simulators, including deductive and concurrent fault simulators, as well as hardware accelerators that also provide significant improvements in fault simulation time. The number of faults to be simulated is a major component in fault simulation time and is a function of the size of the CUT as well as the fault models.

2.2 Fault Models and Detection

To effectively evaluate the quality of a set of tests for a product, as well as to evaluate the effectiveness of a BIST approach in its application to that product, fault models are required for emulation of faults or defects in a simulation environment. The first requirement of a good fault model is that it accurately reflects the behavior of the actual defects that can occur during the fabrication and manufacturing processes as well as the behavior of faults that can occur during the operation of the system. The second requirement of a good fault model, and just as important as the first, is that it must be computationally efficient with respect to the fault simulation environment. As we will see, these two requirements are often in opposition to each other. Currently, the most widely used fault models include gate-level faults, transistor-level faults, bridging faults, and delay faults. All of these fault models can be emulated in a simulation environment and, for the most part, they closely approximate the behavior of actual defects and faults that can occur during the manufacturing process and system operation. In order to detect a given fault model, we must also understand its behavior in the circuit, since the faulty circuit must produce an error at one of its primary outputs in order for the fault to be detected.

2.2.1 Gate-Level Stuck-At Fault Model

The gate-level stuck-at fault model allows the gate inputs and outputs to be either stuck-at-0 (sa0) or stuck-at-1 (sa1) [330]. The typical notation for a stuck-at fault in a gate-level logic diagram places an 'cross' (denoted by '×') at the fault site with sa0 or sa1 adjacent to the '×' to denote the type of fault, as illustrated at the top of Figure 2.2. The faults are emulated by the fault simulator as if the gate input or output has been disconnected from the circuit and tied to either a logic 1 or a logic 0, as illustrated at the bottom of Figure 2.2.

The output response of a gate with a stuck-at fault is compared to that of a fault-free gate in Figure 2.3 for an AND gate, OR gate, and inverter for all possible input patterns. Note that for each possible fault site (each input and the output of the gate) there are two possible faults that can occur, stuck-at-0 and stuck-at-1. The behavior of the gate with these faults is given in the table associated with each gate; note that this faulty behavior table assumes that only one fault is present in the gate. The output responses of the faulty gate that differ from those of the fault-free gate are highlighted in the tables by a gray background. Since every fault illustrated for each of the gates has at least one gray box, every fault can be detected. For those faults with only one gray box, the associated input pattern must be included in the set of test vectors in order to detect that fault, otherwise the fault will not be detected. For the faults that can be detected with multiple patterns (the AND gate output stuck-at-1 or the OR gate output stuck-at-0), only one of those input patterns needs to be included in the set of test vectors. For any *K*-input *elementary logic gate* (AND, NAND, OR, NOR and Inverter), only *K*+1 vectors are needed to detect 100% of the single stuck-at gate-level faults. For an AND or NAND gate, the minimum set of test vectors includes the all 1s pattern and the set of patterns that include a single logic 0 in a field of 1s. For an OR or NOR gate, the minimum set of test vectors includes the all 0s pattern and the set of patterns that include a single logic 1 in a field of 0s. Note that the exclusive-OR gate is not considered to be an elementary logic gate since it is generally constructed from multiple gates such that the set of faults depend on its construction, yet all four test patterns are needed for complete testing regardless of its construction.

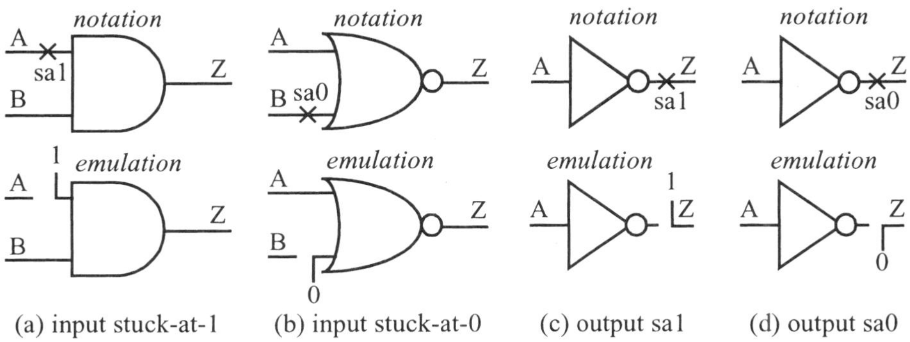

FIGURE 2.2 Gate-level stuck-at fault model.

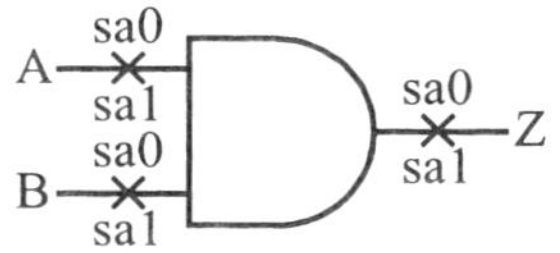

AND		A		B		Z	
AB	Z	sa0	sa1	sa0	sa1	sa0	sa1
00	0	0	0	0	0	0	1
01	0	0	1	0	0	0	1
10	0	0	0	0	1	0	1
11	1	0	1	0	1	0	1

(a) AND gate

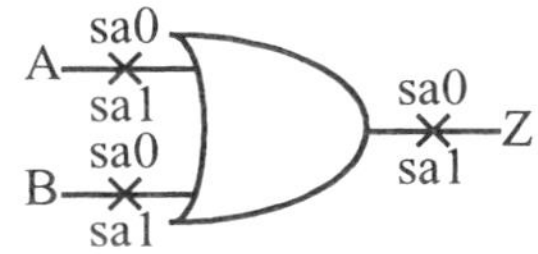

OR		A		B		Z	
AB	Z	sa0	sa1	sa0	sa1	sa0	sa1
00	0	0	1	0	1	0	1
01	1	1	1	0	1	0	1
10	1	0	1	1	1	0	1
11	1	1	1	1	1	0	1

(b) OR gate

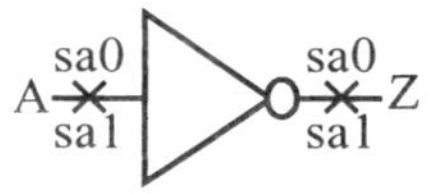

(c) Inverter

NOT		A		Z	
A	Z	sa0	sa1	sa0	sa1
0	1	1	0	0	1
1	0	1	0	0	1

FIGURE 2.3 Gate-level stuck-at fault behavior.

2.2.2 Transistor-Level Stuck Fault Model

Gate-level fault models can accurately reflect the behavior of transistor faults in NMOS circuits, but the advent of CMOS brought with it the need for transistor fault models. The transistor-level stuck fault model allows any transistor to be either stuck-on (also referred to as stuck-short) or stuck-off (also referred to as stuck-open). Figure 2.4 illustrates the emulation of the transistor-level stuck fault model in a 2-input CMOS NOR gate. Figure 2.4a gives the fault-free transistor-level diagram for the 2-input NOR gate. The stuck-on (s-on) transistor-level fault can be emulated by a short between the source and drain of the transistor (the A-input P-channel MOSFET in the example of Figure 2.4b). The stuck-off (s-off) transistor-level fault can be emulated by disconnecting the transistor from the circuit as illustrated in Figure 2.4c (where the A-input N-channel MOSFET has been removed). Alternatively, stuck-on faults can be emulated by disconnecting the gate of the MOSFET from the signal net and connecting it to a logic 1 for N-channel transistors (NFETs) or to a logic 0 for P-channel transistors (PFETs) such that the transistor is always conducting. Similarly, stuck-off faults can be emulated by connecting the gate of the MOSFET to a logic 0 for NFETs or to a logic 1 for PFETs such that the transistor never conducts.

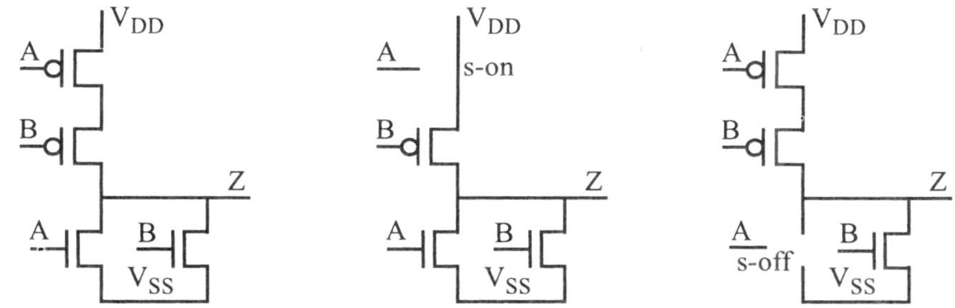

(a) transistor diagram of 2-input NOR gate

(b) A PFET stuck-on

(c) A NFET stuck-off

Fault-Free NOR		A PFET		B PFET		A NFET		B NFET	
AB	**Z**	**s-on**	**s-off**	**s-on**	**s-off**	**s-on**	**s-off**	**s-on**	**s-off**
00	1	1	Z^-	1	Z^-	V_Z/I_{DDQ}	1	V_Z/I_{DDQ}	1
01	0	0	0	V_Z/I_{DDQ}	0	0	0	0	Z^-
10	0	V_Z/I_{DDQ}	0	0	0	0	Z^-	0	0
11	0	0	0	0	0	0	0	0	0

(d) gate-level behavior of transistor-level faults

FIGURE 2.4 Transistor-level stuck fault model and behavior.

The gate-level behavior of transistor-level stuck faults is not as simple as that of the gate-level stuck-at fault model, as illustrated in the table of Figure 2.4d. Note that each possible fault site (transistor) can be stuck-on or stuck-off. The faulty behavior illustrated assumes that only one fault is present in the gate. The output responses that differ from those of the fault-free gate are highlighted in gray. For the transistor stuck-on faults, some input patterns will produce a conducting path from V_{DD} to V_{SS} through the output node during steady-state due to the fault condition. As a result, the voltage at the output node will not be equal to either V_{DD} or V_{SS} but will instead be a function of the voltage divider formed by the channel resistance of the conducting transistors. Therefore, the voltage at the output node, V_Z, would be given by $V_Z=V_{DD}R_N/(R_P+R_N)$, where R_P represents the equivalent resistance of the conducting PFETs (from the output node to V_{DD}) and R_N represents the equivalent resistance of the conducting NFETs (from the output node to V_{SS}). Depending on the ratio of these resistances along with the switching threshold(s) of the gate(s) being driven by the output of this faulty gate, the output voltage of the faulty gate may or may not be interpreted as an incorrect logic value such that the fault may or may not be detected at a primary output. This behavior is further complicated in compound static CMOS

AND-OR-Invert (AOI) or OR-AND-Invert (OAI) gates where different sets of input logic values that expose the faulty behavior will lead to different values of R_N and R_P. The only characteristic of the faulty gate that can be counted on to be different from that of the fault-free gate is that there will be steady-state current flow from V_{DD} to V_{SS} in the faulty circuit (for those input logic values shown in the tables) while in the fault-free circuit there will only be the normal leakage current (primarily due to reversed biased PN-junctions). Therefore, monitoring the steady-state power supply current (I_{DDQ}) has become a popular method for detecting transistor stuck-on faults in static CMOS circuits. This technique is referred to as I_{DDQ} testing [95]. The potential problem with I_{DDQ} testing is that the small currents involved due to a single transistor stuck-on can be difficult to detect when there are millions of transistors in a VLSI device, producing leakage currents from millions of reverse-biased PN-junctions [248]. As a result, new testing techniques which monitor the dynamic power supply current (i_{DDT}) are being explored to overcome this problem [159].

For the transistor stuck-open fault, there will be no path from the output node to either V_{DD} or V_{SS} for some input patterns. As a result, the output node will retain its previous logic value (denoted by Z^-). This creates a situation where a combinational logic gate behaves like a dynamic memory element (more specifically, a dynamic level-sensitive latch). These faults can only be detected when they are in this "storage state", with the opposite logic value from that of the fault-free circuit stored at the faulty gate output. This means that a sequence of vectors is required to get the faulty gate to store a logic value that can be observed as being different from a fault-free circuit in order to detect the fault [117]. For example, consider the four transistor stuck-off faults in the 2-input CMOS NOR gate example of Figure 2.4. In order to detect the A NFET stuck-off, the '00' test vector (ordered for AB) must first be applied to get a logic 1 at the output of the gate. Next, the '10' test pattern must be applied to detect the fault such that the faulty circuit will continue to produce a logic 1 while the fault-free circuit will produce a logic 0. Similarly, to detect the B NFET stuck-off, the '00' test vector must first be applied to get a logic 1 at the output of the gate, followed by the '01' test vector so that the faulty "storage state" is detected. Finally, to detect either the A or B PFET stuck-off, a test vector with at least one logic 1 must be applied to get a logic 0 at the gate output and then the '00' test vector must be applied to detect the difference in the faulty gate and the fault-free gate. While this sequence results in a total of 6 test vectors to detect these four faults, the set of test vectors can be minimized to the sequence {00, 10, 00, 01} for the single fault model. The first '00' test vector sets up the necessary logic 1 at the gate output. The '10' test vector then detects the A NFET stuck-off and sets up a logic 0 at the gate output if the A NFET stuck-off fault is not present. The second '00' test vector detects the A and B PFETs stuck-off and sets up a logic 1 at the gate output if neither PFET stuck-off fault is present. Finally, the '01' test vector detects the B NFET stuck-off.

The same minimum set of test vectors needed to detect the transistor-level stuck-on faults will also detect the gate-level stuck-at faults. This can be seen by comparing the tables in Figure 2.3b and Figure 2.4d. These are also the same test vectors that detect the transistor stuck-off faults. However, the ordering of the test vectors is critical in the case of the stuck-off fault detection. Also critical is the interleaving of the all 0s test pattern for setting up the necessary logic 1 value at the output of the NOR gate. Therefore, for a K-input static CMOS elementary logic gate (NAND or NOR), a minimum of $2K$ vectors are required to detect all of the transistor-level faults. For a NAND gate, this set begins with the all 1s test pattern and alternates this test vector with all possible patterns consisting of a single 0 in a field of 1s. Similarly, for a NOR gate, this set begins with the all 0s test pattern and alternates this test pattern with all possible patterns consisting of a single 1 in a field of 0s. Monitoring the I_{DDQ} current while applying these test vectors will detect any transistors stuck-on, but a stuck-on fault might also be detected at a primary output depending on the value of V_Z.

2.2.3 Bridging Fault Models

Another set of important defects includes opens in wire segments and shorts between wire segments [83]. These faults are common defects that result from over-etching or under-etching during the VLSI or PCB fabrication process. Since an open in a wire prevents the propagation of a signal past the open, inputs to gates and transistors on the other side of the open will remain constant, creating behavior equivalent to the gate-level and transistor-level fault models. As a result, opens can be detected by test vectors for either gate-level or transistor-level fault models. Therefore, only shorts between wires are of interest and they are more commonly referred to as bridging faults. To illustrate the modeling of bridging faults, consider the two wires in Figure 2.5a denoted A and B at the signal source end of the wires and A′ and B′ at the signal destination end of the wires with a resistive short between the two wires. There are two bridging fault models that are the most frequently used in practice today: the wired-AND/wired-OR bridging fault model and the dominant bridging fault model [156]. Both models are illustrated in Figure 2.5. The wired-AND/wire-OR bridging fault model (Figure 2.5b) originated with bipolar integrated circuit technology where the destination ends of the shorted wires receive the logical AND or logical OR of the logic values at the signal sources [7]. The wired-AND model (also denoted as WAND) is sometimes referred to as a 0-dominant bridging fault since a 0 on either wire determines the logic value on both wires just as a logic 0 at any input of an AND gate determines the gate output value. Similarly, the wired-OR (also denoted as WOR) is sometimes referred to as a 1-dominant bridging fault.

The dominant bridging fault model (not to be confused with 1- and 0-dominant bridging faults) was developed to more accurately reflect the behavior of some shorts in

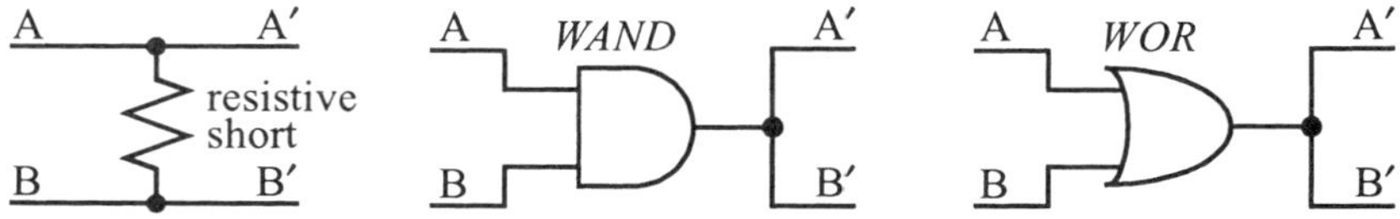

(a) bridging fault (b) wired-AND/wired-OR bridging faults

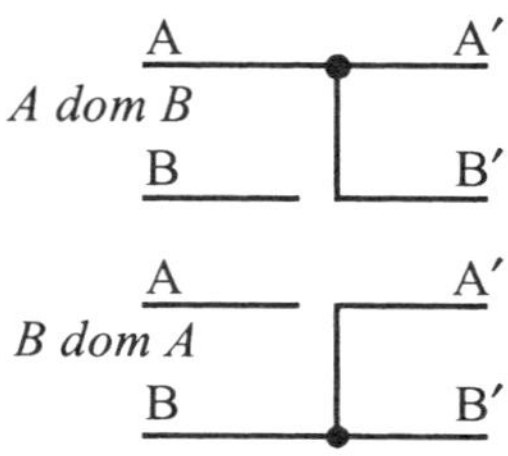

Fault-Free		Wired-AND/OR				Dominant			
		WAND		WOR		A dom B		B dom A	
A	**B**	**A′**	**B′**	**A′**	**B′**	**A′**	**B′**	**A′**	**B′**
0	0	0	0	0	0	0	0	0	0
0	1	0	0	1	1	0	0	1	1
1	0	0	0	1	1	1	1	0	0
1	1	1	1	1	1	1	1	1	1

(c) dominant bridging faults (d) bridging fault behavior

FIGURE 2.5 Wired-AND/OR and dominant bridging fault models.

CMOS circuits where the logic value at the destination end of the shorted wires will be determined by the source gate with the strongest drive capability [84]. As illustrated in Figure 2.5c, the driver for one node "dominates" the driver for the other node; this is typically denoted as "A dom B" or "B dom A" depending on which driver is stronger.

The faulty circuit behavior is given in the table of Figure 2.5d for the wired-AND/OR and dominant bridging fault models where the differences in the faulty circuit from that of the fault-free circuit are highlighted in gray. Given two wires in a circuit, it is difficult to know which driver will be stronger or whether a short will behave as a wired-AND or a wired-OR. Therefore, both faults for a given fault model (wired-AND and wired-OR, or else, "A dom B" and "B dom A") are emulated during fault simulation to determine if a given set of test vectors will detect both faults associated with that particular bridging fault model. This is analogous to emulating both stuck-at-0 and stuck-at-1 for a gate-level fault or stuck-on and stuck-off for a transistor-level fault.

Bridging faults can only be detected by applying opposite logic values to the two wires as can be seen from the table in Figure 2.5d. I_{DDQ} testing techniques can be used to detect bridging faults since, like the transistor stuck-on fault, there will be a conducting path between V_{DD} and V_{SS} during steady-state whenever the two drivers are applying opposite logic values [163]. Similarly, i_{DDT} testing techniques can also be used to detect bridging faults [154]. Using traditional logic value output differences for detection, we can refer to the table in Figure 2.5d. The wired-AND/OR bridging faults can be detected by one of two options: 1) apply either vector (10 or 01) while monitoring both wires at primary outputs, or 2) apply both vectors while monitoring only one of the wires at a primary output. Dominant bridging faults, on the other hand, can only be detected by the first option. Therefore, dominant bridging faults are more difficult to detect than the wired-AND/OR bridging faults but detection of the dominant bridging faults will ensure detection of wired-AND/OR bridging fault behavior. From this standpoint, it is better to use the dominant bridging fault model for fault simulation. In addition, the dominant bridging fault model is the easier of the two models to emulate in a fault simulation environment.

Another bridging fault model is based on the behavior of shorts observed in faulty Application Specific Integrated Circuits (ASICs) and Field Programmable Gate Arrays (FPGAs) [276].[1] The behavior of this bridging fault model is similar to the dominant bridging fault model in that one of the lines is not effected, but the short behaves more like a diode between the shorted wires in terms of its effect on the other wire. For example, when the dominant wire is driven to a logic 0, the other wire can function normally, passing both logic 0s and logic 1s. However, when the dominant wire is driven to a logic 1, the other wire is a logic 1 regardless of the logic value of its driving gate. In other cases, a logic 1 on the dominant wire allows the other wire to function normally while a logic 0 on the dominant wire forces the other wire to a logic 0. This fault model is called the dominant-AND/dominant-OR bridging fault model and assumes that one driving source of the shorted wires is dominant but the dominance is a function of the logic value being driven (see Figure 2.6). There are four

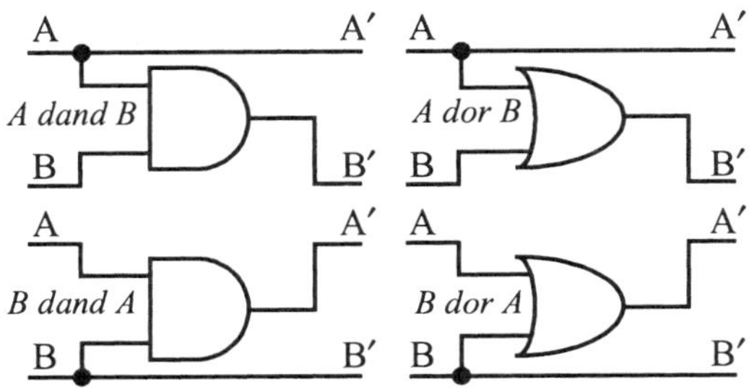

(a) dominant-AND bridging faults

(b) dominant-OR bridging faults

Fault-Free		A dand B		B dand A		A dor B		B dor A	
A	**B**	**A′**	**B′**	**A′**	**B′**	**A′**	**B′**	**A′**	**B′**
0	0	0	0	0	0	0	0	0	0
0	1	0	0	0	1	0	1	1	1
1	0	1	0	0	0	1	1	1	0
1	1	1	1	1	1	1	1	1	1

(c) dominant-AND/OR bridging fault behavior

FIGURE 2.6 Dominant-AND/OR bridging fault models.

1. We first observed this behavior in some of our faulty ASICs at Bell Labs in the mid-1980s.

possible faults to consider at each fault site while there are only two faults in the case of the wired-AND/OR and dominant bridging fault models. To detect these four faults, both vectors ('01' and '10') must be applied while monitoring both wires at the primary outputs, as can be seen from the fault behavior in the table of Figure 2.6c. With this more restrictive condition, detecting all four dominant-AND/OR bridging faults at each fault site ensures detection of all wired-AND/OR and dominant bridging faults.

Bridging faults that create a feedback loop in a combinational logic circuit, where the output of an inverting combinational logic circuit is shorted to its input, may be more difficult to detect. In the case of a feedback bridging fault on a single inverting gate, such as an inverter, the feedback can cause an intermediate voltage at the output of the gate which results in both PFETs and NFETs partially conducting. This results in a current path from V_{DD} to V_{SS} during steady-state that can be detected by I_{DDQ} testing techniques. On the other hand, the logic value may or may not be interpreted as an incorrect logic value at the destination gates on the signal net, similar to transistor stuck-on faults. When there are many gates in the combinational logic loop created by the feedback bridging fault, oscillations may result in the circuit. In order to detect this fault, an incorrect logic value resulting from the oscillation must be observed at a primary output of the CUT since I_{DDQ} testing techniques may not be able to detect the changes in power supply current. Therefore, while transistor-level faults and bridging faults more accurately reflect the behavior of defects in a circuit, they are computationally more difficult to emulate and evaluate in a fault simulator than the traditional gate-level stuck-at fault model.

2.2.4 Delay Faults

Another important class of faults is delay faults, which result in a circuit whose operation is logically correct but does not perform at the required operating frequency [143]. This can result from under- or over-etching (or other defects) during the fabrication process, which produces MOSFETs with channel widths that are much narrower or channel lengths that are much longer than intended. Therefore, some paths through the circuit may not meet performance specifications. The goal of delay fault testing is to find and expose any of these defects that may exist in a fabricated device. While delay faults are much more difficult to model and emulate compared to the fault models discussed previously, the basic objective of delay fault testing is to test paths between flip-flops, primary inputs to flip-flops, and flip-flops to primary outputs through the combinational logic to determine if any path fails to operate at the required system speed. As a result, delay fault testing typically requires a sequential application of two vectors such that the path through the combinational logic is set-up by the first vector while the second vector produces a transition through the path for delay fault detection. This is similar to the 2-vector sequence needed to detect transistor stuck-off faults.

2.2.5 Single vs. Multiple Fault Models

Obviously, multiple defects can be sustained in a given VLSI device or PCB during the manufacturing process or during system operation. However, the sheer number of combinations of possible multiple faults makes the multiple fault model computationally inefficient for simulation. For example, in a logic circuit with a total of N gate inputs and outputs, there would be 3^N-1 different faulty circuits to emulate during fault simulation under the multiple stuck-at gate-level fault model, since any gate input or output can be stuck-at-0, stuck-at-1, or fault-free and there are N fault sites to consider. The one circuit that is not considered (the "-1" in the above expression) is the case where all N fault sites are fault-free; this is the fault-free CUT. The single stuck-at gate-level fault model assumes that only one fault can exist in the circuit at a time such that there would be only $2N$ different faulty circuits to emulate during fault simulation. The same analysis holds true for the transistor-level stuck fault model where each of the N transistors in a device can be either stuck-on or stuck-off. This is also true for N bridging fault sites when we consider the wired-AND/OR or dominant bridging fault models since there are two possible faults for each fault site. For the dominant-AND/OR bridging fault model, there are four possible faults to consider for each of the N fault sites, such that the multiple fault model would result in 5^N-1 faulty circuits to emulate while the single fault model would require emulation of only $4N$ faulty circuits. Therefore, multiple fault simulation is prohibitively expensive in terms of fault simulation time for all but the smallest circuits. Fortunately, it has been shown in a number of studies that high single fault coverage for any of the fault models translates to high multiple fault coverage [113]. As a result, the single fault model is widely used for test development and evaluation. Unless specifically stated otherwise, the single fault model will be assumed in the subsequent sections and chapters.

2.2.6 Equivalent Faults and Fault Collapsing

Under the single fault model, the number of gate-level stuck-at faults associated with each gate is twice the total number of inputs and outputs of the gate since each input or output can be either stuck-at-0 or stuck-at-1. However, from the tables in Figure 2.3 it is evident that some of these faults behave identically and cannot be distinguished from each other as a result of their faulty behavior. For example, any stuck-at-0 input to an AND (NAND) gate is indistinguishable from the output of the gate being stuck-at-0 (stuck-at-1). These faults are *equivalent faults* and detecting any one of these faults is guaranteed to detect all of its equivalent faults [50]. As a result, only one fault in each group of equivalent faults needs to be emulated; this helps to reduce fault simulation time by reducing the number of faults to be emulated. A stuck-at-1 fault at any input of an OR (NOR) gate is equivalent to the output stuck-at-1 (stuck-at-0). In the case of an inverter (buffer), the input stuck-at-0 is equivalent to the output stuck-at-1 (stuck-at-0) while the input stuck-at-1 is equivalent to the output stuck-at-0 (stuck-at-1). Therefore, for any K-input elementary logic gate where $K>1$, the total number of faults that need to be simulated is $K+2$, which includes the output stuck-at-0 and stuck-at-1 along with each input stuck-at-1 for an AND/NAND gate or stuck-at-0 for an OR/NOR gate. For an inverter or buffer, the input faults are equivalent to the output faults such that there are only two faults to be emulated, the output stuck-at-0 and stuck-at-1. The removal of equivalent faults while keeping one from each group for fault simulation is called *fault collapsing* [4].

On a 2-point signal net (with one source and one destination, or *sink*), the output of the source gate stuck-at-1 (stuck-at-0) cannot be distinguished from the input of the sink gate stuck-at-1 (stuck-at-0). Therefore, there are equivalent faults at the structural-level as well as at the gate-level. This is illustrated in the two circuit examples in Figure 2.7. The set of "collapsed" faults for each circuit is indicated in bold italics. In Figure 2.7a, the circuit contains a fan-out stem (at the output of the inverter), which

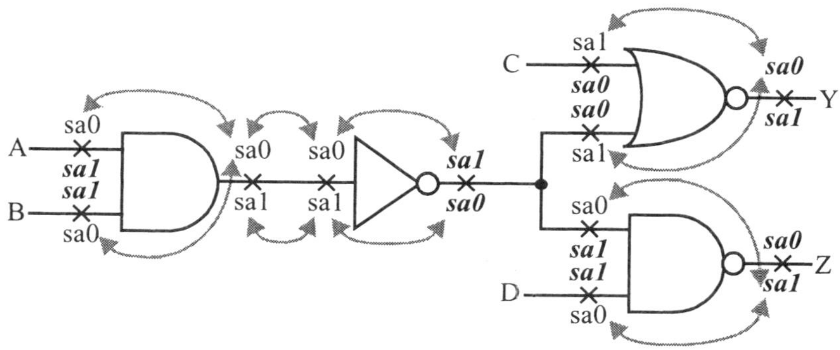

(a) gates with fan-out stem (total collapsed faults = 12)

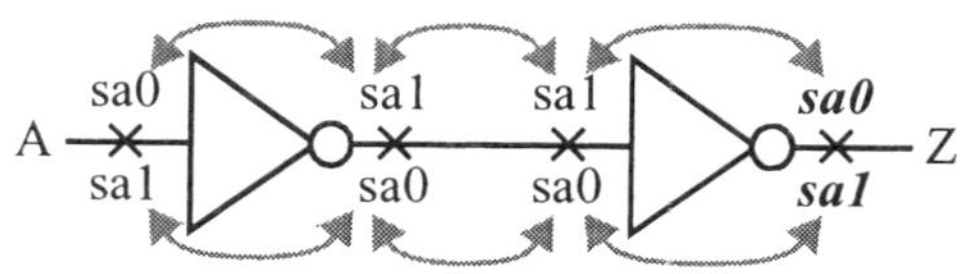

(b) chain of inverters/buffers (total collapsed faults = 2)

FIGURE 2.7 Gate-level stuck-at fault collapsing.

requires special consideration during fault collapsing. The faults at the fan-out stem are not equivalent to the faults at the fan-out branches without violating the single stuck-at fault model assumption that there can be only one fault in the circuit. For example, the stuck-at-1 fault at the fan-out stem would be equivalent to a stuck-at-1 at the input to the NOR gate ***and*** a stuck-at-1 at the input to the NAND gate, violating the single fault model. Therefore, both the stuck-at-0 and stuck-at-1 faults at a fan-out stem must be included in the fault list (similar to a primary output). In the case of chains of series inverters and/or buffers, as illustrated in Figure 2.7b, faults can continue to be collapsed in the absence of fan-out stems resulting in only two faults (the output stuck-at-0 and stuck-at-1) included in the collapsed fault list. Note that any of the faults in a set of equivalent faults can be used to represent all of the faults in the set, but most CAD tools that generate collapsed fault lists tend to favor collapsing in the direction of the primary outputs as shown in the examples of Figure 2.7.

The total number of gate-level faults in an uncollapsed fault list is $2N$ where N is the total number of gate inputs and gate outputs in the CUT. The total number of gate-level faults in the collapsed fault list, F_{CG}, is given by:

$$F_{CG} = 2(P_O + F_O) + G_I - N_I \tag{2.2}$$

where G_I is the total number of gate inputs to all the gates in the circuit, P_O is the number of primary outputs, F_O is the number of fan-out stems, and N_I is the number of inverters (and/or buffers) in the circuit. Gate-level fault collapsing typically reduces the total number of faults to be simulated by about 50% in most circuits, when compared to the number of uncollapsed faults; this results in a significant reduction in fault simulation time.

Transistor-level circuits also have equivalent faults and, hence, fault collapsing can be performed, but not to the extent that can be accomplished with gate-level faults. For example, in the table in Figure 2.4d, the behavior of series transistors stuck-off is indistinguishable, as is also the case with parallel transistors stuck-on. Therefore, for the transistor-level fault model, the total number of uncollapsed faults is $2T$ where T is the total number of transistors in the CUT, while the total number of collapsed faults, F_{CT}, is given by:

$$F_{CT} = 2T - T_S + G_S - T_P + G_P \tag{2.3}$$

where T_S is the total number of series transistors, G_S is the number of independent groups of series transistors, T_P is the total number of parallel transistors, and G_P is the number of independent groups of parallel transistors. This expression is valid for static CMOS circuits for individual gates as well as for the entire CUT. Transistor-level fault collapsing typically leads to about a 25% reduction in the total number of faults to be simulated when compared with the set of uncollapsed faults.

An important question is which set of faults, collapsed or uncollapsed, should be used. For fault simulation the collapsed fault set will obviously reduce the fault simulation time since it will be smaller. However, when considering the quality of a set of manufacturing tests, the uncollapsed fault set provides a more accurate representation of defect coverage since every potential defect site for the given fault model is considered. As a result, the uncollapsed fault set more realistically approximates the complete set of potential defect sites. While we would prefer to use uncollapsed faults for accuracy, we will typically turn to collapsed faults for faster fault simulation time even though there may be several percent difference in fault coverage obtained. Experience has shown that using uncollapsed faults often result in higher fault coverage but the result is circuit dependent.

2.2.7 Bridging Fault Extraction

Unlike gate-level and transistor-level equivalent faults, there is no corresponding fault collapsing for bridging faults. However, a bigger concern is determining the number and location of possible bridging fault sites in the first place. The possible fault sites for gate-level and transistor-level fault models are apparent from the circuit design long before physical layout since we are only concerned with the individual gates and transistors. Bridging faults, on the other hand, deal with pairs of wires and more specifically pairs of wires that are adjacent in the physical layout. Therefore, the possible fault sites to be considered for fault simulation is a concern in bridging fault simulation due to the simulation time, as well as the accuracy of the fault coverage results. For example, if we consider possible shorts between any pair of wires in a circuit with N signal nets, there will be N-choose-2 or $(N^2-N)/2$ possible fault sites to consider, each with 2 or 4 faults to simulate depending on the bridging fault model used. As a result, the fault simulation time will be prohibitive for a circuit of any practical size.

The actual fault coverage obtained when considering shorts between all possible pairs of wires in the circuit is of little, if any, value. For example, two wires on opposite sides of a VLSI device are less likely to have a bridging fault than two wires that are adjacent to each other. If we consider only detection of bridging faults on the two wires that are least likely, the fault coverage is misleading in terms of an accurate indication of the quality of the set of test vectors. If we spend time developing test vectors for bridging faults that are not likely to occur, then we are wasting valuable test development time with little or no value added to the quality of the manufacturing test in terms of fault detection capability. In addition, we are increasing the test time and cost associated with each device or PCB that we test. Therefore, for accurate bridging fault coverage, it is important to consider the physical layout of the circuit in order to consider only those wires that are likely (if not most likely) to sustain a bridging fault.

Realistic potential bridging fault sites can be accurately extracted from the physical design by using existing CAD tools developed for capacitance extraction from the physical design database for post-layout timing simulations [164]. By extracting fringe capacitance we determine potential bridging faults on a given mask layer. Extracting overlap capacitance, on the other hand, yields potential bridging faults between mask layers. Using the parallel plate model, overlap capacitance is given by:

$$C = \frac{\varepsilon A}{d} \tag{2.4}$$

where A is the area of overlap and d is the distance between the two conductors (ε is the inter-layer dielectric constant). Since the probability of a bridging fault occurring between two adjacent conductors is proportional to the area A of adjacency and inversely proportional to the distance d, the capacitance C has a similar relationship to A and d as the probability of a bridging fault occurring between two conductors. By comparing the capacitance between two different pairs of adjacent conductors, we can determine which pair may be more likely to sustain a bridging fault, specifically the pair with the larger extracted capacitance. Differences in ε for various mask layers can be compensated for by dividing by the ε for each layer such that we are left with the A/d ratio to represent the relative probability of occurrence at a given bridging fault site. While the model used to calculate fringe capacitance is more complicated than the parallel plate model, the basic effect is the same in terms of providing a relative likelihood of the occurrence of a bridging fault for two different fault sites on the same mask layer. For example, two conductors that have a longer region of adjacency are more likely to sustain a bridging fault than two conductors with a shorter region of adjacency, even though the distance between the two sets of conductors is the same, as illustrated in Figure 2.8a. Similarly, two conductors that are closer together are more likely to sustain a bridging fault than two conductors that are farther apart even though their lengths are the same, as illustrated in Figure 2.8b. This basic information can be obtained simply by comparing the extracted capacitance of two pairs of adjacent conductors with the higher capacitance indicating the higher probability of sustaining a bridging fault.

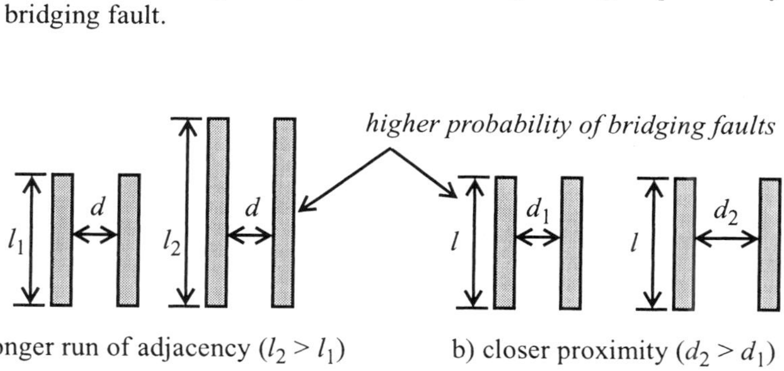

FIGURE 2.8 Capacitance vs. probability of bridging faults.

By ordering the extracted bridging faults according to their capacitance, we obtain an ordered list of bridging fault sites from the most likely to occur (and, therefore, the most important to detect) to the least likely to occur (and the least important if we don't detect them). This facilitates targeting the highest probability bridging fault sites during test development for the most efficient manufacturing testing. Or, if we wish to reduce the number of faults to be simulated, we can remove those potential bridging faults from the list that have the lowest probability of occurrence. Another important aspect of the capacitance extraction is the ability to determine the weighted fault coverage of a given set of manufacturing tests. While the unweighted bridging fault coverage is given by Equation (2.1), the weighted bridging fault coverage, FC_{WBF}, is given by [164]:

$$FC_{WBF} = \frac{\sum_{i=1}^{D} C_i}{\sum_{j=1}^{T} C_j} \tag{2.5}$$

where D is the number of detected faults, T is the total number of faults, and C_i is the capacitance extracted for the ith bridging fault site [164]. As a result, the bridging faults with the higher probability of occurrence have a bigger impact on the weighted fault coverage. For example, when developing manufacturing tests for a production Complex Programmable Logic Device (CPLD), we found that the lowest unweighted bridging fault coverage for a set of test vectors developed specifically for detecting bridging faults was 96.5% while the weighted fault coverage was actually 99.9943% [277]. Therefore, there was no need to continue with test development for higher fault coverage since the faults that were not detected were those faults that were the least likely to occur. On the other hand, we found that a set of existing manufacturing test vectors developed for 100% stuck-at gate-level faults obtained only 50.5% unweighted bridging fault coverage. More importantly, this set of test vectors had a weighted bridging fault coverage of only 29.9% indicating that the bridging faults being detected by this set of tests were those that were the least likely to occur. This case contradicts the notion that high stuck-at gate-level fault coverage implies high bridging fault coverage.

2.2.8 Statistical Fault Sampling

Another important method for reducing the total number of faults to be simulated and the resulting fault simulation time is to simulate a subset of the complete set of faults. This subset of faults is obtained via a random sampling of the fault list. It should be noted that the random selection of faults to simulate is done without replacement;

when a given fault is selected for simulation, it is not considered in subsequent random selections of faults for the sample. The larger the sample, the more accurate the fault coverage obtained from the fault simulation. But if the aim is to reduce fault simulation time, then we would like to know the smallest sample required to obtain a reasonably accurate fault coverage. There are methods to determine the number of random samples required for a given accuracy in the fault coverage obtained from fault simulation as well as techniques to determine the fault coverage accuracy for a given number of randomly sampled faults [50]. However, it has been shown that a minimum random sampling of 1000 faults will give an accuracy within 2% to 3% of the actual fault coverage that would be obtained by simulating the complete set of faults for most circuits, regardless of the circuit size [50]. This can result in a significant reduction in fault simulation time when the complete set of faults is large. This statistical fault sampling technique is the only practical solution to simulation of multiple fault models. In the case of single fault models, the uncollapsed fault list offers the best starting place for the random sampling process such that we obtain a better indication of the defect coverage of the set of test vectors. One drawback of statistical fault sampling is that we loose the ability to target areas of the circuit for further design verification or further test vector development based on the undetected fault list since it is not a complete set of undetected faults. Otherwise, fault sampling provides an excellent method of reducing fault simulation time for very large circuits with minimal loss of accuracy in fault coverage determination.

2.3 Fault Detection and Fault Coverage

Fault coverage in its simplest form is the ratio of the number of faults detected by a given set of test vectors to the total number of possible faults considered, as given by Equation (2.1). However, there are other factors that influence the fault coverage calculation. These include undetectable and potentially detected faults, both of which can lower the overall fault coverage that be obtained for most circuits unless one specifically considers these types of faults and how to deal with them in the fault coverage calculation. As we saw in the case of the gate-level fault model, the individual elementary logic gates are completely and easily testable. Therefore, it is the way in which we interconnect these gates that leads to testability and testing problems.

2.3.1 Controllability and Observability

In order to detect a fault, the test stimuli applied to the CUT must establish conditions within the CUT such that the faulty circuit will produce a different output response than that of the fault-free circuit. Therefore, detection of a given fault (regardless of the fault model considered) involves establishing input stimuli which will produce the appropriate logic value or condition at the suspected fault site to create a contradiction between the fault-free and faulty circuit. Next, this contradiction must be propagated to a primary output of the CUT so that it can be observed. As a result, the testability of a given fault site in a circuit is a function of the *controllability* of the fault site from the primary inputs, as well as the *observability* of the fault site from the primary outputs. Controllability is the ease with which we can control a fault site and observability is the ease with which we can observe a fault site [17]. The more difficult a fault site is to control and/or observe, the more difficult the fault models associated with that fault site will be to detect and, hence, the less testable the fault site. Depending on the fault model, every gate input or output, every transistor, or every pair of adjacent wires in the circuit represents a potential fault site. As a result, the testability of the entire circuit is a function of the controllability and observability of all fault sites within the complete circuit.

2.3.2 Path Sensitization

The controllability and observability of a fault site can most easily be seen through the *path sensitization algorithm* for determining test vectors that will detect gate-level stuck-at faults in a combinational logic circuit with no feedback paths. This algorithm provides the basic idea behind many ATPG programs [4]. The algorithm can also be used to sensitize a path through combinational logic for delay fault testing. The basic algorithm is performed as follows:

Step 1: At the fault site, assign the logic value opposite to that of the stuck-at fault. This step is referred to as *fault sensitization* or *fault activation*.

Step 2: From the fault site, choose a path to a primary output, assigning non-controlling logic values to all other inputs (referred to as *off-path inputs*) to gates in that path (1 = non-controlling logic value for AND/NAND gates and 0 = non-controlling logic value for OR/NOR gates). This step is referred to as *fault propagation* or *path sensitization*.

Step 3: For all assigned values, backtrace to the primary inputs selecting gate input values that will produce the assigned values at the gate outputs. This step is referred to as *fault justification* or *line justification*.

Step 4: If there is a conflict in the logic values to be assigned to a given node, repeat Steps 2 and 3 choosing new paths and/or logic values in Step 3. If no path can be found without producing a conflict, the fault ***may be*** undetectable. On the other hand, if there are no conflicts in the logic values assigned at any of the nodes, the fault is detectable and the logic values at primary inputs form a test vector that will detect the fault.

To illustrate the application of the algorithm, consider the hazard-free multiplexer shown in Figure 2.9. In order to detect the output of the middle AND gate stuck-at-0, we would set the fault site to a logic 1 (the opposite logic value) to activate the fault. To propagate that value to a primary output (this is indicated by '1/0' in the figure to denote fault-free/faulty circuit behavior), we next assign non-controlling logic values (logic 0 in this case since we are propagating through an OR gate, shown in bold italics) to propagate the faulty behavior to the primary output. Finally, we backtrace from all the assigned values, assigning input logic values to the gates as we go. Once we have backtraced from all assigned values (denoted by the normal font 0s and 1s), we have our test vector that will detect the fault. If we encounter conflicts in the logic values assigned to a given node then we need to go back to Steps 2 and 3 to try different choices of gate input values as well as different paths until we have exhausted all possible combinations of paths and logic values. If all possible combinations of logic values and paths are exhausted without finding a conflict-free case, the fault is undetectable, but this is difficult to prove for large circuits.

The path sensitization algorithm also lets us sensitize a given path from the input to the output of a combinational logic circuit. This facilitates toggling the input and observing the output toggle via the path sensitized through the combinational logic circuit. The delay through the path can be measured to determine if faults exist in the path which cause excessive delay. This forms the basic idea behind testing approaches for detecting delay faults [252].

2.3.3 Undetectable Faults

If no test vector or sequence of test vectors exists that can detect a given fault, then that fault is undetectable [170]. This is an interesting situation since the implication is

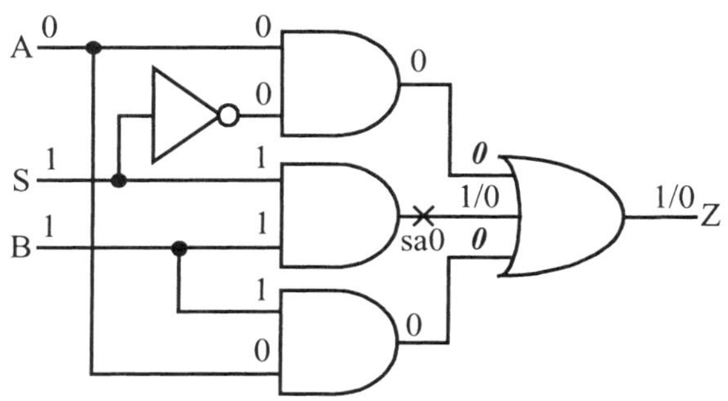

FIGURE 2.9 Application of path sensitization algorithm.

that we cannot tell whether the fault exists or not. As a result, the operation of the system will not be affected by the presence of the fault. Undetectable faults typically result from reconvergent fan-out ***and*** redundancy in the logic; however, the presence of reconvergent fan-out does not always result in undetectable faults. In practice, undetectable faults are difficult to verify. These faults simply show up as undetected faults and degrade the fault simulation time as well as the resulting fault coverage determination for a given set of test vectors. Undetectable faults always slow down fault simulation since the fault is never detected and the entire set of test vectors is simulated for that fault before it is put into the undetected fault list.

An example of an undetectable fault in the hazard-free multiplexer is illustrated in Figure 2.10. The stuck-at-0 fault at the output of the bottom AND gate is undetectable. One conflict resulting from the application of the path sensitization algorithm is shown, but all possible combinations of paths and assigned logic values result in similar conflicts. This undetectable fault is the result of redundancy in conjunction with reconvergent fanout. All three primary inputs (A, B, and S) constitute fanout stems and, in each case, the fanout reconverges at the output OR gate. The redundancy can be seen in the accompanying Karnaugh map were the product term shown in gray (corresponding to the bottom AND gate) is not an essential prime implicant while the other two product terms are essential. Therefore, this additional product term is redundant and provides no essential output function.[1]

The hazard-free multiplexer does have an important place in design applications to prevent a glitch at its output when A=B=1 and S changes from a 1 to a 0. This is par-

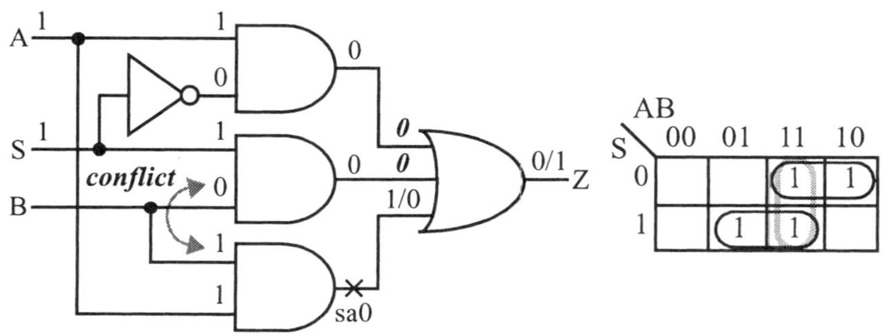

FIGURE 2.10 Example of an undetectable fault.

1. It is interesting to think that while we were being taught the hazard-free multiplexer in our Logic Design 101 course, we were also being taught to design circuits that are difficult to test.

ticularly true when the multiplexer is selecting the source of the system clock for example, otherwise the glitch in the clock could cause problems throughout the system. Therefore, there are cases where system-level considerations require incorporation of circuits with undetectable faults, like the hazard-free multiplexer. But one should use these types of circuits sparingly and only when needed in order to keep track of the known undetectable faults. When a hazard-free multiplexer is desired and this fault exists in the circuit, the glitch-free multiplexer is converted to a "glitchy" multiplexer, yet a proper test for the existence of this fault remains an open problem.

Other types of circuits that can cause undetectable faults include global resets/presets and power-up initialization circuitry. Global resets/presets are good for quick initialization for logic simulation during design verification. Yet, in the absence of global resets/presets, the initialization of the circuit provides a quick testability analysis of the design. The portions of the circuit that are the most difficult to initialize are going to be the portions that are the most difficult to test due to a lack of controllability; therefore, testability is proportional to initializability. While these circuits are sometimes required by system specifications and considerations, designers should minimize the use of these circuits as much as possible.

Logic circuits that are not minimized can also lead to undetectable faults. A common design technique that implies logic is not minimized is tying unused inputs to a logic 1 (V_{DD}) or a logic 0 (V_{SS}). For example, a stuck-at-1 cannot be detected on an input to a gate when that input is tied to V_{DD}, because the input is always a logic 1. Even though this is an intentional design, the fault will not be detected and will remain in the undetected fault list, reducing fault coverage and increasing fault simulation time. Another bad design habit is to fanout an input signal to multiple inputs of a gate when fewer inputs are needed; this creates redundancy and reconvergent fanout. An example is an input signal connected to two inputs of a 3-input NAND gate when only a 2-input NAND function is required. In both of these cases, we are simply wasting area as well as creating undetectable faults that will in turn make fault simulation and fault coverage determination more difficult than necessary.[1] While the first case, tying inputs to logic 1 or logic 0, is a valuable design technique for multi-mode usage and reuse of embedded functions, the second case is of no value. Therefore, it is good design practice to minimize the instances of tying inputs to a logic 1 or 0 and to avoid fanning out a signal to multiple inputs of a gate.

1. At a design review, a novice designer had connected the input of a flip-flop to V_{DD} and the output of the flip-flop then drove other circuitry. I asked the designer what was the point when the first clock cycle after power-up would clock in a logic 1 and it would stay that way leading to many undetectable faults. He sheepishly replied, "That's my synchronous 1s generator!"

2.3.4 Potentially Detected Faults

The feedback typically found in static memory elements of a sequential logic circuit will result in the initialization of every memory element on power-up to some valid logic value in a real circuit. The problem during logic and fault simulation is that we cannot predict the initialization logic value for a given memory element. As a result, the *undefined logic value* (which can represent either a logic 1 or a logic 0 but not a deterministic value) is introduced and used in simulators to represent an un-initialized state for memory elements. When a given memory element is initialized (either through clocking in input data or activation of a preset/reset control), the undefined logic value stored in that memory element is replaced with the known logic value. The notation of the undefined logic value varies with simulators and typical notations include '2', '3', 'U', and 'X'. Note that the 'U' will be used to denote the undefined logic value throughout the rest of this book.

During fault simulation, any fault that prevents initialization of a memory element will result in the undefined logic value being propagated to a primary output when a path from the fault site is sensitized [233]. The fault-free circuit, on the other hand, will produce a valid logic value (either 0 or 1) at the primary output once the memory element is initialized. Since the undefined logic value represents either a logic 0 or a logic 1, there is no definite contradiction at the output of the CUT, only a potential contradiction which is dependent on the actual logic value in the memory element of the real faulty circuit. Therefore, during fault simulation, the simulator cannot determine whether the fault was detected or not when the faulty circuit is producing a 'U' while the fault-free circuit is producing a valid logic value. As a result, the fault simulator marks the fault as potentially detected [4]. These faults are typically included in a separate, potentially detected fault list as well as the undetected fault list. In some cases, a potentially detected fault can be detected with a different set of test vectors but in other cases the fault will remain only potentially detected. An example of this is shown in Figure 2.11 where a stuck-at-0 fault on the active high clock enable to the register will prevent initialization such that output of the register will always be undefined. Similarly, a stuck-at-0 or a stuck-at-1 on the clock input to the rising edge-triggered flip-flop will prevent initialization. As a result, these three faults will only be potentially detected for any single observation of the primary outputs when a path is sensitized from the fault site.

At worst case, we can assume an equal likelihood of a logic 1 or 0 as the actual initialized logic value in the real circuit (during wafer-level testing for example) such that we assume that each potentially detected fault has a 0.5 probability of being detected during manufacturing test. This would translate to only half of all potentially detected faults being detected. In reality, this assumption is true only when the fault-free circuit does not change logic values. However, if the data activity is high at the fault site

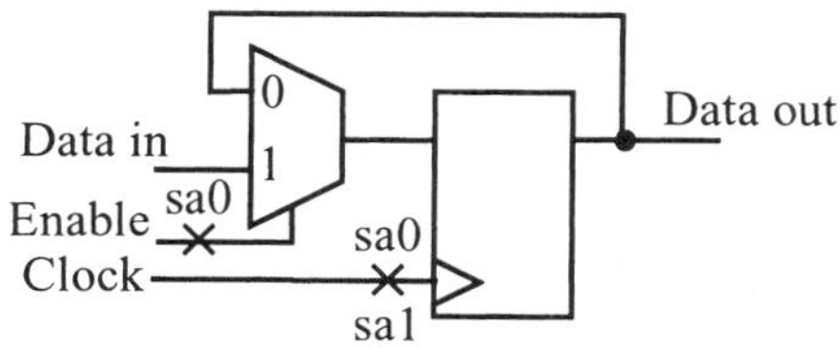

FIGURE 2.11 Example of potentially detected faults in a clock-enabled flip-flop.

and that data is propagated to primary outputs, then we can safely assume that the potentially detected fault is detected as a result of multiple observations of the fault site when the fault-free circuit has propagated both logic values through the fault site. This becomes a judgment call on the part of the designer and/or test engineer based on personal knowledge of the circuit and the test vectors. For example, in the clock-enabled flip-flop example in Figure 2.11, we can safely assume that the three potentially detected faults will be detected if we clock both a logic 0 and a logic 1 through the flip-flop during the set of test vectors and both of these values are observed at primary outputs. These tests are referred to as *multiple-observation* tests and require additional effort during test development and application to ensure fault detection [211].

Another type of fault that falls into the potentially detected category is a fault that causes the circuit to oscillate, such as the feedback bridging fault discussed earlier. Some fault simulators will stop simulation once an oscillation occurs and will place the oscillation faults in a separate fault list. Detection of these oscillation faults depends on the faulty circuit producing an incorrect logic value at the primary outputs in order to produce a mismatch with the expected output response. This will be dependent on the data activity in the fault-free circuit and the propagation of that data to primary outputs as well as the sampling time at the primary outputs. We can safely assume that the probability of detecting a given oscillation fault is 0.5. Therefore, we can treat these faults similar to potentially detected faults.

2.3.5 Fault Coverage

A more complete picture of fault simulation is illustrated in Figure 2.12 where the inputs to the fault simulation environment include the circuit netlist at a level of abstraction sufficient to emulate the desired fault models along with the set of test vectors to be simulated for evaluation of their fault coverage. During the initial fault simulation, the list of faults to be simulated is typically generated by the fault simula-

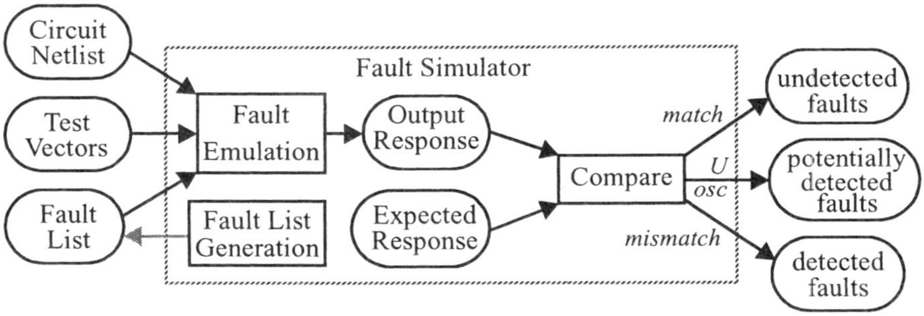

FIGURE 2.12 Fault simulation process.

tor under the direction of the user to obtain the initial set of collapsed or uncollapsed faults. In subsequent fault simulations, the undetected fault list from the previous fault simulation is typically used; in this way, we minimize fault simulation time by not resimulating the previously detected faults. The ability to specify this input fault list is also important for statistical fault simulation where we have produced a set of randomly sampled faults, and for bridging fault simulation where we have produced a set of faults extracted from the physical layout.

At the completion of the fault simulation each fault will be either detected, undetected, or potentially detected by the given set of test vectors. Alternatively, the fault itself may be undetectable, in which case it will reside in the set of undetected faults. However, if we can determine that a fault is undetectable (and no corrective action will be taken to eliminate the source of the undetectable fault, such as logic minimization) then we should remove the undetectable fault from the input fault list to reduce fault simulation time. While a simple fault coverage calculation was given in Equation (2.1), a more realistic fault coverage calculation is given by:

$$FC = \frac{D + xP}{T - U} \tag{2.6}$$

where D is the number of detected faults, P is the number of potentially detected faults (including oscillation faults), T is the total number of faults, U is the number of faults determined to be undetectable, and x is the probability of detection for the potentially detected faults. The value of x can be adjusted in the range $0.0 \leq x \leq 1.0$ based on the data activity of the CUT and based on judgement of the designer and/or test engineer. The fault coverage calculation is conservative (or pessimistic) when x=0.5. While the absolute worst case probability of detection for these faults is 0.0, the assumption that the faulty circuit will initialize to either a logic 0 or logic 1 and remain at that value makes 0.5 a valid conservative value [233]. Alternatively, any potentially detected faults that can safely be assumed to be detected via multiple observations of both logic values can be moved to the detected fault list (from *P* to *D*) for the fault coverage calculation.

Fault simulation is an important part of many BIST implementations to determine the effectiveness of the BIST approach in a given application. Fault simulation time is a concern since, for an *N*-gate circuit, logic simulation time is usually $O(N)$ while fault simulation time is $O(N^2)$ [4]. Fault modeling is also important to BIST implementations since the goal is to make the investment of the overhead of the BIST circuitry, in terms of device area and design effort, pay-off with high quality tests so that few defective parts are shipped to the field. This requires not only an understanding of the various types of fault models used to evaluate manufacturing tests, including BIST approaches, but also their limitations since it is often difficult to efficiently simulate and/or accurately emulate defect behavior.

2.3.6 *N*-Detectability

While the gate-level single stuck-at fault model appears to be the least attractive in terms of it ability to accurately model defects, recent studies of actual faulty VLSI devices have shown that certain sets of test vectors developed using the gate-level fault model are very effective in detecting defects when applied *at-speed*. Specifically, these are test vectors that detect each fault at least *N* times, each time with a different test vector; these sets of test vectors are referred to as *N-detect test sets* [176]. Furthermore, the studies have shown that *N*-detect test sets developed using elementary logic gate representations of the device are the more effective in detecting actual defects when compared to *N*-detect test sets developed using *pin faults* of functional models, such as an exclusive-OR gate, multiplexer, full adder, or flip-flop [54]. A pin fault only considers stuck-at-0 or stuck-at-1 faults for the I/O of the functional model as opposed to the inputs and outputs of the elementary logic gates that would be used to construct the logic function [173]. For example, all pin faults for the functional model of a 2-input exclusive-OR gate can be detected in three test vectors while any elementary logic gate-level model for an exclusive-OR gate requires all four possible input test vectors to detect all gate-level stuck-at faults.

These elementary logic gate, at-speed *N*-detect test sets are as effective in detecting delay faults, bridging faults, and transistor stuck-open faults as test vectors developed specifically for those fault models [181]. The big question is what value of *N*-detect is needed to ensure these results. From the limited data that has been accumulated thus far, it appears that *N* should be a minimum of 3 to 5. The high fault coverage obtained for other fault models using *N*-detect test sets developed for elementary logic gate-level fault models is analogous to the high fault coverage obtained for multiple fault models using single fault model test vectors that provide high fault coverage [176].

These N-detectability results are extremely important in BIST applications since (as we shall see in later chapters) the test patterns produced by the TPGs in most BIST implementations tend to generate many different test vectors. This increases the probability of detecting each fault multiple times. As a result, using the gate-level fault model for fault simulations is an effective approach to evaluating different BIST implementations for a given application. This requires that fault dropping be suppressed during fault simulation such that the N-detectability of the set of test vectors can be determined. This also assumes the fault simulator is capable of indicating the number of times a fault is detected during the set of test vectors. In addition, to increase the value of N-detectability, the BIST sequence will be longer; in fact as much as N to $N\log(N)$ times longer [313]. While both of these factors increase the fault simulation time, N-detectability provides valuable insight into the ability of a given BIST sequence to provide detection of other fault models. Therefore, the additional fault simulation time is a good investment when considering a single fault simulation using the gate-level stuck-at fault model compared to running multiple fault simulations for different fault models. This is because the gate-level fault model can be simulated more efficiently than some of the other fault models we have discussed. More importantly, BIST approaches that provide a higher value of N-detect will be the best candidates for implementation in any given application.

CHAPTER 3 Design for Testability

This chapter presents a basic overview of Design for Testability (DFT) including ad-hoc techniques, full and partial scan design techniques, and Boundary Scan (the IEEE 1149.1 Standard). The intent is to provide the necessary background needed to understand the advantages and disadvantages of BIST compared to the other DFT techniques, as well as specific issues related to the implementation and evaluation of various BIST approaches. For example, some BIST approaches incorporate aspects of the other DFT techniques.

3.1 Overview of Design for Testability

In general, DFT techniques seek to improve the testability of a circuit by including additional circuitry that improves the controllability and observability of the circuit under test (CUT). While the additional DFT circuitry may increase design time, the resultant reduction in test development time usually overcomes the added design time to the extent that a reduction in time-to-market is realized. Some DFT techniques minimize additional design time by providing automatic synthesis of the DFT circuitry in the CUT. In addition, some DFT techniques minimize test development time by providing automatic test pattern generation (ATPG) for all detectable faults in the circuit (also identification of undetectable faults) for near 100% fault coverage. Still other DFT techniques (specifically BIST approaches) facilitate internal test pattern generation and output response compaction within the CUT to further minimize testing time. Collectively, DFT techniques can provide significant reductions in manufacturing testing time and costs, as well as a reduction in testing time and costs over the entire product life-cycle when made accessible for system-level use.

Design methodologies that maximize the effectiveness of DFT select the testing strategy from the very start of the project. In these methodologies, the testing strategy and DFT technique are jointly selected by designers ***and*** test engineers, or the DFT and testing strategy is established for the entire project (even the entire company in some cases). In many successful design methodologies, the designers have been charged with the responsibility of achieving the targeted goals for fault coverage which forces the designers to learn and consider DFT.[1] Conversely, it is typically short-sightedness on the part of designers and managers when testability and testing is not considered early in the product development cycle. The area overhead and performance penalties associated with the incorporation of DFT circuitry are the most frequent excuses, along with the old saying: "Testing is not my problem, that's the test engineer's problem!" This is not to imply that DFT is a "free lunch" because the area overhead and performance penalties are real; but these penalties should be considered as evaluation criteria for the selection of the appropriate DFT approach instead of reasons not to use DFT. In addition to area and performance penalties, one should also consider other issues including package requirements in terms of power dissipation and pins, fault coverage for appropriate fault models, test application time, CAD tool support, automatic test equipment (ATE) costs, and life-cycle usefulness when comparing and evaluating the various DFT techniques. There have been some important findings from case studies done by various companies regarding designs incorporating DFT. Some of these findings include:

1. The cost of conventional test development without DFT is about equal to that of the actual design cost, while the test development cost for designs that incorporate DFT is about 40% of the conventional test development costs for non-DFT designs [75].
2. The reduction in manufacturing costs for products that incorporate DFT is greater than additional design costs required to incorporate DFT [123].
3. ATE costs for ASICs that incorporate BIST are about 1/3 to 1/6 the cost of ATE for ASICs that do not incorporate BIST [335].
4. Projects that are over-budget but on-time lose much less profit than projects that are on-budget but are late with respect to the target market window [51].

1. This design methodology initiated my start in DFT and BIST in 1980. While there is a considerable learning curve associated with this approach, I firmly believe that it is the best design methodology for ensuring testable designs, systems, and products.

While the last finding is not specifically related to DFT, a project can easily fall behind schedule due to testing related issues that could have easily been solved by consideration of DFT early in the product development cycle. Once DFT techniques are incorporated, they often provide additional benefits that were not initially considered or intended. For example, DFT can often assist in yield enhancement since it can help identify fabrication and manufacturing process problems as well as defective components. At the system-level, DFT techniques can improve system diagnosis as well as repair time and costs, resulting in less system down-time. Accurate and efficient test and diagnosis can have a beneficial impact on time-to-market and system life-cycle costs, but this impact is difficult to quantify and is often ignored.

There are three main types of DFT approaches for digital circuits. These include ad-hoc techniques, scan design techniques, and BIST [4]. Ad-hoc and scan design techniques will be discussed in the subsequent sections along with an overview of Boundary Scan. In practice, different DFT techniques may be used in different portions of the CUT or multiple techniques may be combined to improve the effectiveness of DFT. As we shall see in subsequent chapters, ad-hoc techniques, scan design techniques, and Boundary Scan often interface to, or become an integral part of, BIST approaches.

3.2 Ad-Hoc Techniques

The first DFT approach is referred to as *ad-hoc* DFT since it targets only difficult-to-test portions of the CUT in order to improve their controllability and observability. This is usually done by incorporating multiplexers internal to the CUT to create one or more test modes of operation in which the primary inputs and outputs provide access to/from the internal difficult-to-test circuits via the multiplexers. For controllability only, gates can be used to reduce the area overhead and performance penalty from that incurred with a multiplexer. For example, the N-bit counter in Figure 3.1a requires at least 2^N clock cycles during testing in order to see that the counter goes through its full count sequence and produces a carry-out as the counter rolls-over. By partitioning the counter into two $N/2$-bit counters and inserting a multiplexer, as shown in Figure 3.1b, we can test the counter in just over $2^{N/2}$ clock cycles by applying an active carry-in signal to the second counter via the multiplexer. This is a reduction in testing time on the order of $2^{N/2}$ at the cost of a multiplexer and extra primary inputs to control the test mode and provide test data to the internal node. Alternatively, an OR gate with an active high test mode input can be used to force a logic 1 on the internal node that is difficult to control otherwise (in this case, the carry-in to the second counter assuming the carry-in is active high as shown in Figure 3.1c). In

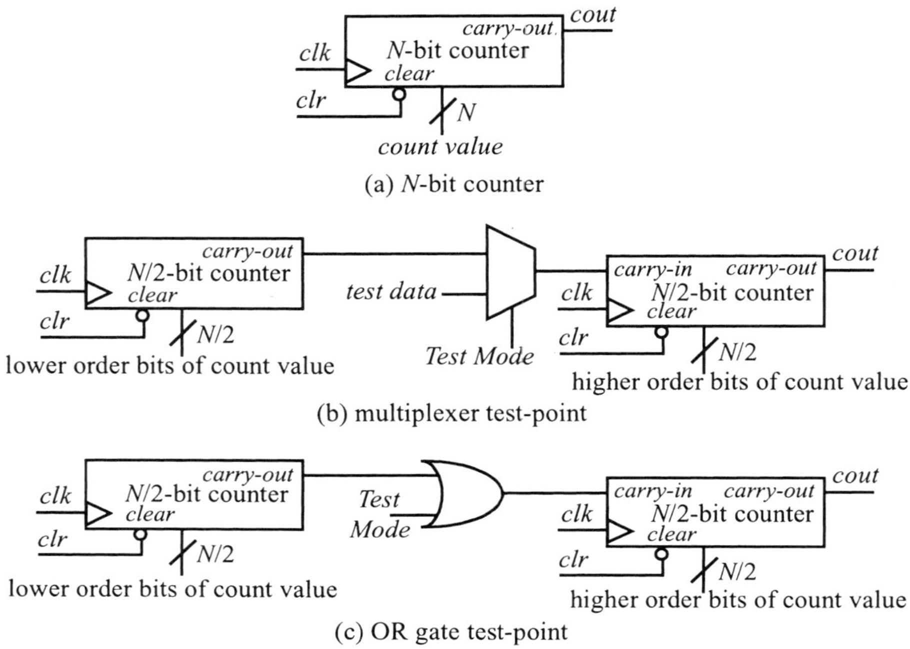

FIGURE 3.1 Examples of ad-hoc DFT.

this example, the insertion of multiplexers or gates (also referred to as *test point insertion*) improves the controllability of the CUT. Multiplexer based test-points can also be used to improve the observability of the CUT by inserting the multiplexers at primary outputs such that internal nodes in the CUT can be observed directly at the primary outputs during the test mode. Gate-based test-points results in lower area and performance penalties compared to multiplexers, but they are typically only useful for improving controllability. As illustrated in Figure 3.2, exclusive-OR gates can be used to linearly combine the logic values at multiple internal nodes that can, in turn, be observed at a primary output via a test point multiplexer.

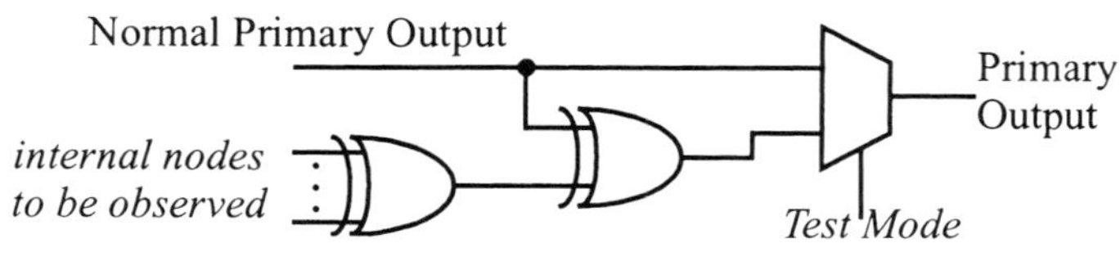

FIGURE 3.2 Observability test-point.

Since test points are inserted only where needed, control of the area overhead can be maintained and one can trade-off area overhead for fault coverage. Critical timing paths in the CUT can be avoided for test point insertion to eliminate performance penalties. Creative ad-hoc testability solutions applied on a case-by-case basis to the difficult-to-test portions of the CUT can yield significant increases in fault coverage, along with significant reductions in test time. The down side of ad-hoc DFT is the number of additional I/O pins required for test modes, and the fact that we must determine the best place to insert test points on a case-by-case basis. Since there are few CAD tools to assist in test point insertion, the ad-hoc DFT process is usually a manual one and can take considerable time. The process typically consists of inserting test-points, developing test vectors, running fault simulations, evaluating fault coverage, and repeating the entire process until the desired fault coverage is obtained.

Typical targets for test point insertion include feedback loops, large counters (as illustrated in the example of Figure 3.1), embedded core logic, asynchronous logic, clock generation circuits, memory initialization inputs, and intentional redundant logic for system purposes [4]. Ad-hoc DFT techniques can provide a number of benefits to the design verification process, including providing easy initialization of the design as well as partitioning the design to provide access to embedded circuits, as illustrated in Figure 3.3. In the latter case, existing test vectors previously developed for the core logic can then be used to test the embedded core. While the example in Figure 3.3 may not be very realistic from the standpoint of many of the applications we encounter, it does illustrate some system-level considerations that must be addressed. As long as there are more primary inputs and outputs than inputs to and outputs from the embedded core logic, only one additional input pin is needed for the *Test Mode* input. However, we are inserting multiplexers between the input logic and the embedded core which will add a multiplexer delay of performance penalty. In addition, we are

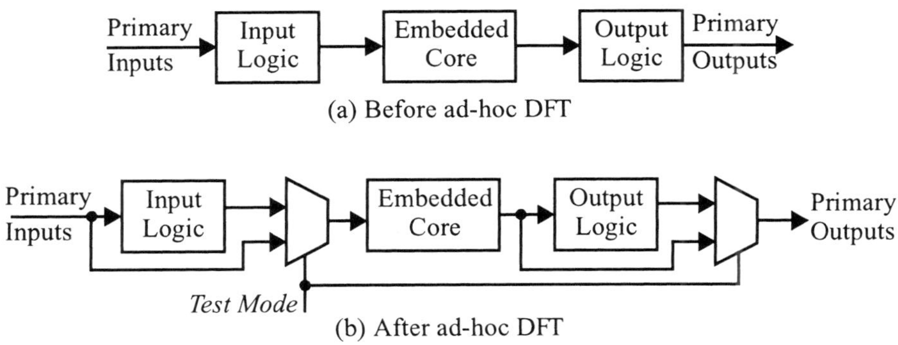

FIGURE 3.3 Ad-hoc DFT for embedded core logic.

adding multiplexers at the primary outputs which will add a multiplexer delay to the clock-to-output delay to the overall circuit. Therefore, we should avoid those primary outputs with the most critical timing requirements. The performance penalty at the primary inputs is not as serious since we are only increasing the loading, but we should start with the primary inputs that are in the least critical timing paths.

3.3 Scan Design Techniques

The next major DFT approach is referred to as *scan design* [50]. It is also frequently referred to as Level Sensitive Scan Design (LSSD) based on its original implementation using level sensitive latches [80]. In scan design-based DFT, the flip-flops of a sequential logic circuit are transformed into a shift register by adding a multiplexer to the input of each flip-flop. The transformation of a general sequential logic circuit to a scan design is illustrated in Figure 3.4. One input to the multiplexer is connected to the normal system logic and the other input to the multiplexer is connected to the output of another "scan" flip-flop. The select input to the multiplexer then controls the mode of operation: either system mode or shift register mode (usually referred to as scan mode or test mode). The *Scan In* input to the multiplexer of the first flip-flop in

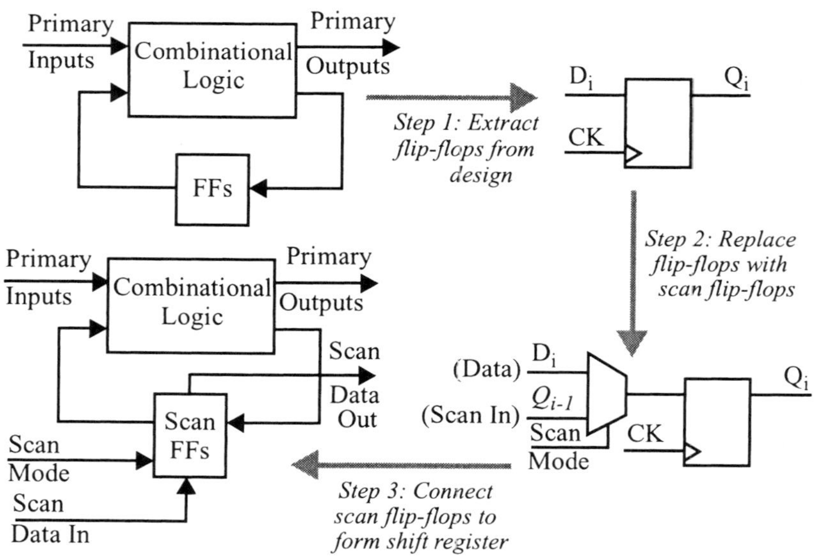

FIGURE 3.4 Implementation of scan design-based DFT.

the shift register chain is connected to a primary input to "scan in" test data. The output of the last flip-flop in the shift register chain is connected to a primary output to "scan out" test results. As a result of this scan chain, all flip-flops are easily controllable and easily observable. Therefore, the problem of testing a sequential logic circuit is reduced to simply that of testing the combinational logic. The excellent controllability and observability of the flip-flops in the scan chain also provide good controllability and observability of the embedded combinational logic of the CUT.

There are many CAD tools available to perform ATPG for combinational logic that can achieve 100% single stuck-at gate-level fault coverage for all detectable faults. Some CAD tools will even identify any combinationally redundant faults in the logic circuit to avoid undetectable faults through further logic minimization. Scan design-based DFT can be completely automated, including insertion of the scan flip-flops, interconnection of the scan chain, generation of the test vectors, and analysis of the fault coverage (with no fault simulation of the sequential circuit required). As a result, scan design-based DFT is probably the best overall DFT approach ever developed.

The most optimized approach to scan design automation, in terms of minimizing area overhead and performance penalties, is a two step synthesis approach [14][50]. Once the gate-level design of the VLSI device is complete, a pre-layout step replaces regular flip-flops with scan-based flip-flops. Connections are made to the clock input, the Q and $\overline{Q}$ outputs, and the normal system data input to the scan-based flip-flop. However, the *Scan Mode* and scan chain connections between the flip-flops are not made at this time. Physical layout is then performed giving all timing considerations to the system function and not the scan chain. In a post-layout step, the scan chain connections are made by daisy-chaining the scan-based flip-flops in each row of cells (here we are assuming standard cell or gate array implementation media) such that only one routing track for each pair of rows of cells is required for the scan chain connections. Similarly the *Scan Mode* connections to all flip-flops are made using only one routing track for each pair of rows of cells. As a result of this approach, performance penalty and area overhead are at their absolute minimum. Test vectors are then automatically generated by ATPG software based on the post-layout scan chain connections and ordering.

A test vector for scan design is a test vector for the combinational logic of the CUT and consists of two parts: the portion of the test vector to be applied by the flip-flops of the scan chain and the portion of the test vector to be applied at the primary inputs to the CUT. Similarly, the output response of each test vector consists of a portion that drives the *D*-inputs to the flip-flops and a portion that drives the primary outputs of the CUT. The basic idea of the testing process with scan design-based DFT is to put the CUT into the scan mode of operation and shift in the flip-flop portion of each test vector. Once the test vector is aligned in the scan flip-flops, the primary input portion of the test vector is applied while the system mode of operation is activated for one clock cycle. This applies the complete test vector to the combinational logic while the output responses from the combinational logic appear at the primary outputs and are clocked into the scan flip-flops (replacing the original test vector). The CUT is then placed back into the scan mode and the test results are shifted out for comparison with the expected responses of the fault-free circuit. While the output responses are being shifted out, the flip-flop portion of the next test vector can be shifted in. This process continues until all test vectors for the combinational logic have been applied and the associated output responses retrieved (all via the scan chain).

The flip-flops are completely tested (with the exception of flip-flop reset and/or preset control lines) by applying a *toggle sequence* of alternating pairs of 0s and 1s (00110011...) through the scan chain until the first eight bits of the toggle sequence are observed on the *Scan Data Out* output [50]. An 8-bit toggle sequence can be prepended to the first test vector to test the flip-flops while the flip-flop portion of the first test vector is being shifted into the scan chain. As a result, the flip-flops are completely tested prior to application of the first test vector to the combinational logic. The reset and preset inputs to the flip-flops can also be easily tested. For example, the reset capability can be tested by shifting an all 1s test pattern into the flip-flops via the scan chain, applying the reset signal, and shifting out the response. The test time in terms of the number of clock cycles, N_{CC}, required to completely test the CUT is given by:

$$N_{CC} = N_{TV} \times (N_{FF} + 1) + N_{FF} + 8 \tag{3.1}$$

where N_{TV} is the number of test vectors needed to test the combinational logic of the CUT, and N_{FF} is the total number of flip-flops in the CUT. The additional 8 clock cycles are for application of the toggle sequence to test the flip-flops of the scan chain. The test sequence for scan design is summarized as follows:

Step 1: Apply a *toggle sequence* to test the scan chain flip-flops (*Scan Mode* = 1).

Step 2: Scan (shift) in the flip-flop portion of the test vector (*Scan Mode* = 1).

Step 3: Apply the remaining primary input portion of the test vector to the primary inputs and observe combinational logic output responses at primary outputs.

Step 4: Apply one clock cycle while in the system mode of operation to capture the combinational logic output responses in the scan chain (*Scan Mode* = 0) and observe combinational logic output responses at primary outputs.

Step 5: Scan (shift) out the results of the current test vector and scan in the flip-flop portion of the next test vector (*Scan Mode* = 1).

Step 6: Go to Step 3 and continue until all test vectors have been processed.

The area overhead and performance penalties associated with scan design are due to the multiplexers added to the inputs of each flip-flop. There is a 2-to-1 multiplexer delay (one or two gate delays depending on the construction of the multiplexer) in every path between flip-flops. The area overhead is dependent on the combinational logic to flip-flop ratio of the CUT. There is also routing overhead for the scan chain connections and the *Scan Mode* control signal. Typical area overhead values are on the order of a 2% to 10% increase in the total chip area [50]. While there are three new pins introduced for the scan design implementation (*Scan Data In*, *Scan Mode* control, and *Scan Data Out*), practical applications typically require only one additional pin for *Scan Mode* since *Scan Data In* and *Scan Data Out* can usually be shared with existing primary inputs and outputs.

Scan design-based DFT does have drawbacks, including long test application time due to the serial application of the test vectors and retrieval of test results. This test time can become significant for PCB-level testing where hundreds of devices with thousands of flip-flops in many different chips are daisy-chained together to form a single, and very long, scan chain. For system-level use of scan design-based DFT, the test patterns and expected responses must be stored in the system. Another problem at system-level testing is the difficulty of applying the tests at the system operating frequency. Since the *Scan Mode* select signal has as many loads as the system clock signal, another "clock" buffer must be designed and incorporated to handle the loading on Scan Mode so that it can be switched between test mode and normal operation at the system operating frequencies for test vector application and expected response capture [69]. Another problem is the result of test patterns shifting through the scan chain between the application of each vector. This makes detection of transistor stuck-off faults and delay faults extremely difficult, if not impossible. Transistor stuck-off fault models require a two vector sequence for detection, but the shifting data in the scan chain between vectors destroys the condition set-up by the first vector for detection of the fault by the second vector. This can be overcome by implementing a double-latched scan chain (similar to the Boundary Scan cell discussed in the next section), but the area overhead can be significant [162]. Therefore, scan design-based DFT is most frequently used for gate-level stuck-at fault and bridging fault detection.

There are a number of restrictions placed on circuits that are candidates for scan design implementation [50]. The circuit must use edge-triggered, D-type flip-flops that are clocked from primary inputs. These restrictions are required in order to implement and correctly operate the scan chain shift register in both normal and scan modes of operation. This usually means that clocks cannot be gated [50]. However, the restrictions or "rules" placed on circuits for scan design implementations represent good design practices for sequential logic design. Some scan design CAD tools provide audit software to determine any rule violations that will prevent proper implementation and operation of this DFT approach. Despite its drawbacks, scan design can be used at all levels of testing and can be applied hierarchically to chips, PCBs, and systems.

3.3.1 Multiple Scan Chains

Testing time can be reduced by creating multiple scan chains of equal length such that there are fewer clock cycles required to scan in test vectors and scan out test results between application of the test vectors to the CUT. Multiple scan chains also provide a good solution to circuits with multiple clocks and/or multiple edges of the same clock since the basic scan design approach is oriented toward single-clock, single-edge, synchronous design. In this case, one or more scan chains would be associated with each clock and/or clock edge. As long as the clocks can be manipulated independently during manufacturing testing, the multiple scan chains can be used to test the combinational logic. The multiple scan chains will also require multiple *Scan Data In* and *Scan Data Out* pins, a separate pair for each scan chain (usually obtained by using existing primary inputs and outputs), but in general a single *Scan Mode* pin can be used to service all scan chains.

3.3.2 Partial Scan Design

The area overhead and performance penalties can be reduced by using *partial scan design* techniques where only a subset of the flip-flops in the CUT are replaced with scan flip-flops that are connected to form the scan chain [59]. As a result, scan flip-flops can be avoided in critical timing paths to reduce or eliminate the performance penalty introduced by the multiplexer. Some of the techniques used to select which flip-flops should be replaced with scan-based flip-flops include: 1) structural (select flip-flops to cut feedback loops) [59], 2) ATPG-based (select flip-flops that are useful for ATPG) [96], and 3) testability-based (select flip-flops to maximize fault coverage) [92]. Most partial scan approaches typically require sequential ATPG software since the non-scan portion of the CUT is sequential rather than combinational as in full scan. However, there are partial scan techniques that avoid the need for sequential ATPG [96]. In addition, partial scan design trades fault coverage for reduced area overhead which is often on the order of 50% to 70% of full scan [50]. Since there are fewer flip-flops in the scan chain, it would seem that testing time would be reduced as well, but partial scan usually requires more test patterns than full scan (due to the sequential logic to be tested as opposed to combinational logic in full scan). Other issues associated with partial scan include limited CAD tool support for synthesis and test pattern generation. The big question in partial scan implementation is which flip-flops should be incorporated into the scan chain for maximum fault coverage with minimum area and performance penalties; only a few CAD tools have been developed to do this analysis automatically.

3.4 Boundary Scan

The success of scan design led to the application of scan design-based DFT techniques for testing interconnect and solder joints on surface mount PCBs. This became known as Boundary Scan. Following a proposal by the European Joint Test Action Group (EJTAG) in the mid 1980's, representatives from all over the world quickly joined the renamed Joint Test Action Group (JTAG - a term sometimes used interchangeably with Boundary Scan today) [162]. Boundary Scan eventually became the IEEE 1149.1 Standard and now provides a generic test interface, not only for interconnect testing but also for access to DFT features and capabilities (including BIST) within the core of a chip (see Figure 3.5a) [258]. The standard Boundary Scan interface includes four I/O pins for *Test Clock* (TCK), *Test Mode Select* (TMS), *Test Data In* (TDI), and *Test Data Out* (TDO). A *Test Access Port* (TAP) controller is included to access the Boundary Scan chain and any other internal features designed into the device, such as access to internal scan chains or, in the case of FPGAs and CPLDs, access to the configuration memory. The TAP controller is a 16-state finite state machine (FSM) with the standardized state diagram given in Figure 3.5b where all state transitions occur on the rising edge of TCK based on the value of TMS [161]. This facilitates common access to any device that incorporates the 1149.1 Standard interface. As a result, Boundary Scan is supported by many manufacturers, CAD tools, and ATE vendors.

The Boundary Scan cells used for testing the interconnect on a PCB are illustrated in Figure 3.6 where the basic design of a Boundary Scan cell and its functional operation are given in Figure 3.6a. A more complex construction of the triple Boundary Scan cell for a bi-directional I/O buffer is illustrated in Figure 3.6b. The bi-directional buffer illustrates the need for the double-latched scan chain design of the basic Boundary Scan cell. This double-latching action of the Boundary Scan cell prevents back driving of other bi-directional buffers on a PCB as a result of shifting in test patterns and shifting out the test results. The Update flip-flop holds all values (including the tri-state control value) stable at the pads during the shifting process. Once the test pattern is shifted into the Boundary Scan chain via the Capture flip-flops, the Update Data Register (DR) state of the TAP controller (see state diagram in Figure 3.5b) transfers the test patterns to the Update flip-flops for external application to the PCB interconnect. A similar double latching system of scan chains would be required for complete transistor-level testing and delay fault testing using scan design-based DFT.

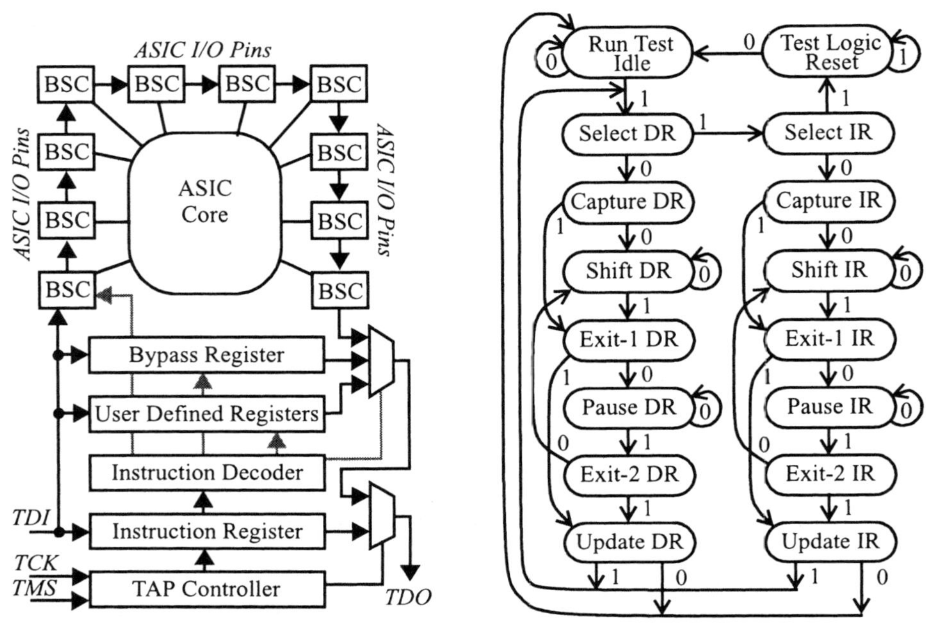

(a) Boundary Scan implementation (b) TAP controller state diagram

FIGURE 3.5 Boundary Scan architecture and state diagram.

It is important to realize that Boundary Scan by itself is primarily a board-level DFT technique for testing interconnect on PCBs and not necessarily an internal chip DFT technique.[1] But the board-level testing problems that Boundary Scan solves (including components mounted on both sides of the board as well as floating vias and micro-vias) are significant. The Boundary Scan interface and circuitry can be used at all levels of testing and Boundary Scan can be hierarchically applied to the entire system. In addition, Boundary Scan provides a standardized interface for maintenance and testing functions, including access to BIST through the implementation of User Defined Registers shown in Figure 3.5a. While the TAP controller does not need to be modified for these registers, the Instruction Register, Instruction Decoder, and output multiplexer for TDO will require application specific design.

1. Many times I have been asked the question, "If we include Boundary Scan, our chip will test itself, right?" Wrong! This is a misconception based on an optional RUNBIST instruction in the IEEE 1149.1 Standard; you have to implement BIST in order to "RUNBIST".

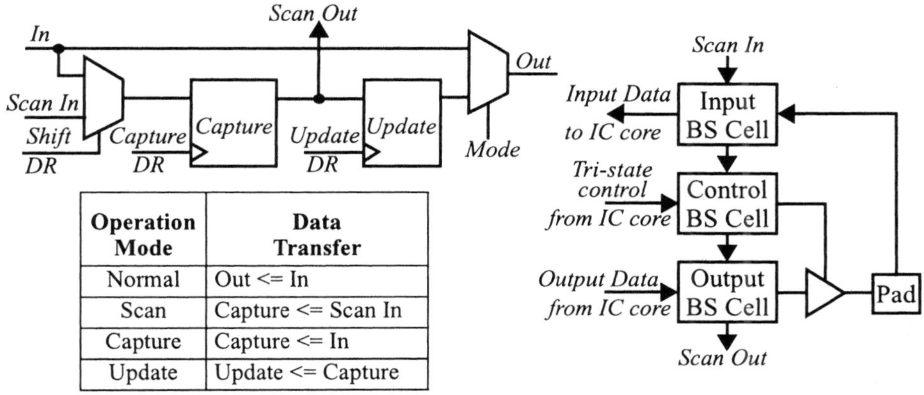

Operation Mode	Data Transfer
Normal	Out <= In
Scan	Capture <= Scan In
Capture	Capture <= In
Update	Update <= Capture

(a) Basic Boundary Scan cell

(b) Bi-directional buffers require multiple Boundary Scan cells

FIGURE 3.6 Boundary Scan cells.

The design of the basic Boundary Scan cell shown in Figure 3.6a also facilitates application of test patterns at the input buffers to the internal core of the VLSI device, as well as capture of output responses from the internal core logic at the output buffers. This internal test, or INTEST as the instruction is referred to in the IEEE 1149.1 Standard, is a valuable instruction (particularly for system-level use of BIST as we shall discuss in subsequent chapters), although it is not always implemented in a effort to reduce the area overhead and performance penalty associated with Boundary Scan. The external test for PCB interconnect, referred to as EXTEST in the IEEE 1149.1 Standard, is a mandatory part of Boundary Scan in order to test the PCB interconnect for which Boundary Scan is intended. Another mandatory instruction and feature in the IEEE 1149.1 Standard is the BYPASS instruction which uses the Bypass Register shown in Figure 3.5a. The Bypass Register consists of a single flip-flop and allows the entire Boundary Scan chain to be bypassed in order to provide faster access (in terms of the number of clock cycles required to shift data through the serial chain) to other devices on the board with daisy-chained Boundary Scan interfaces.

The area overhead associated with Boundary Scan can be significant as implied by Figure 3.5a and Figure 3.6. For example, for each I/O pin we add two flip-flops and two multiplexers. When adding the four I/O pins and their associated buffers and pads to the device for the 4-wire interface, these I/O pins do not incorporate Boundary Scan cells so that they can be manipulated independently of the test vectors being shifted and applied. The TAP controller, Instruction Register, Instruction Decoder,

Bypass Register, and output multiplexer also have considerable impact on the area overhead. Performance penalties of one multiplexer delay are incurred at every input and output pin. While this performance penalty and area overhead for each I/O pin can be reduced by removing logic that will support the INTEST instruction, this circuitry can be valuable in many BIST approaches and saves area overhead for those implementations. With additional area overhead, Boundary Scan can facilitate (via the User Defined Registers) other DFT techniques, including scan design-based DFT (but only single scan chains can be implemented) and BIST. While Boundary Scan cannot run at system speed, we can access BIST which can run at system speed.[1]

3.5 BIST

The final DFT approach is BIST, which incorporates test pattern generation and output response analysis capabilities in the chip (or on the PCB) itself. Other than initiating the BIST sequence, and reading the BIST results at the end of the sequence, there is no need for external manipulation or monitoring of the device during the testing sequence, as is the case in all of the other DFT techniques. While this external intervention for the other DFT approaches prevents testing at the system operating frequency in many cases, BIST can easily facilitate testing at system speed (at-speed testing) if implemented properly [244]. BIST can also be used at all levels of testing, from wafer-level to system-level (vertical testability). These are the main advantages BIST has over the other DFT techniques. It was the limitations of ad-hoc and scan design-based DFT that led to the quest for BIST, circa 1980 [245]. Other advantages include the ability to provide efficient and cost effective burn-in testing, faster testing time, less expensive ATE at device and board-level testing, and system-level diagnostic resolution to the device or subcircuit. While the test patterns for ad-hoc and scan design primarily use deterministic vectors developed to detect specific faults (essentially a 1-detect test set), BIST approaches typically generate many test patterns which lead to N-detect test sets where $N > 1$. As a result, BIST is more likely to provide high fault coverage for a wide variety of fault models due to the multiple-detect test sets in conjunction with testing performed at-speed [176].

DFT techniques are often categorized into structured and non-structured techniques. Scan design and Boundary Scan fall into the structured category with ad-hoc techniques obviously falling into the non-structured category. Some people refer to BIST

1. There is an optional fifth pin, TRST, which is an input that resets the TAP controller to the Test-Logic Reset state. However, this state can easily be reached from anywhere in the state diagram by applying five TCK cycles while holding TMS=1. Hence, TRST is rarely included.

as an unstructured (or ad-hoc) technique since the choice of BIST implementation for a given CUT must be selected on a case-by-case basis. Other people refer to BIST as a structured technique since many of the BIST approaches developed have been successfully automated in a manner similar to scan design. In a sense, both schools of thought are correct. Different types of circuit structures require different types of BIST approaches, but this can be viewed as another advantage to BIST. For example, scan design-based DFT is great for general sequential logic in VLSI devices but cannot be efficiently used in structures like RAMs. While there are different BIST approaches to test general sequential logic and RAMs, BIST in general does cover more types of CUTs than the other DFT techniques. But BIST is an off-spring of the ad-hoc and scan design-based DFT approaches (just as Boundary Scan was an off-spring of scan design) and, as a result, many BIST approaches incorporate some of the best features of these other DFT approaches. The principle disadvantage of BIST compared to the other DFT approaches is that full scan design is fully automated and can ensure near 100% fault coverage without the need for sequential fault simulation, while many BIST approaches require varying amounts of design effort, and fault simulation is usually a necessary and essential part of the evaluation of the effectiveness of many BIST approaches.

3.6 Evaluation Criteria for DFT Approaches

The most obvious criteria for evaluating different DFT approaches, including BIST, are area overhead and performance penalties [198]. Various methods for calculating area overhead are summarized in Table 3.1. The first method is the increase in chip area as a result of the incorporation of DFT or BIST and represents area overhead in its truest sense. This is not the way area overhead is typically calculated since this would require physical layout of the device both with and without DFT or BIST. While the increase in chip area does influence the cost, a pad-limited chip may not experience any increase in chip area as a result of DFT, but the additional DFT circuitry will increase the active area of the device which, in turn, increases the area of the device that is susceptible to defects. A more realistic method for expressing the area overhead is given by the second method in Table 3.1. This method is based on the total number of gates in the design with and without DFT. The most frequently used method for calculating and reporting area overhead is given by the fourth method, based on the number of gates added to a design for DFT and the total number of gates in the design with DFT.

Since gates vary in size in terms of the number of transistors and the resultant area they require for implementation, a more meaningful method for determining area overhead is to use the number of gate I/O. In static CMOS implementations, the number of transistors is twice the number of inputs to the gate. In NMOS implementations, on the other hand, the number of transistors is equal to the number of inputs, plus the load transistor. Therefore, the total number of gate inputs gives an accurate relationship to the area of a CMOS implementation while the total numbers of gate inputs and outputs gives an accurate relationship to the area of an NMOS implementation.[1] The area overhead calculation would then be determined by either of the last two methods in Table 3.1 depending on the type of implementation (CMOS or NMOS).

TABLE 3.1 Methods for calculating area overhead.

Method	Equation
chip area	$AO = \dfrac{\text{area of chip with DFT}}{\text{area of chip without DFT}}$
number of gates	$AO_{G1} = \dfrac{\text{total gates with DFT}}{\text{total gates without DFT}}$
a frequently used method	$AO_{G2} = \dfrac{\text{total gates for DFT}}{\text{total gates without DFT}}$
most frequently used method	$AO_{G3} = \dfrac{\text{total gates for DFT}}{\text{total gates with DFT}}$
number of gate inputs	$AO_{GI} = \dfrac{\text{total gate inputs for DFT}}{\text{total gate inputs with DFT}}$
number of gate I/O	$AO_{GIO} = \dfrac{\text{total gate I/O for DFT}}{\text{total gate I/O with DFT}}$

The size of multiplexers and exclusive-OR gates is sometimes difficult to determine in area overhead calculations because they vary in size depending on their construction. For example, exclusive-OR gates can vary between 6 and 16 transistors while 2-to-1 multiplexers can vary between 4 and 12 transistors, depending on how they are implemented in terms of gates and/or transistors. They are often considered as a single gate but this becomes misleading when considering area overhead. Therefore, in the remaining chapters, area overhead and performance penalty considerations will be broken down into multiplexers, exclusive-OR gates, and elementary logic gates in an effort to account for these variations in construction.

1. While we think of current VLSI implementations as being CMOS, NMOS-type designs continue to be alive and well since dynamic CMOS circuits and many regular structures such as RAMs, PLAs, and large multiplexers (like those used in the programmable interconnection networks in FPGAs and CPLDs) tend to be structured as NMOS circuits. This is done in order to minimize area.

Performance penalty is an important consideration, particularly in the case of "catalog" or "off-the-shelf" devices where speed sorting takes place and the chips are priced based on the performance results. The addition of gates for DFT in any critical timing path will reduce the maximum operating frequency of the device. In addition, the increased area of the device will tend to lengthen wires causing additional delay penalties. ASICs, on the other hand, are typically designed to meet some specified system clock frequency. As long as the insertion of DFT circuitry in the critical timing paths does not prevent system operation at the specified frequency, one may consider the DFT implementation to have no performance penalty.

The CAD tool support for a given DFT approach is a serious consideration with respect to the efforts that will be required during the design and test development processes. The ability to automatically insert a DFT approach can save considerable design time as well as reduce the risk of introducing a design error as a result of manual implementation of the DFT technique. In this case, full scan design and Boundary Scan have the upper hand. A useful CAD tool that has been developed for scan design-based DFT techniques is an audit of the circuit to determine if there are any design features (such as gated clocks, internal tri-state busses, or asynchronous circuitry) that could prevent the proper implementation and operation of scan design. Any such audit or analysis program can save considerable design time. ATPG is also important from the standpoint of test development time for some DFT approaches. Consideration must still be given to how long ATPG will need to run for the given design and DFT technique.

Fault simulation is a major concern with many DFT approaches since most will require fault simulation in order to evaluate the quality of the test vectors. As a result, we must consider the availability and capabilities of the fault simulation tools including fault simulation time, ability to suppress fault dropping and determine N-detectability, fault models supported, and support of different HDLs. A related issue is the consideration of the various fault models to be simulated and their impact on fault simulation time. Finally, we must consider the fault coverage that can be obtained with candidate DFT techniques for the various fault models. All of these considerations can have a major impact on test development time and cost.

Test application time and the cost of ATE that will be needed to test the fabricated device or PCB are very important when evaluating DFT techniques. Both of these factors can be impacted by the memory requirements that will result from the complete set of test vectors that will be applied to the device or PCB. The test vectors that must be applied are a direct result of the DFT technique chosen for implementation (or the lack of DFT implementation). The ability of a DFT approach to be used for other areas of testing such as burn-in testing, unit or system-level testing and diagnosis, and the PCB repair process are also important considerations when evaluating different DFT approaches. These issues directly impact the manufacturing testing costs.

Different DFT approaches can have varying impact on package requirements, including power dissipation and the number of additional pins that may be required. When some of the pins on a package are unused, we can use these pins for DFT without having to change the package to a more expensive one with a higher I/O pin count. However, another consideration is that the additional pins used will also require additional routing on the PCB. Power dissipation resulting from some DFT implementations, particularly BIST, is often ignored but can be a serious problem if the data activity during the testing sequence is much higher than the normal expected activity during system operation. As a result, the DFT technique can be the determining factor in the temperature rating for the device package, heat sinks associated with the package, the physical location of the device on the PCB, power supplies in the unit or system, and even in ambient temperature requirements that determine if we need forced air or air conditioning in the system.

An interesting but important consideration is "How easy is the DFT circuitry to test?" Scan design-based DFT techniques are well tested during the process of testing the combinational logic in the CUT. Some BIST structures, on the other hand, are not completely tested during the BIST sequence itself so that additional test vectors must be applied to test the BIST circuitry (examples of these structures will be given in Chapter 5). In these cases, additional DFT approaches, such as ad-hoc techniques, must be incorporated in order to completely test the BIST circuitry. If one has to incorporate extra DFT circuitry and/or write extra test vectors for the BIST circuitry, then serious consideration should be given to whether that particular BIST approach is worth implementing or if there is a better BIST approach for the application.

Considerations of the life-cycle usefulness of a DFT approach including system-level testing, field support, maintenance, diagnostics, and repair are extremely important. The effects of these considerations are hard to quantify and are often ignored. This is particularly true when the end product for a company or a division is a device rather than a system since it is even more difficult to know what the system requirements will be for that device over its life-cycle in different products. In these situations, manufacturing testing tends to be the only concern for DFT considerations with even more emphasis on minimizing area overhead and performance penalties. But these devices must be tested in the system in many applications. Incorporating DFT approaches that are undocumented and not accessible by the user at the system-level may be counterproductive to sales of the device and the customer's perception of its quality. Life-cycle usefulness is the area where BIST has the most potential advantage over the other DFT techniques as will be discussed in Chapter 6.

CHAPTER 4

Test Pattern Generation

We begin our detailed look at BIST with one of its most important components, the test pattern generator (TPG). The fault coverage that we obtain for various fault models is a direct function of the test patterns produced by the TPG and applied to the CUT. This chapter presents some of the basic TPG design and implementation techniques used in BIST approaches.

4.1 Types of Test Patterns

There are several classes of test patterns. As a result, TPGs are sometimes classified according to the class of test patterns they produce [4]. These classes of test patterns are described below along with examples of the types of circuits that can be used in a BIST environment to generate these test patterns.

- **Deterministic test patterns** are developed to detect specific faults and/or structural defects for a given CUT. An example of hardware for applying deterministic vectors would include a ROM with a counter for addressing the ROM, as was used in the simple BIST design in Figure 1.5. As pointed out in Chapter 1, this type of approach has limited applicability for BIST. This approach is often referred to as "stored test patterns" in the context of BIST applications.
- **Algorithmic test patterns** are similar to deterministic patterns in that they are specific to a given CUT and are developed to detect specific fault models in the

CUT. However, because of the repetition and/or sequence typically associated with algorithmic test patterns, the hardware for generating algorithmic vectors is usually a finite state machine (FSM). There is considerable applicability of this test pattern generation approach to BIST for regular structures such as RAMs.

- **Exhaustive test patterns** produce every possible combination of input test patterns. This is easy to see in the case of an N-input combinational logic circuit where an N-bit counter produces all possible 2^N test patterns and will detect all detectable gate-level stuck-at faults as well as all detectable wired-AND/OR and dominant bridging faults in the combinational logic circuit without the need for fault simulation. This counter-based approach will not detect all possible transistor-level faults or delay faults since those faults require specific ordering of vectors as well as the ability to repeat certain test vectors within the vector set. As a result, fault simulation will be required to determine the fault coverage for transistor and delay faults. Exhaustive test patterns are not practical for large N.
- **Pseudo-exhaustive test patterns** are an alternative to exhaustive test patterns. In this case, each partitioned combinational logic subcircuit will be exhaustively tested; each K-input subcircuit receives all 2^K possible patterns, where $K < N$, while the outputs of the subcircuit are observed during the testing sequence [172]. As a result, this approach is more practical for large N as long as K is not large. Like exhaustive test patterns all detectable gate-level stuck-at faults and bridging faults within the combinational logic subcircuits are guaranteed to be detected without the need for fault simulation. Bridging faults between the subcircuits are not guaranteed detection, along with transistor and delay faults, requiring fault simulation to determine fault coverage. Candidate hardware for this type of test pattern generation for BIST includes counters, Linear Feedback Shift Registers (LFSRs), and Cellular Automata (CA).
- **Pseudo-random test patterns** are the most commonly produced patterns by TPG hardware found in BIST applications. The primary hardware for producing these test patterns are LFSRs and CA. Pseudo-random test patterns have properties similar to those of random pattern sequences but the sequences are repeatable.
- **Weighted pseudo-random test patterns** are good for circuits that contain random pattern resistant faults. This type of test pattern generation uses an LFSR or CA to generate pseudo-random test patterns and then filters the patterns with combinations of AND/NAND gates or OR/NOR gates to produce more logic 0s or logic 1s in the test patterns applied to the CUT.
- **Random test patterns** have frequently been used for external functional testing of microprocessors [136] as well as in ATPG software [4]. But the generation of truly random patterns for a BIST application is of little value since these test patterns cannot be repeated and since the fault coverage obtained would be different from one execution of the BIST sequence to the next.

The types of test patterns described above are not mutually exclusive. For example, pseudo-random test patterns may also be pseudo-exhaustive. In some BIST applications, multiple types of test patterns are used; pseudo-random test patterns may be used in conjunction with deterministic test patterns. The subsequent sections address test pattern generation from the hardware standpoint, in the context of BIST, with emphasis on minimizing area overhead and maximizing performance. In just about all TPG implementations, a very important capability is initialization of the TPG to some known starting place in the test pattern sequence; unless otherwise specified, we will assume that a reset or preset capability is incorporated into all of the TPGs discussed in the subsequent sections.

4.2 Counters

An N-bit binary counter will generate all 2^N possible test patterns which is sufficient to completely test an N-input combinational logic circuit with no feedback for all detectable gate-level stuck-at faults and bridging faults. However, this type of TPG is only practical for N less than 22 to 25 since N=20 requires over 1 million clock cycles. In addition, the fact that the least significant bit (LSB) toggles every clock cycle while the most significant bit (MSB) only toggles half way through and at the end of the count sequence, makes these two bits of the counter less than desirable for many sequential CUTs. As a result, one must pay attention to the assignment of counter bits to the inputs of the CUT to maximize fault coverage. Alternatively, there are other types of counters that can be considered for TPG applications, such as grey-code counters (to minimize the number of bits that change with each count) and constant-weight counters (to produce an N-out-of-M code, for example 3 bits out of 8 are always logic 1s with the remaining bits being logic 0s). These types of counters would be desirable only for specific types of CUTs that have special test pattern requirements for fault detection. Therefore, a counter by itself is rarely used for the implementation of a TPG in practice. Instead, counters are more commonly used for the test controller or in conjunction with other circuitry for the generation of algorithmic test patterns. Since counters are common components in many system functions, using an existing counter for a TPG helps reduce area overhead of a BIST approach.

4.3 Finite State Machines

Finite state machines (FSMs) are a common TPG implementation for algorithmic test pattern generation. FSMs are also used in conjunction with counters for TPGs. For example, random access memories (RAMs) typically use algorithmic test patterns and there are many RAM testing algorithms [319]. One of these algorithms, the March Y algorithm, is given in Figure 4.1a. During the course of the March Y test, we cycle through all addresses in the RAM four times, writing to and reading from the memory locations as indicated by the algorithm. Note that the address ordering (either ascending or descending addresses) in the first and last loops is a "don't care"; this can be used to help minimize the logic of the FSM (ascending order was used in the pseudocode description of Figure 4.1a). An example TPG implementation using an FSM in conjunction with a counter is given in Figure 4.1b. In this example, the RAM has an active high write enable, N address bits, and M data bits with separate ports for *Data In* and *Data Out*; this is a typical RAM structure used in ASICs and supports pipelined write and read operations. The TPG implementation uses an N+2-bit binary up-counter with an active high count enable (*cen*). The N least significant bits (q_{N-1} to q_0) are used for moving through the 2^N address space. The two most significant bits are used to control which of the four address loops is being executed in the March Y algorithm. In addition, a 2-bit FSM with active high synchronous reset (*clr*) is used to control the read and write operations within the second and third loops of the algorithm. The state sequence for this 2-bit FSM is f_1f_0={00, 01, 10, 00, ...} and can be easily implemented using one-hot-zero encoding. Prior to execution of the BIST sequence, both the counter and FSM would be initialized to all 0s; while not shown in Figure 4.1b, this initialization logic is considered in area overhead and fault coverage

```
for address = 0 to 2^N-1 loop *
    write 0s
end loop
for address = 0 to 2^N-1 loop
    read (expect 0s)
    write 1s
    read (expect 1s)
end loop
for address = 2^N-1 to 0 loop
    read (expect 1s)
    write 0s
    read (expect 0s)
end loop
for address = 2^N-1 to 0 loop *
    read (expect 0s)
end loop
```

* *address can be ascending or descending for these steps*

(a) March Y algorithm

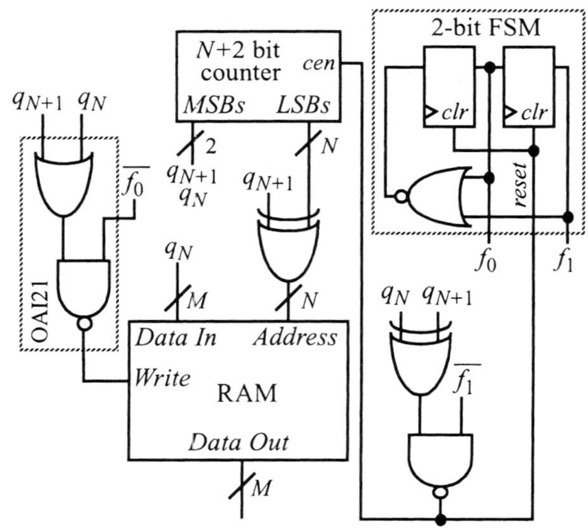

(b) FSM-based TPG for March Y algorithm

FIGURE 4.1 Example of an FSM-based TPG.

analysis discussed at the end of this chapter. During the first and last loops through the RAM addresses, the 2-bit FSM is held in the reset state. During the second and third passes through the RAM addresses, the 2-bit FSM is enabled to cycle such that the TPG performs the read-write-read operations specified by the algorithm.

4.4 Linear Feedback Shift Registers

The LFSR is one of the most frequently used TPG implementations in BIST applications [32]. One reason for this is that an LFSR is more area efficient than a counter, requiring less combinational logic per flip-flop. There are two basic types of LFSR implementations, the internal feedback and external feedback LFSRs illustrated in Figure 4.2.[1] The external feedback LFSR in Figure 4.2b best illustrates the origin of the name of the circuit: a shift register with feedback paths that are linearly combined via the exclusive-OR gates. Internal and external feedback LFSRs are duals of each other [32]. Both implementations require the same amount of logic in terms of exclusive-OR gates and flip-flops. In the external feedback LFSR in Figure 4.2b, there are two exclusive-OR gates in the worst case path from the output of the last flip-flop in the shift register to the input of the first flip-flop in the shift register. On the other hand, the internal feedback LFSR has, at most, one exclusive-OR gate in any path between flip-flops. Therefore, the internal feedback LFSR provides the implementation with the highest maximum operating frequency for use in high performance applications. The main advantage of external feedback LFSRs is the uniformity of the shift register; hence, there are some applications where external feedback is preferred.

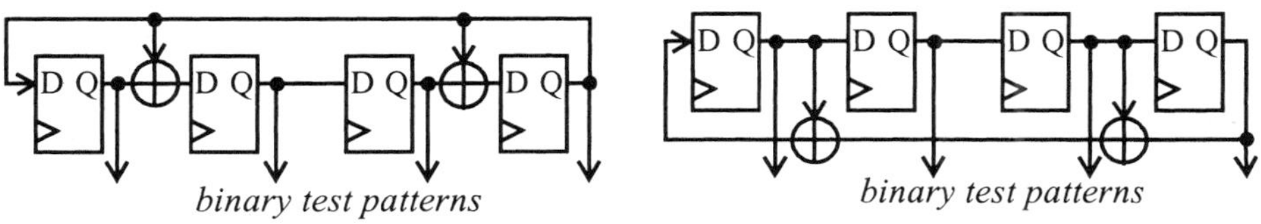

(a) internal feedback LFSR (b) external feedback LFSR

FIGURE 4.2 LFSR implementations.

1. Internal feedback LFSRs are sometimes referred to as Type 1 LFSRs while external feedback LFSRs are referred to as Type 2 LFSRs [32]. One problem with this terminology is remembering which is which, but a more serious problem is that the designations are often reversed in some papers where the external feedback LFSR is referred to as Type 1 and the internal feedback LFSR is referred to as Type 2 [218]. The internal and external feedback designations are unambiguous, avoid confusion and, more importantly, avoid design errors.

The sequence of test patterns produced at the outputs of the flip-flops in an LFSR is a function of the placement of the exclusive-OR gates in the feedback network. This is illustrated by the state diagrams associated with the two LFSR implementations in Figure 4.3. Both LFSRs are internal feedback type implementations but when initialized to any state other than the all 0s state the LFSR in Figure 4.3b goes through all possible states except the all 0s state before repeating the sequence. The test patterns produced by the LFSR in Figure 4.3a are a function of the initial state of the LFSR, where the longest sequence is only six patterns before repeating. Obviously, the longest possible sequence of test patterns is of most interest for TPG implementations in BIST applications. The longest possible sequence of patterns for any n-bit LFSR will be 2^n-1 since any LFSR initialized to the all 0s state will remain in that state. An LFSR that generates a sequence of 2^n-1 unique patterns before repeating is referred to as a *maximum length sequence* (also referred to as a maximal length sequence or as an m-sequence) LFSR [94]. The placement of exclusive-OR gates with respect to the flip-flops of the LFSR is defined by the *characteristic polynomial* of the LFSR. The characteristic polynomials, denoted as $P(x)$, are given for the two LFSRs with each flip-flop labeled according to its position in the polynomial. Note that the characteristic polynomial for the external feedback LFSR in Figure 4.2b is the same as that of the internal feedback LFSR in Figure 4.3a. Each non-zero coefficient in the characteristic polynomial represents an exclusive-OR in the feedback network; the exceptions are the x^n and x^0 coefficients which are always non-zero but do not represent the incorporation of an exclusive-OR gate, although they do represent connections to

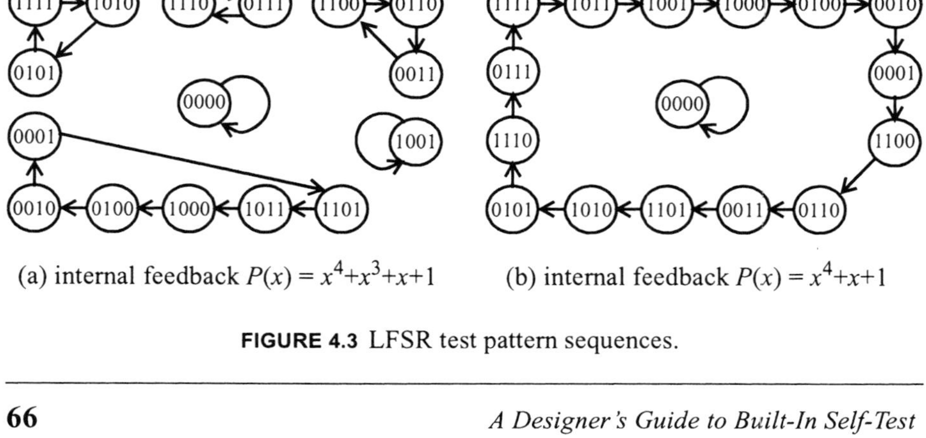

(a) internal feedback $P(x) = x^4+x^3+x+1$ (b) internal feedback $P(x) = x^4+x+1$

FIGURE 4.3 LFSR test pattern sequences.

exclusive-OR gates. Note that $x^0=1$ such that the "1" in the characteristic polynomial represents the x^0 term.

For LFSRs, n is referred to as the *degree of the polynomial* and results in an n-bit LFSR. For example the LFSRs in Figure 4.2 and Figure 4.3 all have characteristic polynomials of degree 4. Therefore, the characteristic polynomial completely defines the construction of the LFSR (for either internal or external feedback implementations) where the degree of the polynomial gives the number of flip-flops and the number of non-zero coefficients (not including x^n and x^0) gives the number of exclusive-OR gates. The LFSR is much more area efficient than a counter which has a minimum of one exclusive-OR gate and one AND gate per flip-flop. This gate count is based on a ripple-carry counter whose critical timing path is through the ripple-carry chain with N gate delays for an N-bit counter, while the internal feedback LFSR has only one exclusive-OR gate delay in its critical timing path, regardless of the size of the LFSR.

4.4.1 Primitive Polynomials

Polynomials that result in a maximum length sequence are called *primitive polynomials* and polynomials that do not produce a maximum length sequence are called *non-primitive polynomials* [32].[1] Primitive polynomials produce a maximum length sequence regardless of whether the implementation is of the internal or external feedback type. However, the actual sequence of patterns will differ depending on the implementation as illustrated in Figure 4.4 for an external feedback implementation of $P(x) = x^4+x+1$, whose sequence can be compared to that of internal feedback implementation of the same polynomial in Figure 4.3b. Therefore, the characteristic

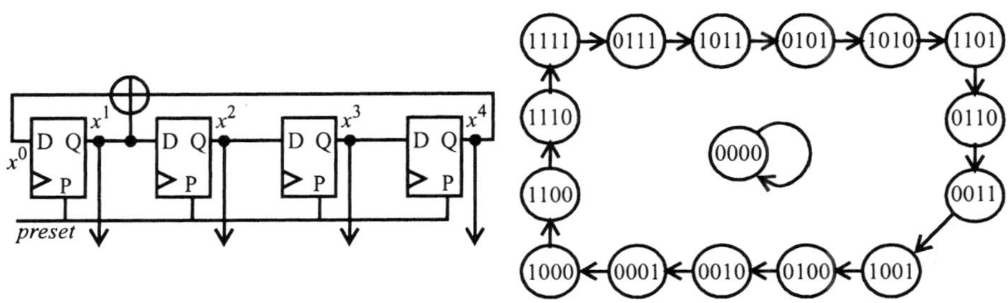

FIGURE 4.4 LFSR pattern sequence for external feedback primitive polynomial.

1. A primitive polynomial is like a prime number in that it is only divisible by itself and by 1.

polynomial of the LFSRs in Figure 4.3b and Figure 4.4 is a primitive polynomial while the characteristic polynomials of the LFSRs in Figure 4.2 and Figure 4.3a are non-primitive. Primitive polynomials with a minimum number of non-zero coefficients are the desired characteristic polynomials for LFSRs to be used for TPGs in BIST applications. Table 4.1 gives primitive polynomials with minimum non-zero coefficients (and, therefore, minimum exclusive-OR gates) for degrees $2 \le n \le 74$.[1] As can be seen from the table, the minimum number of exclusive-OR gates required to implement any of these primitive polynomials is between 1 and 3.

TABLE 4.1 Primitive polynomials through degree 74.

Degree (n)	Polynomial	Degree (n)	Polynomial
2, 3, 4, 6, 7, 15, 22, 60, 63	$x^n + x + 1$	12	$x^n + x^6 + x^4 + x^3 + 1$
5, 11, 21, 29, 35	$x^n + x^2 + 1$	33	$x^n + x^{13} + 1$
8, 19, 38, 43	$x^n + x^6 + x^5 + x + 1$	34	$x^n + x^{15} + x^{14} + x + 1$
9, 39	$x^n + x^4 + 1$	36	$x^n + x^{11} + 1$
10, 17, 20, 25, 28, 31, 41, 52	$x^n + x^3 + 1$	37	$x^n + x^{12} + x^{10} + x^2 + 1$
13, 24, 45, 64	$x^n + x^4 + x^3 + x + 1$	40	$x^n + x^{21} + x^{19} + x^2 + 1$
14, 16	$x^n + x^5 + x^4 + x^3 + 1$	42	$x^n + x^{23} + x^{22} + x + 1$
18, 57	$x^n + x^7 + 1$	46	$x^n + x^{21} + x^{20} + x + 1$
23, 47	$x^n + x^5 + 1$	54	$x^n + x^{37} + x^{36} + x + 1$
26, 27	$x^n + x^{12} + x^{11} + x + 1$	55	$x^n + x^{24} + 1$
30, 51, 53, 61, 70	$x^n + x^{16} + x^{15} + x + 1$	58	$x^n + x^{19} + 1$
32, 48	$x^n + x^{28} + x^{27} + x + 1$	65	$x^n + x^{18} + 1$
44, 50	$x^n + x^{27} + x^{26} + x + 1$	69	$x^n + x^{29} + x^{27} + x^2 + 1$
49, 68	$x^n + x^9 + 1$	71	$x^n + x^6 + 1$
56, 59	$x^n + x^{22} + x^{21} + x + 1$	72	$x^n + x^{53} + x^{47} + x^6 + 1$
66, 67, 74	$x^n + x^{10} + x^9 + x + 1$	73	$x^n + x^{25} + 1$

Like the counter, we are limited to values of n around 22 to 25 if we intend to run the full maximum length sequence in any reasonable amount of time. However, LFSRs are also good for output response analysis circuits (discussed in Chapter 5) which do not have the same size constraints. LFSRs with primitive characteristic polynomials are also the basic component in Cyclic Redundancy Check (CRC) circuits that are commonly used as on-line concurrent fault detection circuits, as will be discussed in Chapter 15 [122].

1. This number should be sufficient for most BIST applications encountered in practice but primitive polynomials for larger degrees can be found in [50] for $n \le 100$ and in [32] for $n \le 300$.

We are usually interested in minimizing the area overhead of the TPG and, therefore, implement a specific primitive polynomial for the LFSR based on the number of bits (or degree of the polynomial). However, there are applications where a generic LFSR is desired so that we can change the characteristic polynomial in the system. Examples of generic (or generalized) 4-bit internal and external feedback LFSRs are illustrated in Figure 4.5. The coefficients of the desired characteristic polynomial are applied to the inputs of the AND gates where a logic 1 for a non-zero coefficient will enable the feedback, while a logic 0 for a zero coefficient will disable the feedback.

Primitive polynomials make the initialization of LFSRs a simpler task since any non-zero state guarantees that all non-zero states will be visited in the maximum length sequence. The presets to each flip-flop in the LFSRs of Figure 4.3 and Figure 4.4 are used to initialize the LFSRs. Initializing to the all 1s state allows all flip-flops to have the same implementation (with a synchronous preset). In addition, the beginning and end of the sequence are easy to decode using an AND or NAND gate. As can be seen in the state diagram of Figure 4.3a, initializing to the all 1s state for the non-primitive polynomial does not lead to the longest sequence of patterns in this particular case. We have to either determine the state diagram to find the longest sequence or else we have to incorporate additional circuitry to preload the LFSR to a number of different starting states; this is another good argument for using primitive polynomials. For large LFSRs where n is greater than 20 to 22, the ability to preload the LFSR to different states allows the LFSR to visit various subsets of the state diagram. While initializing the LFSR to a given value is often referred to as *seeding* the LFSR, preloading different starting values is generally referred to as *reseeding* the LFSR.

Primitive polynomials also exhibit an important feature in that the sequence of test patterns produced is pseudo-random. The test patterns have many properties of random patterns when one looks at any single bit in the LFSR. First, the number of 1s is one greater than the number of 0s (this is due to the fact that the LFSR never enters the all 0s state, which is an isolated state). Second, the number of *runs* of 1s is equal

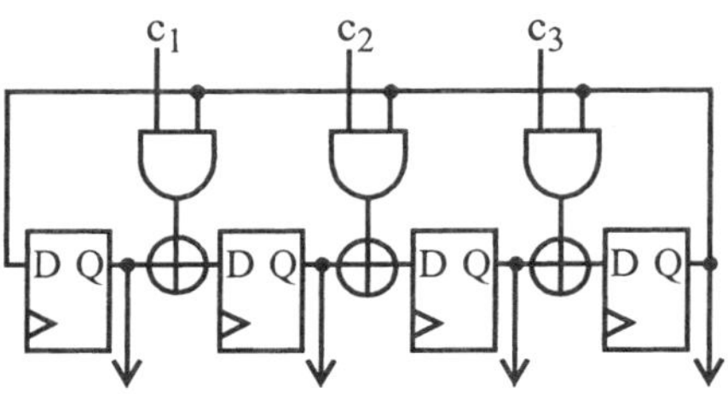

(a) internal feedback LFSR

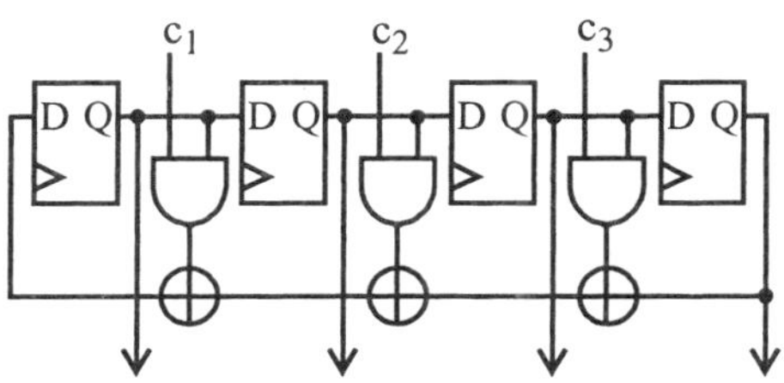

(b) external feedback LFSR

FIGURE 4.5 Generic LFSR implementations.

to the number of runs of 0s (where a *run* is a sequence of consecutive 1s or consecutive 0s) and the total number of runs is 2^{n-1} (which is also the total number of transitions from $1 \rightarrow 0$ and $0 \rightarrow 1$). Finally, the lengths of the runs are distributed as follows: half of the runs will have a length of 1, a quarter of the runs will have length 2, an eighth of the runs will have length 3, a sixteenth of the runs will have length 4, and so on. This pattern breaks down once the fraction no longer results in an integral number of runs [4]. For example, consider the left most bit of the state diagram in Figure 4.3b. There are eight 1s and seven 0s. Out of the total eight runs, four are runs of 0s and four are runs of 1s. Finally, four runs are of length 1, two are of length 2, one is of length 3, and one is of length 4. This is independent of whether the LFSR is constructed with an internal or external feedback implementation as can be seen by the bit values in the state diagram in Figure 4.4. An LFSR is sometimes referred to as a pseudo-random pattern generator (PRPG).

4.4.2 Producing the All 0s Pattern

The LFSR can be modified to produce the all 0s pattern during the pseudo-random sequence by adding extra logic. Specifically, the logic values of all stages except for x^n are logically NORed with the output of the NOR exclusive-ORed with the feedback value. As a result, the feedback value is modified when the LFSR is in the 000...001 state to produce the all 0s state on the next clock cycle, after which the LFSR continues to proceed through its normal pseudo-random sequence of states for a total of 2^n patterns before the sequence begins to repeat. This sequence is sometimes referred to as a *de Bruijn sequence* and the modified LFSR is sometimes called a *de Bruijn counter* [32]; it is also called a Complete Feedback Shift Register (CFSR) [331]. The LFSR modification for the all 0s pattern is illustrated in Figure 4.6 for both internal and external feedback implementations of $P(x) = x^4+x+1$ along with their resulting state diagrams.

The area overhead required for producing the all 0s pattern in a N-bit LFSR includes an N-1-input NOR gate and an exclusive-OR gate. While this is not a large overhead, it can be significant for large LFSRs given the typical fan-in limitations of static CMOS gates, particularly when one considers that only one additional test pattern is produced by the modified LFSR. The performance penalty is even more significant where an internal feedback LFSR increases from one exclusive-OR gate between flip-flops to two exclusive-OR gates plus the delay through the NOR function which may consist of two gate delays for $4<n<17$ and three gate delays for $16<n<65$ to construct NAND-NOR and NOR-NAND-NOR networks, respectively. Alternatively, one could increase the size of the LFSR by one bit to ensure that the n-bit all 0s is produced during the course of the 2^{n+1}-1 bit sequence produced by an n+1-bit LFSR; this approach results in lower area overhead and performance penalty but longer test time.

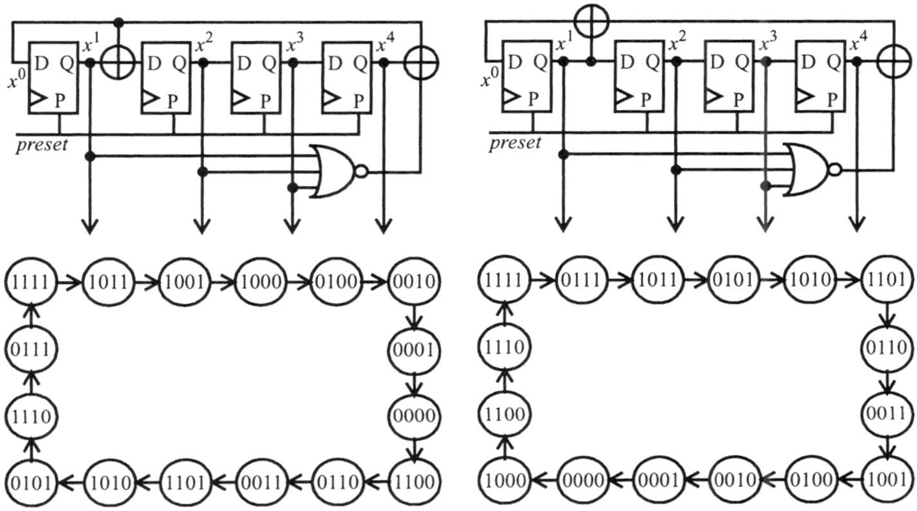

(b) internal feedback LFSR (b) external feedback LFSR

FIGURE 4.6 LFSR implementations to produce all 0s pattern.

4.4.3 Reciprocal Polynomial

One last attribute of polynomials and LFSRs worth noting is referred to as the *reciprocal polynomial* [32]. The reciprocal polynomial of a primitive polynomial is also primitive, while the reciprocal polynomial of a non-primitive polynomial is non-primitive. The reciprocal polynomial, $P^*(x)$, of a polynomial, $P(x)$, is obtained by:

$$P^*(x) = x^n P(1/x) \qquad \textbf{(4.1)}$$

As an example, consider the polynomial for degree n=8 from Table 4.1, $P(x) = x^8 + x^6 + x^5 + x + 1$. Then the reciprocal polynomial, $P^*(x) = x^8 (x^{-8} + x^{-6} + x^{-5} + x^{-1} + 1) = 1 + x^2 + x^3 + x^7 + x^8 = x^8 + x^7 + x^3 + x^2 + 1$. The test pattern sequence produced by an internal feedback LFSR implementing the reciprocal polynomial is in reverse order with a reversal of the bits within each test pattern when compared to that of the original polynomial, $P(x)$. This property can be useful in some BIST applications. LFSRs implementing reciprocal polynomials are sometimes referred to as a reverse-order pseudo-random pattern generators [319]. Note that primitive polynomials with small degrees (n=2 and n=3) do not have a reciprocal polynomial since in these cases $P^*(x) = P(x)$ [184]. An interesting use of the reciprocal polynomial will be discussed in the next chapter.

4.5 Cellular Automata

The Cellular Automata (CA) is similar to the LFSR in that it generates pseudo-random pattern sequences [56]. The big advantage of CA over the LFSR is that we do not observe the effect of the bits shifting through a CA register like we do in the LFSR, particularly in the external feedback LFSR. This is illustrated in Figure 4.7 for the first 23 test patterns produced by 8-bit internal and external feedback LFSRs and an 8-bit CA, started from the all 1s state. Note that a logic 1 is shown as a gray box while a logic 0 is shown as a white box in the figure. The LFSRs are constructed with a primitive characteristic polynomial taken from Table 4.1. As can be seen from the figure, test patterns produced by the CA appear to be more random than those produced by the LFSRs. However, the construction of the CA register (sometimes referred to as a CAR) is not as simple as the LFSR and, as a result, CA registers require considerably more exclusive-OR gates than an LFSR. Like LFSRs, the exclusive-OR gates of the CA require initialization of the register to a non-zero state.

Construction of a CA register is based on the logical relationship of each flip-flop to its two nearest neighbors [325]. These relationships are referred to as "rules" and two of the most popular rules (rule 90 and rule 150) are illustrated in Table 4.2. Because most CA registers are constructed from a combination of rules (different rules for different stages of the CA register), a CA register is sometimes referred to as a Linear Hybrid Cellular Automata (LHCA) or a Linear Cellular Automata Register (LCAR).

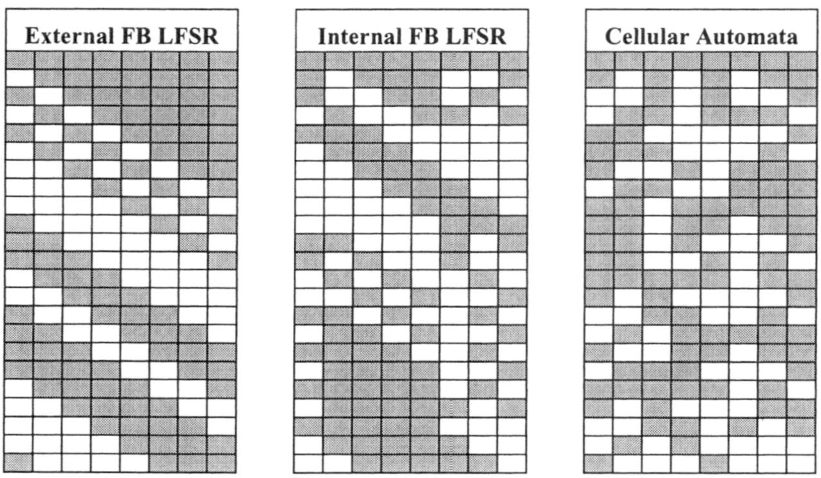

FIGURE 4.7 Test pattern comparison of LFSR & CA based TPGs.

The name of a given rule is obtained from the decimal value of the binary code generated for the next state of flip-flop x_i as a result of it current state and the current states of its two nearest neighbors, x_{i-1} and x_{i+1}. The logic for the next state of a given rule can be obtained from the Karnaugh map of $x_i(t+1)$ with respect to $x_{i-1}(t)$, $x_i(t)$, and $x_{i+1}(t)$. For example, rule 128 is the logical AND of $x_{i-1}(t)$, $x_i(t)$, and $x_{i+1}(t)$ while rule 254 is the logical OR of $x_{i-1}(t)$, $x_i(t)$, and $x_{i+1}(t)$.

TABLE 4.2 Rules 90 and 150 for Cellular Automata implementations.

		7	6	5	4	3	2	1	0
	$x_{i-1}(t)\ x_i(t)\ x_{i+1}(t)$	111	110	101	100	011	010	001	000
rule	$x_i(t+1)$	0	1	0	1	1	0	1	0
90	value = 90		2^6		2^4	2^3		2^1	
rule	$x_i(t+1)$	1	0	0	1	0	1	1	0
150	value = 150	2^7			2^4		2^2	2^1	

This next state for rules 90 and 150 is generated by the exclusive-OR of combinations of the three current state values. Rule 90 is constructed with a flip-flop receiving the exclusive-OR of its two neighboring flip-flops; $x_i(t+1) = x_{i-1}(t) \oplus x_{i+1}(t)$ as illustrated in Figure 4.8a. Rule 150 is constructed with a flip-flop receiving the exclusive-OR of itself and its two neighboring flip-flops; $x_i(t+1) = x_{i-1}(t) \oplus x_i(t) \oplus x_{i+1}(t)$ as

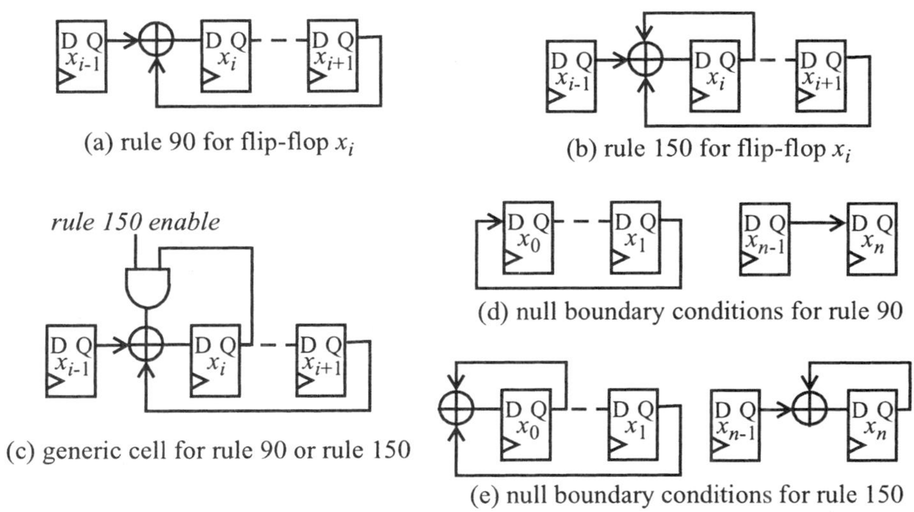

(a) rule 90 for flip-flop x_i

(b) rule 150 for flip-flop x_i

(c) generic cell for rule 90 or rule 150

(d) null boundary conditions for rule 90

(e) null boundary conditions for rule 150

FIGURE 4.8 CA rule 90 and rule 150 implementations.

illustrated in Figure 4.8b. A generic cell which is capable of implementing either rule 90 or rule 150 (when a logic 1 is applied to the *rule 150 enable* input) is shown in Figure 4.8c. There are 256 possible rules but rules 90 and 150 are the two most frequently combined rules for obtaining maximum length pseudo-random sequences. For the flip-flops at the ends of the CA register (referred to as *boundary conditions*) there are two options for implementation: *null* and *cyclic* boundary conditions. For null boundary conditions, the missing flip-flops are assumed to supply a logic 0. The null boundary conditions for rules 90 and 150 are illustrated in Figure 4.8d and Figure 4.8e, respectively. Cyclic boundary conditions assume that the two end flip-flops are adjacent and supply logic values to each other; as a result, there is not reduction in the number of exclusive-OR gates.

An example CA register implementation that generates a maximum length sequence is illustrated in Figure 4.9. The CA uses a combination of rules taken from n=6 in Table 4.3 (rule 150 for 'X' entry in the left most flip-flop and rule 90 for the remaining flip-flops) using null boundary conditions. Since rule 150 implies two exclusive-OR gates for non-boundary cells, maximizing the use of rule 90 minimizes area overhead of CA-based TPG implementations. The total number of exclusive-OR gates, N_{XOR}, required for an n-bit CA register with null boundary conditions is given by:

$$N_{XOR} = n + R_{150} - 2 \quad \textbf{(4.2)}$$

where R_{150} is the number of bits using rule 150. The reduction by two from the total number of exclusive-OR gates is due to the null boundary conditions. When cyclic boundary conditions are used, the '-2' can be removed from the expression to accurately predict the number of exclusive-OR gates.

Like primitive polynomials for LFSR, the designer needs to know the combinations of rules that will result in maximum length sequences for a given number of bits in order to design the CA register [93]. Table 4.3 gives a set of combinations of rules 90 and 150 for register sizes in the range $2 \le n \le 28$, which give a maximum length sequence of 2^n-1 when used with null boundary conditions [108]. The advantage of null boundary conditions over cyclic boundary conditions is a reduction in area over-

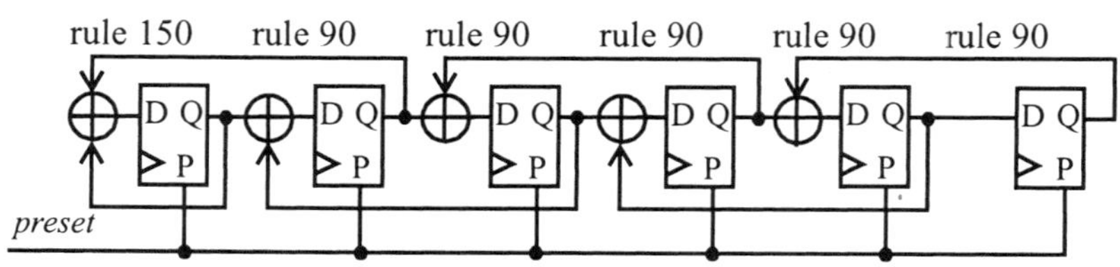

FIGURE 4.9 Example 6-bit cellular automata implementation.

head along with avoidance of the potential performance penalty due to feedback paths between the end registers. Note that the 'X' in a given bit position in Table 4.3 can represent either rule 90 or rule 150 with the absence of an 'X' representing the opposite rule. In addition, the sequence of rules can be reversed (from left to right or vice versa) so that many of the rows in the table represent 4 different CA implementations.

TABLE 4.3 Rule 90 and 150 combinations for maximum length sequence CARs.

n	**Rule Combinations (with null boundary conditions)**																										
2	X																										
3	X																										
4		X		X																							
5	X	X			X																						
6	X																										
7	X	X		X		X																					
8	X	X		X		X		X																			
9	X	X			X		X		X																		
10		X		X		X		X		X																	
11	X	X		X		X		X		X																	
12		X		X		X		X		X		X															
13	X	X			X		X		X		X																
14		X	X	X	X	X		X	X	X	X	X	X														
15	X			X				X		X					X												
16	X	X		X		X		X		X		X		X		X											
17		X	X	X	X	X		X	X	X	X	X	X			X	X										
18		X		X		X		X		X		X		X		X		X									
19		X	X		X			X	X		X	X				X			X								
20	X	X	X	X			X	X	X		X	X		X	X	X	X	X	X	X							
21		X	X	X	X			X	X						X	X	X	X		X	X						
22		X		X		X		X		X		X		X		X		X		X		X					
23	X	X		X		X	X	X			X	X	X		X				X	X		X					
24	X	X	X	X	X	X		X			X		X	X		X		X		X		X	X				
25	X		X	X	X	X		X		X		X		X			X	X	X	X			X				
26		X		X	X		X		X	X		X				X		X	X	X		X	X				
27					X	X	X	X	X						X	X			X					X	X		X	
28		X		X		X		X		X		X		X		X		X		X		X		X		X		X

4.6 Weighted Pseudo-Random Test Patterns

A CUT may exhibit low fault coverage when tested with pseudo-random test patterns. The typical situations where this occurs include circuits with large input AND

(NAND) logic functions or large input OR (NOR) logic functions. A large AND (NAND) function will produce a logic 1 (logic 0) infrequently due to the equal likelihood of logic 0s and logic 1s in each bit of the pseudo-random test patterns. Similarly, a large OR (NOR) function will produce a logic 0 (logic 1) infrequently. Faults that would require many logic 1s generated at the output of a large AND function or logic 0s at the output of a large OR function are not easily detected and are referred to as *random pattern resistant* faults [38].

Another important example of a practical application for weighted pseudo-random test patterns is a CUT that incorporates a global reset or preset to the flip-flops. Frequent resetting of flip-flops by pseudo-random test patterns will clear the test data propagated into the flip-flops and prevent internal faults from being detected. In a study of pseudo-random test patterns applied to the 1989 International Symposium on Circuits and Systems (ISCAS'89) sequential benchmark circuits [45], it was found that single stuck-at gate-level fault coverage was as low as 11% to 15% for circuits that had global resets or presets to the flip-flops due to this fault detection blocking effect of pseudo-random test patterns [234]. When the reset/preset signal is controlled by other means, which can include the use of weighted pseudo-random test patterns, the fault coverage climbs to greater than 90% for these circuits.

A solution to this problem is to take the equal likelihood of 1s and 0s in pseudo-random sequences of an LFSR and use additional logic to create weighted pseudo-random patterns [32]. For example, more frequent logic 1s can be generated by a logical NAND of two or more bits of the LFSR while more frequent logic 0s can be generated by a logical NOR of two or more bits of the LFSR. Given that the probability of a given bit in the LFSR being a logic 0 is approximately 0.5 (denoted $p_0 \approx 0.5$), NANDing two bits of the LFSR will produce a bit that has $p_0 \approx 0.25$; NANDing three bits will result in $p_0 \approx 0.125$, NANDing four bits results in $p_0 \approx 0.0625$, and so on. An example of a weighted-LFSR based TPG is illustrated in Figure 4.10 where each of the normal LFSR outputs has $p_0 \approx 0.5$ for a given test pattern while the weighted out-

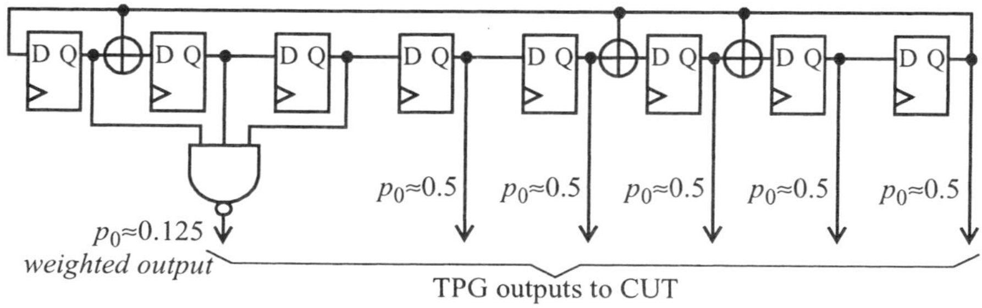

FIGURE 4.10 Example weighted-LFSR implementation.

put has $p_0 \approx 0.125$. Note that the preset to the weighted-LFSR has not been shown in Figure 4.10. If the weighted output was driving an active low global reset or preset in the CUT, then initialization of the TPG to the all 1s state would also generate a global reset/preset during the first test vector for initialization of the CUT.

The probabilities (or weights) can be controlled for higher fault coverage test pattern generation and/or shorter length BIST sequences for a given CUT [189]. Of course the weights and the number of CUT inputs that need to be "weighted" has to be determined on a case-by-case basis since this is a function of the CUT. In calculating the weight for generating a global reset/preset signal to the CUT, a rule of thumb is to make the reset signal weight 2^{-m}, where m is chosen to be greater than the sequential depth of the CUT. The sequential depth of the CUT is the number of flip-flops in the longest path between primary inputs and outputs. This requires an m-input NAND or NOR gate but helps to ensure that test patterns have sufficient clock cycles to propagate through the CUT before a reset occurs.

The weighting requires extra hardware in terms of the NAND/NOR gates as well as additional LFSR bits since multiple bits are logically combined to produce a single test pattern bit. The increased area overhead along with the increased design time are a limitation of this TPG approach. However, the weighted-LFSR (or WLFSR, as it is sometimes referred to) is a powerful technique for increasing fault coverage in circuits with random-pattern resistant faults. Similar weighting can be applied to CA-based TPG implementations as well. However, the pseudo-random test patterns produced by a CA register can also be weighted by the choice of CA rules [192].

4.7 Other TPGs

There are a couple of other types of TPGs worth noting at this point. The first uses an LFSR or CA register with a shift register added to produce a TPG with more bits than that of the LFSR or CA register. This is illustrated in Figure 4.11a where an N-bit LFSR feeds an M-bit shift register to produce an N+M-bit TPG. While this facilities making a larger TPG from a smaller LFSR, one needs to be careful about the test patterns produced by this type of TPG. Depending on the initial value in the shift register at the beginning of the LFSR sequence, there will be at most 2^N+M different patterns produced by the TPG during its first cycle of test patterns. However, there will only be 2^N-1 unique test patterns produced after the first cycle of the pattern sequence of the LFSR. In addition, the properties of the test patterns, such as the number of runs, is determined by the LFSR and its characteristic polynomial. As a result, the test patterns will contain additional properties, called *linear dependencies* [32] that can pre-

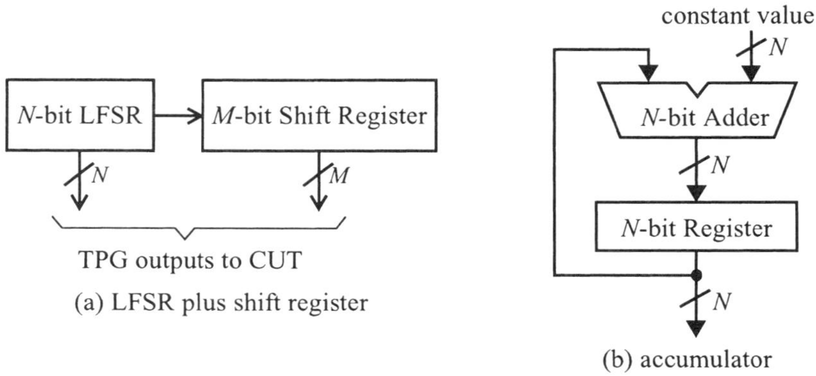

FIGURE 4.11 Other types of TPGs.

vent high fault coverage in some circuits, as will be discussed in more detail in Chapter 10. A related type of TPG is an LFSR that drives a network of exclusive-OR gates which, in turn, drives the CUT. This type of TPG can overcome the problems of linear dependencies and will also be discussed in Chapter 10.

Adders are sometimes common in VLSI implementations and can be used in conjunction with flip-flops to construct an accumulator that can function as a TPG. In some VLSI implementations, accumulators are a part of the normal system function. Accumulators will be discussed in more detail in the next chapter but they basically consist of an N-bit adder and an N-bit register, as illustrated in Figure 4.11b. By initializing the register to the appropriate value and adding an appropriate constant value at the input, the accumulator will progress through all 2^N possible combinations of patterns [98]. A simple example of this is a counter where the constant to be added each clock cycle is 1; this increments the N-bit counter through all possible 2^N combinations. The difficult problem with this type of TPG is determining the initial value for the register and the constant to be added to produce all possible 2^N test patterns. Of course the constant value of 1 can always be used with any initial value for the register to ensure that the accumulator increments through all combinations of test patterns, but a simple counter would be more area efficient in this case. As discussed previously, the test patterns produced by counters can have undesirable properties such as the high and low toggle rates of the least and most significant bits, respectively. Therefore, the initial value and constant that are of most interest would be those that produce test pattern sequences that are more similar to the pseudo-random patterns of an LFSR or a CA register. Other arithmetic functions such as multipliers have also been investigated as TPGs for BIST applications [219].

4.8 Comparing TPGs

Some of the TPGs discussed in this chapter are compared in Table 4.4 in terms of the number of flip-flops, exclusive-OR gates, and elementary logic gates needed to construct the TPG. Note that the number of gate I/O does not include the flip-flops since, in most BIST applications, area overhead is reduced by using existing flip-flops in the CUT to implement the TPG function. All implementations include circuitry for reset or preset capabilities of the flip-flops. The LFSR is an 8-bit implementation using the primitive polynomial for degree n=8 from Table 4.1. The weighted-LFSR (WLFSR) uses a 12-bit LFSR with an additional 3-input NAND gate and 3-input NOR gate to produce bits with a probability of 0.125 of being a logic 0 and logic 1, respectively; using the remaining six pseudo-random bits of the 12-bit LFSR, we obtain an 8-bit weighted pseudo-random TPG. The CA is an 8-bit design using a combination of rules 90 and 150 from Table 4.3 (rule 90 was used for each position with an 'X') with null boundary conditions. The counter is an 8-bit binary up-counter with active high count enable but with no carry-out. The March Y FSM is the counter-based TPG shown in Figure 4.1b designed to test a small RAM with a 5-bit address.

TABLE 4.4 Comparison of TPG implementations.

TPG	# Flip-Flops	# Exclusive-ORs	# Gates	# Gate I/O
LFSR	8	3	8	33
Weighted LFSR	12	3	15	51
Cellular Automata	8	9	8	51
Counter	8	8	15	69
March Y FSM	9	9	19	94

The question of how well the TPG portion of the BIST circuitry tests itself is addressed in the graph in Figure 4.12, in terms of single stuck-at gate-level fault coverage as a function of the number of clock cycles in the BIST sequence. As can be seen from the graph, 100% fault coverage was obtained for all TPG implementations. However, the number of clock cycles required for 100% fault coverage gives an indication of how easy the TPG is to test. An easier to test TPG approaches 100% fault coverage more rapidly than a harder to test TPG. The time to obtain 100% fault coverage correlates with the area overhead of the TPG implementations. Therefore, the LFSR is not only the most area efficient design, but is also easiest to test. There is very little difference in the fault coverage curves for the internal feedback LFSR and CA implementations. However, the external feedback LFSR implementing the same polynomial takes a few more clock cycles to achieve 100% fault coverage than the internal feedback LFSR or the CA. As a result, the internal feedback LFSR with primitive characteristic polynomial is the best TPG in terms of area overhead, performance penalty, and the ability to test itself. It is easy to implement as long as one has access to a list of primitive polynomials, such as the ones given in Table 4.1. The LFSR is also a key component in the design of output response analysis circuits. However, the pseudo-random patterns produced by the LFSR are not good for testing all CUTs. Therefore, it is important to keep the other TPGs in one's BIST tool kit.

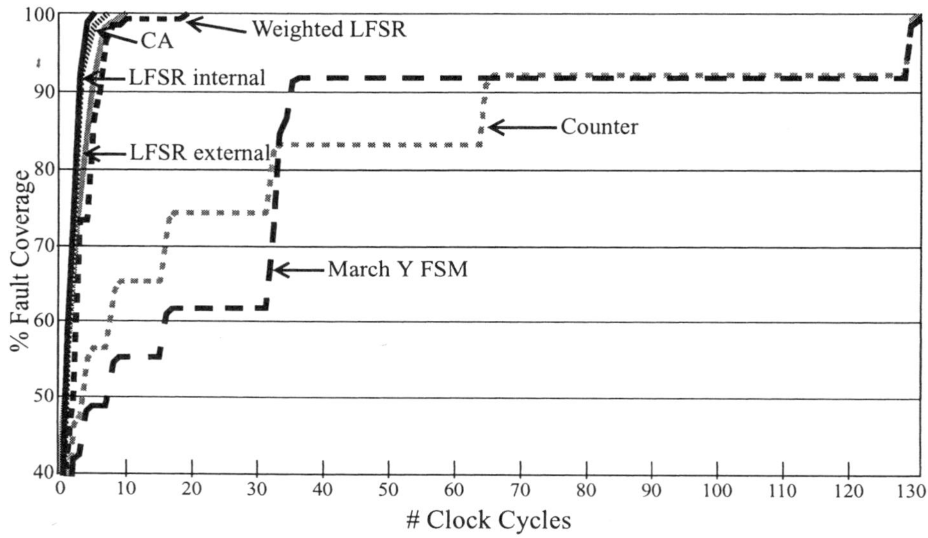

FIGURE 4.12 Fault coverage comparison of TPGs.

CHAPTER 5

Output Response Analysis

We continue our look at BIST with another important component, the output response analyzer (ORA). The ORA compacts the output responses of the CUT to the many test patterns produced by the TPG into a single *Pass/Fail* indication (usually a multiple-bit "signature"). The ORA is sometimes referred to as an output data compaction (ODC) circuit [4]. The significance of the ORA is that there is no need to compare every output response from the CUT with the expected output response external to the device. Only the final *Pass/Fail* indication needs to be checked at the end of the BIST sequence in order to determine the fault-free/faulty status of the CUT. This chapter presents the basic ORA design and implementation techniques along with their capabilities and limitations for use in BIST applications.

5.1 Types of Output Data Compaction

In most ORA techniques, the output responses of the CUT to the test patterns are "compacted" into a "signature" which is compared to the expected signature for the fault-free circuit. As we shall see, the *Pass/Fail* indication is often composed of a set of data bits that contain the signature of the BIST sequence, rather than a single *Pass/Fail* bit. Note that the term *compaction* is used rather than *compression* since compression implies no loss of information; since most ORA techniques incur some loss of information, compaction is a more accurate term [171]. There are several classes of

ORA circuits based on the manner in which the output data compaction takes place. Each of these approaches is discussed in more detail in the subsequent sections.

- **Concentration** typically performs output data compaction of multiple outputs from a CUT into a single output data stream. While this technique is not considered to be a typical class of ORA, it is often used with other ORA techniques such as counting techniques to reduce the area overhead of the overall ORA implementation. The typical hardware for concentration includes a tree of exclusive-OR gates connected in a manner similar to a parity circuit.
- **Comparison-based** ORAs use a comparator to detect mismatches in the fault-free and faulty circuits. An example of a comparator-based approach is the simple BIST design in Figure 1.5, where the expected responses were stored in a ROM and compared on a vector-by-vector basis to the output responses of the CUT. While this particular BIST technique is not typically used due to the large area overhead, there are other comparator-based ORA implementations that are area efficient and, as a result, are found in practical BIST implementations. However, a comparator by itself will require monitoring the output of the comparator during every test vector unless we incorporate an RS latch to retain any mismatches encountered during the test sequence. Therefore, typical hardware includes a comparator and an RS latch for the *Pass/Fail* indication.
- **Counting techniques** count the number of 0s, 1s, or transitions in the output response of the CUT with the resultant count value of the specific attribute at the end of the test sequence providing the "signature" for the *Pass/Fail* indication. At the end of the BIST sequence, we compare the resultant count value(s) to the expected count value(s) for the fault-free circuit to determine the faulty/fault-free status of the CUT. As a result, typical hardware includes counters for each output of the CUT.
- **Signature analysis** is the most commonly used technique for ORAs in BIST implementations [15]. Signature analysis uses an LFSR as the primary component of the ORA implementation. The basic idea behind signature analysis is to divide the polynomial representing the output response of the CUT by the characteristic polynomial of the LFSR used to implement the ORA [40]. The resultant "signature" is the remainder of the polynomial division and is compared to the signature for the fault-free circuit at the end of the BIST sequence. The typical hardware is a single LFSR with additional exclusive-OR gates to facilitate single or multiple output CUTs.
- **Accumulators** sum the output responses of the CUT by treating each output response as a binary magnitude. These are the same circuits that are used for checksum-based concurrent fault detection circuits [145]. At the end of the BIST sequence, the accumulated value (or checksum) is compared to that obtained for

the fault-free circuit to determine the pass/fail status of the CUT. The typical hardware consists of a full adder and a register to hold the accumulated sums.[1]

- **Parity check** techniques can also be used for output response compaction. The basic idea is to calculate parity for the output response data and to check for the correct parity at the end of the BIST sequence. As a result, the typical hardware consists of exclusive-OR gates for the parity calculation for each output response along with a flip-flop to calculate the parity over the entire sequence of output responses.

Since all ORAs perform data compaction, there will be some loss of information. When the lost information contains fault detection data, it is possible for the BIST results to indicate that a faulty CUT is fault-free. This is often referred to as *fault masking* or, in the cases of some ORA techniques, *signature aliasing* [40]. Each type of ORA has its own fault masking characteristics as well as techniques that can be used to reduce the probability of fault masking. Therefore, fault masking is an important concern when considering a candidate ORA for a BIST implementation.

As was the case with TPGs, multiple types of ORAs can be used in a BIST application. The subsequent sections address output response analysis from the hardware standpoint, in the context of BIST, with emphasis on minimizing area overhead and maximizing performance. In most ORA implementations, an important requirement includes initialization of the ORA to some known value; unless otherwise specified, we will assume that a reset capability is incorporated into the ORAs discussed in the subsequent sections. Another important requirement is the need to enable data compaction only after complete initialization of the CUT as well as the ability to disable the data compaction and hold the final "signature" until it can be retrieved for comparison with the expected results for the fault-free circuit (the importance of this enable/disable feature will be discussed in more detail in Chapter 6).

5.2 Concentrators

Concentrators are valuable in reducing the total number of outputs of the CUT that must be monitored during the test sequence. However, they do not alleviate the need to monitor the output during every clock cycle of the test sequence. As a result, concentrators do not comprise ORAs in their own right, but they are very effective in

1. For more information on concurrent fault detection circuits (CFDCs) including parity, CRC, checksum, Berger and Bose-Lin codes, refer to the overview of CFDCs in Chapter 15.

reducing the total amount of hardware when combined with other ORA techniques, thus reducing the area overhead of the BIST approach. The typical approach to concentration is to linearly combine a multiple output response into a single bit using a tree of exclusive-OR gates, as is typically used in parity-based concurrent fault detection circuits for parallel data. For an N-bit output response, N-1 exclusive-OR gates are needed to perform the linear combination to a single bit, as illustrated in Figure 5.1 for an 8-bit output response. The concentrator output will be a logic 1 whenever there exists an odd number of logic 1s at the inputs to the exclusive-OR tree. While this is a reasonably efficient way to compact the output response, there is a possibility of fault masking in exclusive-OR tree-based concentrators. A double-bit error in a given output response of the CUT will cancel each other and fault masking will occur [32]. In fact, any even number of bit errors in the output response of the CUT will result in the same concentrator response as that of the fault-free CUT.

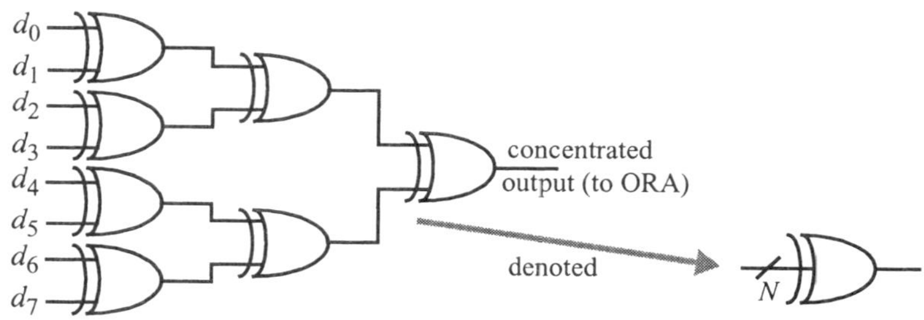

FIGURE 5.1 Example of 8-bit concentrator.

5.3 Comparators

Comparators are attractive for ORA implementations due to their low area overhead; an N-bit comparator requires N exclusive-OR gates and an N-input OR/NOR gate. When a CUT is composed of many identical circuits, the test patterns can be applied to all of those circuits simultaneously. Since these circuits are identical in their design and receive identical test patterns (and are initialized to identical states in the case of sequential circuits), their output responses will be identical for the fault-free case. Therefore, a comparator can be incorporated at the outputs of these identical circuits to monitor and compare their outputs. Any disagreement in the output responses of these circuits would indicate that a fault is present and, therefore, a comparator will suffice for the fault detection. The only limitation to this approach is the case where the identical circuits have equivalent faults such that the erroneous output responses do not mismatch, leading to fault masking [8]. While this is a situation that may have a low probability of occurrence, it is still a situation that should not be ignored when considering a comparator-based ORA. The more identical circuits that can be compared with each other, the lower the probability of fault masking due to equivalent faults in all of the circuits being compared.

Another good candidate application for comparator-based ORAs is when coupled with algorithmic test pattern generation. For example, consider the March Y algorithm in Figure 4.1a. The general implementation for a comparator-based ORA that would work with the March Y TPG implementation (from Figure 4.1b) is illustrated in Figure 5.2. Assuming that we do not know what the data output of the RAM will be during write operations, the mismatch signal to the RS latch is forced to the non-mismatch logic value during write operations (the *Set* input to the synchronous RS latch is assumed to be active high in this example). Note that it is important to use an edge-triggered flip-flop for the RS latch to avoid setting the latch to a *Fail* indication due to glitches at the output of the comparator that can easily result in a fault-free circuit. The RS latch is reset to a *Pass* indication (logic 0 in this example) prior to the execution of the BIST sequence. This can be accomplished with the *BIST Start* signal (assuming the *Reset* input to the RS latch is active low). Since the expected output response is known and is a simple sequence of patterns, it can easily be generated by the FSM for the TPG and used as the expected response for the comparator, as illustrated in Figure 5.2 for the March Y algorithm TPG in Figure 4.1b. Recall that f_1 is the most significant bit of the FSM that controls the read/write sequence to the RAM while q_N and q_{N+1} are the most significant bits of the N+2-bit counter. It should be noted that fault masking can occur in this approach if the TPG fails to produce the correct sequence of test patterns.

One of the main problems with comparator-based ORAs is that the ORA is not completely tested. To illustrate this, consider a fault at the output of the comparator that prevents a mismatch from propagating to the RS latch and setting the *Fail* indication (a stuck-at-1 at the output of the N-bit comparator would be such a fault in the ORA of Figure 5.7, for example). In order to detect this fault, we must produce a mismatch in the comparator, but this will never happen in the fault-free circuit since the RAM

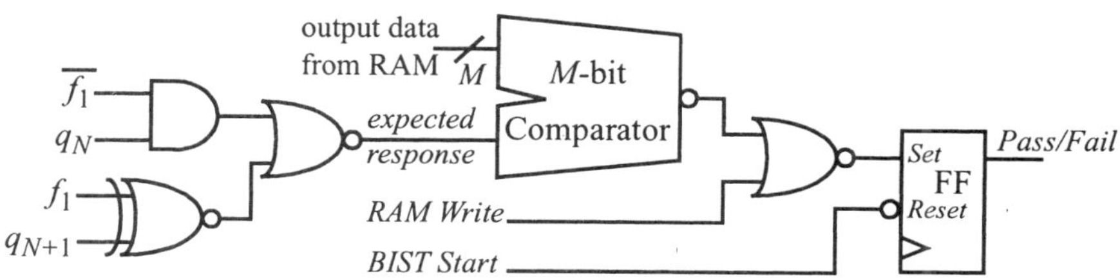

FIGURE 5.2 Example of comparator-based ORA for March Y RAM BIST.

will produce the expected response. There are many such faults in a comparator-based ORA. Fault simulation of this ORA during the March Y algorithm for a fault-free RAM yielded only 67.8% fault coverage for single stuck-at gate-level faults. This means that additional test vectors must be applied outside the BIST sequence in order to determine that the ORA is fault-free and to ensure that we are not getting *Pass* indications for faulty circuits (fault masking). This also means that we would have to apply some type of ad-hoc DFT technique to give access to the ORA during a test mode in order to apply these additional test vectors. Alternatively, we could design another BIST circuit just for the comparator-based ORA, but now we are adding more circuitry that could end up needing additional testing, just as the comparator-based ORA needs.

5.4 Counting Techniques

Counting techniques compact the output response of the CUT by counting the number of 1s or 0s in the output response [33]. This requires a counter on each output of the CUT with each counter having a reset capability for initialization prior to the BIST sequence. In addition, each counter must have an active high or active low synchronous count enable to allow the counter to increment each time a logic 1 or logic 0 is encountered, respectively. At the end of the BIST sequence, the count value for each counter associated with a CUT output is read and compared with the correct count value for that output for the fault-free circuit. If any of the count values disagree with the expected count values, the CUT is determined to be faulty. The choice of counting 1s or 0s is arbitrary. However, a special case of 1s counting, referred to as *syndrome analysis*, has been studied in detail and it has been shown that when exhaustive test patterns are applied to a combinational logic CUT, any single stuck-at gate-level fault will be detected with syndrome analysis [32].

Another type of counting technique is counting the transitions in the output response of the CUT. While this seems a simple approach with the output of the CUT connected to the clock input of a counter, glitches that are common to combinational logic outputs will cause major problems for this type of BIST implementation, including metastability [88]. Since the final count of the number of transitions determines the pass/fail status of the CUT, glitches will cause additional increments of the counters and cause a fault-free circuit to produce a failing indication in the BIST results.[1] Therefore, one must be careful to include a digital one-shot circuit when

1. The last thing we want to do is to throw away fault-free circuits and add to the expense of manufacturing and system repair.

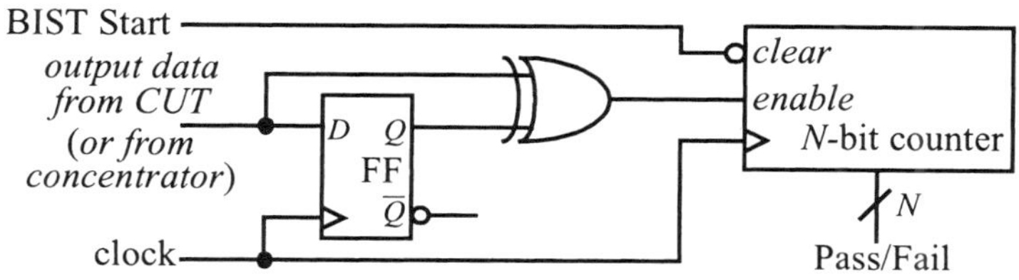

FIGURE 5.3 Transition counting with digital 1-shot.

implementing transition counting, as illustrated in Figure 5.3 for counting transitions, both rising and falling, in the output response of the CUT. While most textbooks suggest that both rising and falling transitions are counted, there have been few reports of actual applications of counting techniques in practical BIST applications (probably due to the many counters needed for a multiple output CUT). Since fault simulation is required to determine the actual fault coverage obtained, it is reasonable to experiment with rising-edge or falling-edge transition counting to see what provides the best solution for a potential application. The only difference in those circuits compared to that of Figure 5.3 would be replacing the exclusive-OR gate with an AND gate (for rising-edge transition counting) or a NOR gate (for falling-edge transition counting) while taking the output from the $\overline{Q}$ output of the flip-flop in either case.

There is the possibility of fault masking with counting techniques. For example, if a fault causes errors in the output response of the CUT but does not change the number of 1s, 0s, or transitions (whichever attribute is being counted), the final count value for the faulty CUT will be the same as that of the fault-free CUT. Given a test sequence of length T and an attribute count value K, it has been shown that the probability of fault masking increases as K approaches $T/2$ and decreases as K approaches 0 or T [32].

Another issue in using counting techniques for an ORA implementation is the size of the counter. The maximum sized counter needed is $\lceil \log_2(T) \rceil$. This is analogous to Berger code used in concurrent fault detection circuits. From the fault simulation results for the 8-bit counter in Figure 4.12, it can be seen that the single stuck-at gate-level fault coverage for the counter does not reach 100% until 129 clock cycles, which corresponds to the most significant bit of the counter toggling. Therefore, given an attribute (1s, 0s, or transitions) count for the fault-free circuit of K, the size of the counter should not exceed $\lceil \log_2(K) \rceil$ in order for the counter to be completely tested during the BIST sequence. Smaller counters reduce the area overhead of the BIST approach but will also increase the probability of fault masking. This is analogous to Bose-Lin code used in concurrent fault detection circuits. When $2^N < K$, where N is the number of bits in the counter, then the signature, $S(K)$, of the BIST count that is used to determine the pass/fail status of the CUT is given by $S(K) = K$ modulo 2^N. As N decreases, there are more faulty CUT count values, K', that will match the fault-free CUT signature $S(K)$ such that $S(K)=S(K')$ and the probability of fault masking increases.

Choosing the size of the counter for an ORA brings up an interesting issue in that one must know the number of 1s, 0s, or transitions that are to be counted in the output response of the CUT. This will be a function of not only the CUT but also of the TPG. Once both CUT and TPG are designed, logic simulation can be used to determine the size of the ORA counter(s). Any subsequent changes to the design of the CUT or TPG may require a change in the ORA counter size since the number of 1s, 0s, or transitions may change significantly. Neglecting the influence of such design changes could result in a reduction in fault coverage or an increase in the probability of fault masking. With multiple counters for a multiple output CUT, a relatively small design change to the CUT or the TPG could result in significant design changes in the ORA in terms of the sizes of the counters. Concentration can be used at the multiple outputs of the CUT to reduce the number of counters to a single counter as indicated in Figure 5.3. Initialization of counters used for ORAs is typically to the all 0s state. The initialization control can be performed using the *BIST Start* signal, also illustrated in Figure 5.3 where the *BIST Start* is assumed to be active high.

5.5 Signature Analysis

Signature analysis uses the LFSR as its basic component [94]. While the LFSR used for a TPG implementation is a closed system (once initialized), the LFSR used for signature analysis needs input data, specifically the output response of the CUT. The output response of the CUT can be represented by a polynomial, called the *data polynomial.* The basic idea behind signature analysis is to divide the data polynomial by the characteristic polynomial of the LFSR. The remainder of this polynomial division is the signature used to determine the fault-free/faulty status of the CUT at the end of the BIST sequence. This process is illustrated in Figure 5.4 for a 4-bit signature analysis register (SAR) constructed from an internal feedback LFSR with characteristic polynomial from Table 4.1. Since the last bit in the output response of the CUT to enter the SAR in time represents the coefficient for x^0, the data polynomial of the output response of the CUT can be determined by counting backward from the last bit to enter the SAR to the first. Thus the data polynomial for this example is given by $K(x)$, as illustrated in Figure 5.4a. The contents of the SAR for each clock cycle of the output response from the CUT are illustrated in Figure 5.4b along with the input data, $K(x)$, shifting into the SAR on the left hand side and the data shifting out the end of

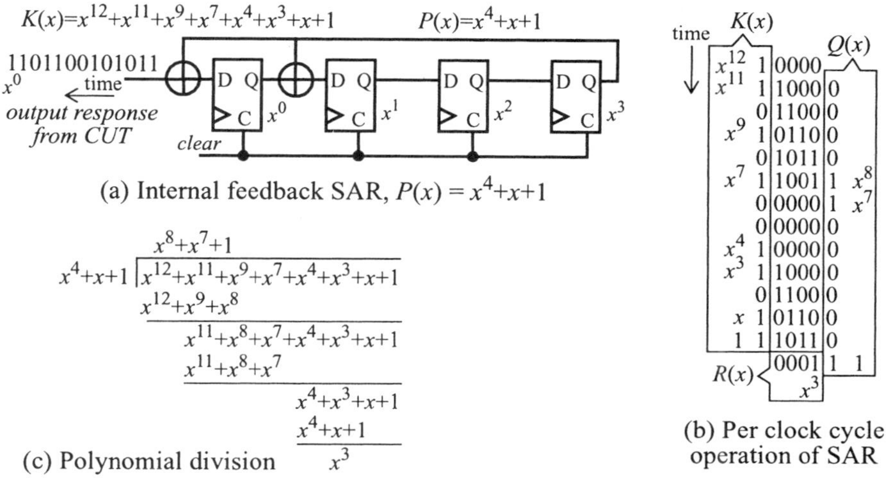

FIGURE 5.4 Example of signature analysis.

the SAR, $Q(x)$, on the right-hand side. The signature contained in the SAR at the end of the BIST sequence is shown at the bottom of Figure 5.4b and is denoted $R(x)$. The polynomial division process is illustrated in Figure 5.4c where the division of the CUT output data polynomial, $K(x)$, by the LFSR characteristic polynomial, $P(x)$, results in a quotient, $Q(x)$, which is shifted out of the right end of the SAR, and a remainder, $R(x)$, which is contained in the SAR at the end of the BIST sequence.[1]

Signature analysis is based on the same principle (and the circuit design is the same) as Cyclic Redundancy Check (CRC) codes [207]. There are a number of interesting properties of signature analysis that are worth understanding. The polynomial division process of the SAR is given by:

$$K(x) = Q(x)P(x) + R(x) \tag{5.1}$$

This is more often denoted as:

---

1. Note that in polynomial division, $x^n + x^n = x^n - x^n = 0$; therefore, we are performing the exclusive-OR of the two terms during the subtraction process. Also note that for the signature, $R(x)$, the notation of the flip-flops in the SAR has changed with respect to that used for the implementation of the characteristic polynomial of the LFSR such that the data contained in the left-most bit of the SAR at the end of the signature analysis process represents the x^0 coefficient for the signature, $R(x)$.

$$R(x) = K(x) \bmod P(x) \tag{5.2}$$

Once the fault-free signature has been established, fault detection results when the signature of the faulty circuit is different than that of the fault-free circuit. Given a faulty CUT response, $K'(x)$, the signature resulting from signature analysis, $R'(x)$, would be given by:

$$R'(x) = K'(x) \bmod P(x) \tag{5.3}$$

The relationship of the faulty output response, $K'(x)$, to the fault-free output response, $K(x)$, is given by:

$$K'(x) = K(x) + E(x) \tag{5.4}$$

where $E(x)$ is referred to as the *error polynomial* since it denotes the erroneous bits in the output response. The difference between the signature of the faulty circuit, $R'(x)$, and the fault-free circuit, $R(x)$, is given by:

$$R_E(x) = R(x) + R'(x) = E(x) \bmod P(x) \tag{5.5}$$

where $R_E(x)$ is the signature of the error polynomial, $E(x)$. Therefore, a faulty CUT will be identified anytime $R_E(x) \neq 0$. When $R_E(x) = 0$, the faulty CUT escapes detection and is referred to as *signature aliasing* [338]. This happens whenever $E(x)$ is a non-zero multiple of $P(x)$ (the zero multiple represents the fault-free circuit).

Fault detection is illustrated in Figure 5.5a where a single bit error ($E(x) = x^4$) has been introduced in the output response of the CUT, $K'(x)$. In this case, $R'(x) \neq R(x)$ and the fault causing the error is detected. However, like in any division process, there are multiple dividends that can be divided by the same divisor to yield the same remainder (while the quotients will differ), which leads to signature aliasing. An example of this is illustrated in Figure 5.5b where three errors ($E(x) = x^4 + x + 1$) in the output polynomial of the CUT, $K''(x)$, result in $R''(x) = R(x)$ such that the fault(s) causing the errors fails to be detected. Note that $E(x) = P(x)$ in this example such that $E(x)$ is a non-zero (one) multiple of $P(x)$. Also note that in both examples, the quotient, $Q(x)$, was different from that of the fault-free circuit. Therefore, the example of signature aliasing would result in detection of the faulty CUT if the quotient were saved and compared with that of the fault-free CUT. However, this would require considerable memory (one bit of memory per clock-cycle of the BIST sequence) if stored on-chip, otherwise, each bit of the quotient would have to be monitored on a per clock-cycle basis which eliminates some of the major advantages of BIST (similar to monitoring outputs using a concentrator or comparator).

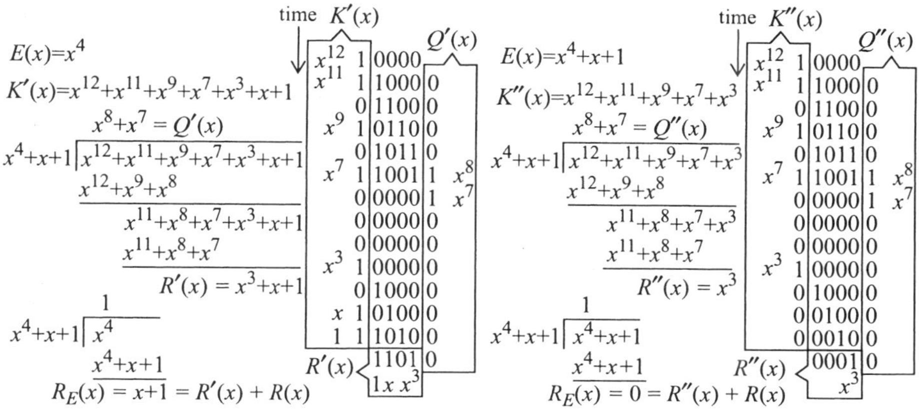

(a) Fault detection example (b) Signature aliasing example

FIGURE 5.5 Examples of fault detection and signature aliasing.

For a multiple output CUT, a concentrator, as illustrated in Figure 5.1, can be used in conjunction with an SAR. However, the more common approach is to construct a multiple input signature register (MISR) by adding exclusive-OR gates between the inputs to the flip-flops of the SAR for each output of the CUT. This is illustrated in Figure 5.6 for a 4-output CUT with an internal feedback LFSR-based MISR implementing a characteristic polynomial $P(x)=x^4+x+1$. This implementation requires five exclusive-OR gates, one for the characteristic polynomial and one for each output of the CUT. Note that the three inputs to the exclusive-OR symbol at In_2 implies 2 gates.

Aside from signature aliasing, MISRs are susceptible to one other type of fault masking usually referred to as *error cancellation* [32]. This occurs when a bit error coming

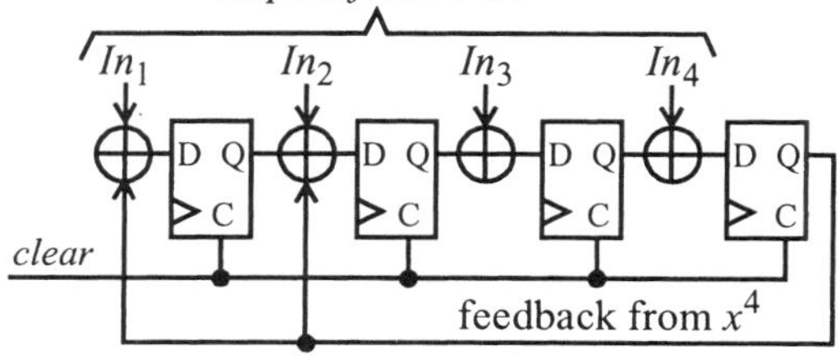

FIGURE 5.6 Example of a MISR.

into the MISR is cancelled by a second bit error as the first bit error shifts down the MISR. For example, assume that a bit error arrives at input *In*3 and in the next clock cycle a second bit error arrives at input *In*4, then the two error bits will cancel each other as they enter the last flip-flop in the MISR. While the initial bit error resulted in an incorrect signature once it was clocked into the MISR, the bit error during the next clock-cycle sets the signature back to that of the fault-free circuit. Another way to look at this situation is that a exclusive-OR tree concentrator driving an SAR is the equivalent to a MISR, but with a time sequence change in the most significant bits of data. An even number of bit errors in a given output response of the CUT will cancel each other in the exclusive-OR tree and the SAR cannot detect the fault. A similar situation holds true for the MISR if the bit errors are spaced appropriately in time.

SARs and MISRs are capable of detecting any single bit errors. They are also capable of detecting any burst errors (with the exception of error cancellation in a MISR) as long as the burst of error bits is contained within no more than n consecutive bits, where n is the degree of the characteristic polynomial of the LFSR [32]. In addition, SARs and MISRs are capable of single bit error correction when using primitive polynomials for the LFSR as long as the degree of the data polynomial, $K(x)$, is less than 2^n-1 [169]. In fact, characteristic polynomials can be constructed in such a way as to detect and correct multiple bit errors [71]. The ability to identify error bits in BIST leads to the potential for diagnosis of the faulty portion of the CUT. However, the algorithms for error identification are complex and can take considerable time and processing power. Therefore, it is the fault detection capabilities of signature analysis that are of primary interest for BIST implementations.

While signature aliasing cannot be avoided, there are a number of steps that can be taken in the design of the BIST circuitry to reduce the probability of aliasing. The classical approximation for the probability of aliasing is 2^{-n} where n is the degree of the characteristic polynomial of the LFSR. While this is not realistic since it assumes that a probability of 0.5 for any given bit in the output of the CUT to be in error, it does show the first step in reducing the aliasing problem - use large LFSRs for signature analysis. Studies have shown that primitive polynomials help to reduce signature aliasing (compared to non-primitive polynomials), particularly for short test sequences [339]. Therefore, 16-bit SARs and MISRs with primitive characteristic polynomials are a safe place to start, with larger SARs/MISRs providing further reductions in the probability of signature aliasing. Repeating the test sequence while performing signature analysis with a different characteristic polynomial reduces the probability of aliasing [104]. Similarly, repeating the test sequence with a different ordering of test patterns (preferably in reverse order) also helps. Observing intermediate signatures will also help, as can be seen in the signature aliasing example of Figure 5.5b, where observing any of the last four signatures prior to the final signature would have resulted in fault detection. These last three techniques help to reduce the probability of error cancellation in MISRs. Finally, signature aliasing is also a function of the length of the test sequence. Most signature aliasing studies have shown that the probability of aliasing is higher for short test sequences of less than several hundred clock cycles [338].

SAR and MISR implementations can be constructed with internal or external feedback LFSRs. However, the internal feedback implementation of SARs and MISRs provides the higher performance compared to external feedback initialization since the worst case delay path will have no more than two exclusive-OR gates in any path between flip-flops in a MISR and only one exclusive-OR for a SAR. In addition, the polynomial division used in Figure 5.4 and Figure 5.5 to calculate the resultant signatures only works for the internal feedback implementation [32]. For some data polynomials, the external feedback SAR will not always yield the signature that is calculated by polynomial division; the quotient is always correct but not the remainder. But there are applications where an external feedback LFSR is advantageous; for example, if a shift register already exists in the system function, it can more easily be modified to operate as an external feedback LFSR in a test mode.

Unlike LFSRs for TPGs, SARs and MISRs are typically initialized to all 0s, but one should be careful that the fault-free circuit signature does not result in an all 0s signature. Otherwise, the SAR or MISR fails to be self-testing; there are faults, such as the data input to the SAR stuck-at-0, that would prevent the compaction of the output response data while the final signature appears to be correct. Therefore, it is a good practice to always look for a final signature that is different from the SAR or MISR initialization value. The following procedure provides a technique that will facilitate determination of the initial value (or seed) for the SAR or MISR required to give a desired final signature, given the output response of the fault-free CUT [168]:[1]

Step 1: Initialize the SAR or MISR to the all 0s state and run a logic simulation to determine the final signature, $R(x)$, for the fault-free CUT output response.

Step 2: Perform a bit-wise exclusive-OR of the final signature, $R(x)$, and the desired signature, $R_d(x)$, to obtain a seed value, $S_1(x)$.

Step 3: Reversing the order of the bits, use the seed value, $S_1(x)$, to initialize an LFSR constructed with the reciprocal polynomial, $P^*(x)$, of the characteristic polynomial of the SAR or MISR. Run a logic simulation to clock the LFSR the same number of clock cycles as was used to compact the CUT output responses in Step 1. The resultant contents of the LFSR at the end of this sequence of clock cycles represent a second seed value, $S_2(x)$.

1. I have used this technique to make the final signature result in my first two initials and the year I designed the chip (for example, "CE89") or just the year for a 16-bit signature register.

Step 4: Reversing the order of the bits, use the second seed value, $S_2(x)$, to initialize the SAR or MISR at the beginning of the BIST sequence. The desired signature, $R_d(x)$, will be the resultant fault-free circuit signature.

This procedure can also be used to obtain an all 0s or all 1s signature which are the most efficient when on-chip determination of the pass/fail status is desired. A simple NOR or NAND function can be used with an all 0s or all 1s final signature, respectively. It is surprising that this procedure only requires the CUT output responses to the test vectors for determining the initial signature in Step 1 while Steps 2 through 4 are independent of the output responses.

Signature analysis can also be performed with CA registers [249]. This is accomplished by adding an exclusive-OR gates to the CA register in the same manner as done to an LFSR to create an SAR. Similarly, an exclusive-OR gate can be added to each stage of the CA register in the same manner as for a MISR, this is sometimes referred to as a Multiple Input Cellular Automata (MICA) [297].

5.6 Accumulators

Accumulators are commonly used for checksum circuits that perform concurrent fault detection in ROMs or on large blocks of transmitted data [122]. There are four basic types of accumulators used for checksum. Three of these are illustrated in Figure 5.7 and the fourth accumulator is the Honeywell accumulator which is of little value in BIST applications. The simplest accumulator is the single precision accumulator and consists of an N-bit adder which sums the incoming data with the current contents of the N-bit accumulator register and stores the result back into the accumulator register. The overflow (carry-out) of the adder is ignored such that the sum is modulo 2^N. The residue accumulator is similar to the single precision accumulator but the carry-out of the N-bit adder is delayed by one clock cycle and used as the carry-in to the adder during the subsequent summing operation. As a result, the overflow information is retained but is reduced in significance by moving it to the least significant bit. The double precision accumulator, on the other hand, accumulates a $2N$-bit value such that the carry-out information is not reduced in magnitude. While this can be viewed as a $2N$-bit single precision accumulator with the upper N-bits to the adder tied to logic 0s, an equivalent implementation is to simply use an N-bit counter with an active high count enable input (*cen*) that is driven by the carry-out output of the N-bit adder. The double precision accumulation is the sum modulo 2^{2N}. Like the other ORA circuits discussed thus far, accumulators are typically reset to all 0s prior to the BIST sequence (note that the mechanism for resetting the accumulator registers, counter,

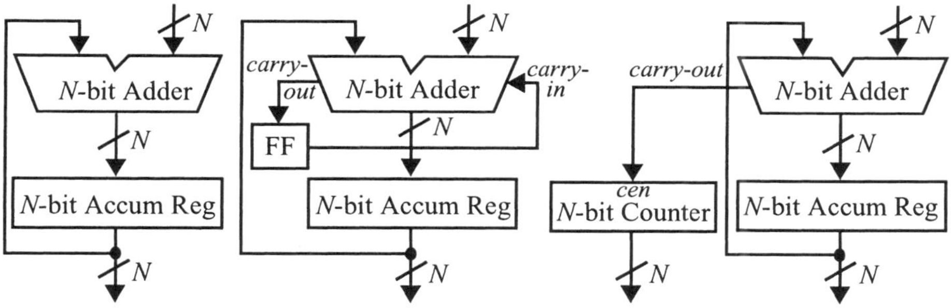

(a) Single precision (b) Residue (c) Double precision

FIGURE 5.7 Three types of accumulators.

and flip-flop is not shown in Figure 5.7). The single precision accumulator has been shown to have the highest probability of signature aliasing of the three accumulators [261].

The advantage of an accumulator-based ORA is particularly important in BIST applications for asynchronous digital systems and in mixed-signal systems where the fault-free output response sequence can vary clock cycle by clock cycle from one execution of the BIST sequence to the next. In mixed-signal systems, for example, this is caused by acceptable variations in the analog component parameters, environmental factors including voltage and temperature, and quantization noise in the digital-to-analog and analog-to-digital converters [279]. Accumulators will allow a range of resultant values to be considered for the fault-free CUT, while any resultant values that fall outside this range would indicate a faulty CUT. This will be discussed in more detail in Chapter 14.

5.7 Parity Check

Parity is a common error detection code used in concurrent fault detection circuits for system-level, on-line fault detection. The 8-bit concentrator shown in Figure 5.1 is a parallel parity circuit and will generate a logic 1 if there are an odd number of 1s in the 8-bit input data. The parity check that is sometimes proposed for use in output response compaction for BIST architectures is a serial parity check circuit as illustrated in Figure 5.8 [19]. Parallel output data from the CUT can be compacted by add-

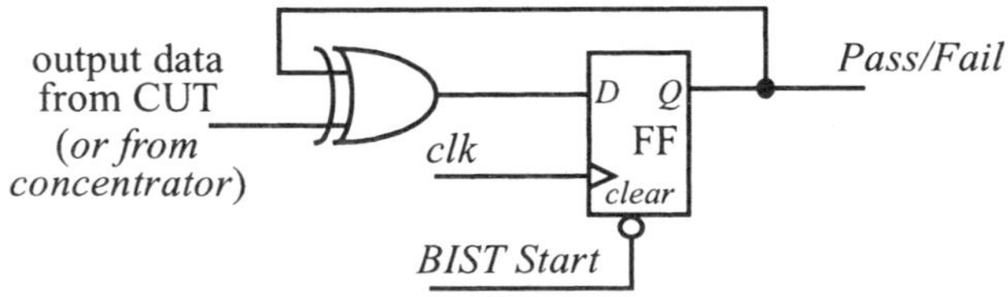

FIGURE 5.8 Parity check compaction.

ing a serial parity check circuit to each CUT output or by using a concentrator and a single serial parity check compaction circuit; note that this is equivalent to a serial/ parallel parity check circuit [274].

Prior to the BIST sequence, the single flip-flop is reset via the active low *clear* input to the flip-flop using the active high *BIST Start* control signal. During the BIST sequence the parity check circuit calculates the parity over the serial output data response of the CUT (or from the concentrated multiple output response). At the completion of the BIST sequence, the parity bit is checked for the correct result to determine the faulty/fault-free status of the CUT. The serial parity check circuit is equivalent to a one's counting compaction technique with a 1-bit counter. The serial parity check technique is also equivalent to a 1-bit single precision accumulator. Similarly, the serial parity check technique is also equivalent to a 1-bit SAR with characteristic polynomial $P(x) = x + 1$. As a result, the probability of aliasing with the parity check compaction technique is $2^{-1}=0.5$ and this is the reason that parity check compaction is rarely used in BIST implementations. However, since parity is a commonly used concurrent fault detection circuit, it is often included as part of the system function and can be considered for use in the BIST architecture where the probability of signature aliasing can be reduced by incorporating methods that will be discussed in Chapter 15.

5.8 Fault Simulation Considerations

As described in Chapter 2, once a fault has been detected by a fault simulator, it is put in a detected fault list, simulation ceases for that fault, and the fault simulator continues to simulate other faults in the list. The basic idea is that once a fault is detected, there is no need to waste additional fault simulation time on that fault, hence the fault is discarded. This stopping of simulation and discarding of detected faults is also referred to as *fault dropping* [4]. A similar strategy exists in manufacturing wafer and device-level testing; as soon as a chip is determined to be faulty, the test machine stops testing it and moves on to the next chip. This is often referred to as "trip-on-first-fail" in test machine terminology. While fault dropping improves fault simulation time (and reduces testing time in manufacturing), it presents a special problem in BIST applications.

The ORA function of BIST removes the need to continuously monitor the output response of the CUT during testing since the only the *Pass/Fail* indication needs to be checked at the end of the BIST sequence. Since all ORA techniques suffer from fault masking, inaccurate fault coverage is obtained when continuously monitoring the BIST results and allowing fault dropping. To achieve a true fault coverage determination from fault simulations, the entire BIST sequence would be executed during fault simulation prior to monitoring the pass/fail indication for fault detection. This expense, in terms of fault simulation time, can be avoided by periodically monitoring the pass/fail status of the BIST sequence during fault simulation and allowing fault dropping to occur. However, the fault coverage will only be accurate if this same technique is used in manufacturing testing in order to reduce testing time by not running the entire BIST sequence when a fault is detected early in the sequence. This technique also reduces the probability of fault masking as in the case of signature analysis when observing intermediate signatures.

5.9 Comparing ORAs

The area overhead for the various ORAs discussed in this chapter is summarized in Table 5.1 for 8-bit implementations in terms of the number of flip-flops, exclusive-OR gates and elementary logic gates. The area overhead includes circuitry to initialize the ORA at the beginning of the BIST sequence but it does not include the circuitry that facilitates holding the BIST results until they can be read for the determination of the pass/fail status of the CUT during system-level use of BIST. The only circuit that does not have 8 inputs is the SAR without concentration; this circuit has a single input. While the comparator and parity check have the fewest flip-flops, the SAR without concentration requires the fewest gates as well as the fewest gate I/O of any of the implementations; of course, it can only handle single output CUTs. Note that the SAR with 8-bit concentration has the same number of gates and gate I/O as the 8-bit MISR. The most area inefficient circuit is the 1s counting ORA without concentration. Therefore, from an area overhead standpoint, it is much better to use concentration when using counting techniques (unless there happen to be a lot of counters in the system function that can double as ORAs during the BIST mode).

TABLE 5.1 Comparison of ORA implementations.

ORA	# Flip-Flops	# Exclusive-ORs	# Gates	# Gate I/O
Comparator	1	8	3	45
1s Count with concentration	8	15	15	92
1s Count w/o concentration	64	64	120	554
SAR w/o concentration (single input)	8	4	8	36
SAR with concentration	8	11	8	57
MISR	8	11	8	57
Single Precision Accumulator	8	15	27	130
Residue Accumulator	9	16	33	149
Double Precision Accumulator	16	24	48	206
Parity check with concentration	1	8	1	69
Parity check w/o concentration	8	8	8	48

The single stuck-at gate-level fault coverage of various ORAs is shown in Figure 5.9 to give an indication of how easily the ORAs are tested as part of the BIST sequence. The fault simulations for all ORAs used the same set of 400 random test patterns. As can be seen, the SAR and MISR are easily tested within a few clock cycles from the

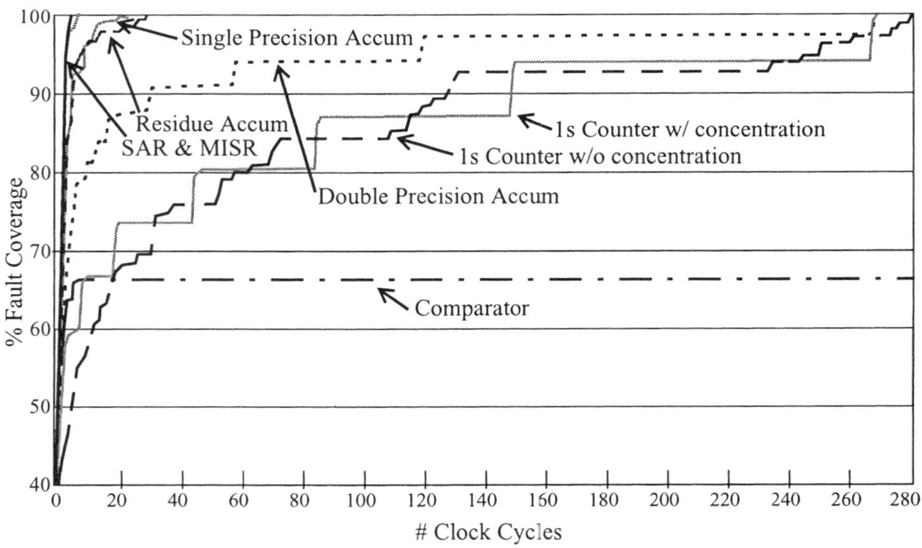

FIGURE 5.9 Fault coverage comparison of ORAs.

start of the BIST sequence (again, the LFSR leads the BIST options). The parity check circuits (with and without concentration) follow the fault coverage curves for the SAR and MISR very closely and, as a result, are not shown in Figure 5.9. The single precision and residue accumulators are also easily tested as part of the BIST sequence as seen by their fault coverage curves. The counter-based implementations, including the double precision accumulator where the MSBs of the accumulator circuit functions as a counter, are not as easily tested.

As pointed out previously, the comparator-based ORA cannot be completely tested as part of the BIST sequence as can be seen in the graph where a maximum fault coverage of about 68% was obtained. This means that additional test vectors would have to be applied, external to the BIST sequence, to completely test the comparator-based ORA. This brings up an interesting situation in that some designers advocate comparing the BIST results with the fault-free signature on-chip [168]. However, this leads to BIST circuitry that is not completely tested during the self-test, requiring additional DFT circuitry to test the comparator that would be needed to compare the signatures. This also requires the development and application of additional vectors to completely test the comparator, not to mention the possibility of undetectable faults if one were to tie the inputs to the comparator to logic 1s and 0s once the fault-free signature is known. Any design modifications can lead to a new fault-free signature and more design modifications just to get the BIST to work correctly. Therefore, it is best to compare the signatures off-chip; after all, what is the difference in comparing one *Pass/Fail* bit to comparing a sixteen or thirty-two bit signature?

In order to reduce the area overhead of a BIST implementation, existing flip-flops are typically used from the system function to double as the TPG and ORA functions during the BIST mode of operation [16]. As a result, only the additional combinational logic required to make the existing flip-flops function as TPGs and ORAs is needed along with multiplexers to switch modes of operation. The choice of which existing flip-flops to use for the BIST TPG and ORA functions depends on the level of access of the BIST capability (system-level and/or manufacturing-level testing) as will be discussed in some of the subsequent chapters.

It is interesting to note that most ORA techniques have analogies to error detection codes in concurrent fault detection circuits for on-line testing [122]. Concentration uses parity trees, counting techniques are similar to Berger and Bose-Lin codes, signature analysis uses the same LFSRs as CRC, and accumulators use the same circuits as checksum. Therefore, when these concurrent fault detection circuits are implemented and existing as part of the normal system function, they can often be used for ORAs during BIST mode to reduce area overhead. This will be discussed in more detail in Chapter 15.

5.10 Summary of Methods to Reduce Aliasing

All ORAs suffer from aliasing (fault masking) due to the compaction of the output response data. The undesirable result of aliasing is that faulty circuits can produce a fault-free circuit indication at the completion of the BIST sequence. The one possible exception is the comparator, but even the comparator is not immune to aliasing in cases where the comparator is used to detect faults in identical circuits that have equivalent faults. Therefore, it is important to consider BIST design guidelines that reduce the probability of aliasing. While these were discussed for signature analysis, they are summarized here in order of effectiveness in the context of how they pertain to other ORA designs including CA registers, counting techniques, and accumulators as well as LFSRs:

1. Use large registers based on a general rule of thumb that the probability of aliasing for an n-bit register will be approximately 2^{-n}.
2. Use primitive polynomials (or maximum length sequence design) when designing LFSR or CAR-based ORAs.
3. Repeat the BIST sequence when possible using a different primitive polynomial (or maximum length sequence design) for LFSR or CAR-based ORAs. In signature analysis for example, this reduces the probability that error polynomials will be a non-zero multiple of both of the characteristic polynomials, the condition which leads to aliasing.
4. Apply a long BIST sequence, more than a few hundred clock cycles, but less than 2^n clock cycles if possible.
5. Observe intermediate BIST results when possible and when reproducible results can be obtained.
6. Repeat the BIST sequence with the test pattern sequence in reverse order if possible.

While implementing all of these techniques is seldom possible and often impractical in terms of test time and cost, not to mention area overhead and performance penalty, the first two techniques are the most effective. Therefore, it is a good idea to observe these when designing ORAs. Some of these techniques, such as observing intermediate BIST results, are more practically implemented during manufacturing testing with no impact to area, performance, or testing time. When using comparator-based ORAs with identical circuits, it is important to compare as many different CUTs with each other as possible to reduce the probability of equivalent faults escaping detection.

CHAPTER 6

Manufacturing and System-Level Use of BIST

We continue our look at BIST with two final components that are important for system-level use of BIST: the test controller and input isolation. In order to better understand the significance of these components, we will first look at the use of BIST during manufacturing testing and the potential benefits that can be obtained. Next, the requirements for ensuring proper operation of BIST during system-level testing will be discussed along with the need for a test controller and input isolation. This will give insight into design decisions that must be made when implementing BIST.

6.1 Manufacturing Use of BIST

Since the test vectors are generated and the output responses are compacted internally by the BIST circuitry in the VLSI device or on the PCB, only power and the clock input(s) need to be applied once the BIST sequence has been initiated. When using BIST during manufacturing testing, there are a number of benefits that can be taken advantage of to reduce manufacturing testing costs.

6.1.1 Collapsed Vectors Sets

Most test machines provide the ability to specify a single instruction that will result in many clock cycles being generated by the test machine and applied to the CUT. During this clocking sequence, the input logic values cannot be changed. Similarly, outputs of the CUT cannot be monitored and compared with expected output responses since the expected output responses cannot be stored in the instruction specifying the generation of the sequence of clock cycles. However, the test patterns being produced by the TPG and the output response compaction performed by the ORA allow the primary inputs to be held constant and allow the primary outputs to be ignored while the BIST sequence is being executed. Hence, we only need to retrieve the resultant signature from the ORA at the end of the BIST sequence to determine the faulty/fault-free status of the CUT. This ability to apply many clock cycles with a single test machine instruction significantly reduces the number of test vectors that must be stored in ATE. This reduced set of test vectors is usually referred to as a *collapsed vector set.* BIST facilitates collapsing vector sets with dramatic results.[1] A collapsed vector set for BIST execution consists of a short set of vectors to initiate the BIST sequence, a single test machine instruction to supply sufficient clock cycles for the BIST sequence to complete, and a short set of vectors to retrieve the BIST results for comparison with the expected results. While collapsed vector sets reduce the ATE memory requirements for storing the vectors, the test time is still a function of the number of clock cycles applied during the BIST sequence.

For long BIST sequences, overall manufacturing testing time can be reduced by retrieving intermediate BIST results during the course of the BIST sequence. This allows many faulty chips to be detected earlier in the testing process and the test machine can move onto the next chip. How early in the BIST sequence to retrieve intermediate results is the big question, but this can be determined through fault simulations. If the fault simulator used to evaluate the BIST implementation has the ability to profile the fault coverage as a function of the number of test vectors or clock cycles, this information can be used to determine at what point(s) in the BIST sequence the fault coverage is sufficient to warrant checking intermediate BIST results. This assumes that fault dropping is being used to speed up the fault simulations. Alternatively, short sets of test vectors can be developed that do not execute the entire BIST sequence but do incorporate retrieval of intermediate BIST results near the end of each vector set. These vector sets can be simulated to determine the fault coverage as a function of the number of clock cycles elapsed in the BIST sequence. From these fault simulation results, a determination can be made of the best point(s), in terms of fault coverage, for intermediate BIST results retrieval during manufacturing testing.

---

1. My first 100,000 transistor chip had a collapsed vector set of only 300 test vectors that obtained over 98% single stuck-at gate-level fault coverage. The actual number of clock cycles applied during the test sequence was well over 32,000 with the vast majority of these clock cycles used for the BIST sequence.

Taking intermediate BIST results also provides a good opportunity to change the primary input logic values during subsequent clock cycle sequences. Otherwise, the primary inputs to the device are held constant during the application of the sequence of clock cycles for the collapsed vector set. The fault coverage obtained for a single test machine instruction BIST sequence may not be the maximum that could result if the logic values at the primary inputs can be changed. Therefore, intermediate BIST results also facilitate maximizing the fault coverage during manufacturing testing.

6.1.2 Low-Cost ATE

The opportunity arises in some cases for using a low-speed, low-cost test machine during manufacturing testing, since only clock and power must be supplied to the CUT during the BIST sequence. To illustrate this, we will consider the example of a BIST approach that was incorporated in a set of VLSI devices for a high-speed, self-routing, broadband switching system [266]. The system software interface was low-speed (around 2 MHz) and operated asynchronously to the main data processing core of the devices that ran at a higher clock frequency (around 184 MHz). The low-speed software interface and control circuitry accounted for less than 10% of the total circuitry in most of the VLSI devices in the set, leaving more than 90% of the circuitry that needed to be tested at high-speed. During manufacturing device-level package testing, the BIST sequence was initiated and the results were retrieved via the low-speed interface using a low-speed test machine with the high-speed clock supplied from an external oscillator mounted on the test fixture. While this approach worked well for the internal high-speed core of the VLSI devices, there were still high-speed I/O that needed to be tested.

The BIST architecture used for these devices also facilitated system-level testing of the high-speed interfaces between the VLSI devices of the self-routing switching system.[1] Taking advantage of this feature, a test fixture was constructed with the primary outputs of the VLSI device routed to the primary inputs such that the same low-speed test machine could be used to test the primary I/O at the system operating frequency, as illustrated in Figure 6.2. In addition, the packaged device could be tested at higher frequencies (for speed sorting) by adjusting the external oscillator on the test fixture. This technique also significantly reduced the total number of device I/O pins that needed to be serviced by the test machine. Therefore, in some applications, BIST can enable the use of lower speed test machines as well as reduce the number of pins and vectors that must be serviced, both of which have the potential for providing significant savings in ATE costs.

1. The details of this specific BIST architecture are presented in Chapter 11.

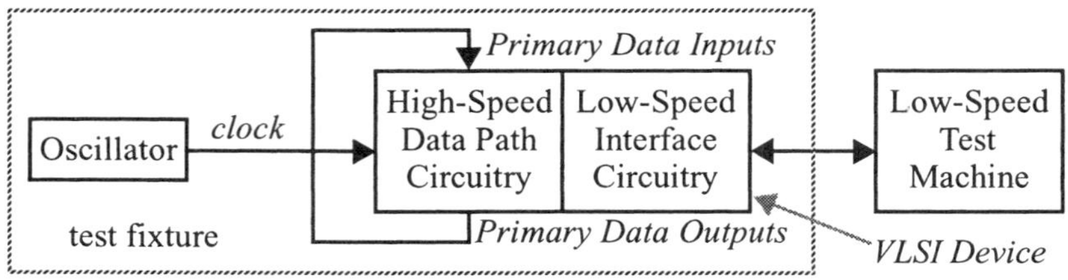

FIGURE 6.1 Using low-speed ATE for high-speed testing with BIST.

6.1.3 Burn-In Testing

The data activity and resultant power dissipation of many BIST implementations is higher than that of normal system data [90]. This provides a good environment for burn-in testing to determine if the device fails due to "infant mortality" defects. Many devices can be tested in parallel with test fixtures that supply power and clock to the devices, while BIST is executing to stress the parts. BIST can also be used with the more conventional high voltage and temperature stress used during burn-in testing. To facilitate the use of BIST for burn-in as well as to keep the burn-in test fixtures simple and inexpensive, the initiation of the BIST sequence and retrieval of BIST results should be made as simple as possible. For example, a single *BIST Start* control input (as was illustrated in Figure 1.5) allows for easy initialization and initiation of the BIST sequence. Retrieval of the BIST results is usually more difficult unless an on-chip comparator is used with the *Pass/Fail* indication driving a primary output of the chip. However, the devices can be re-tested on a test machine after the burn-in cycle to determine which devices have failed. Therefore, it is the stress of the power dissipation due to the high data activity of the BIST sequence that is most important such that simple control of the BIST sequence is the primary consideration for easy and inexpensive manufacturing burn-in testing.

Since burn-in testing takes place for extended periods of time, there are additional design considerations with respect to the TPG and test controller portions of the BIST architecture. First, it is important that the TPG continues to operate for the duration of the burn-in period in order to fully stress the part. Therefore, it is best to design the TPG and test controller such that the TPG can continue to produce and apply test patterns to the CUT after the normal BIST sequence has completed. A second consideration is that the power dissipation as a result of BIST does not cause the temperature rating of the package to be exceeded, since burn-in testing is normally performed on packaged devices. This continuous operation can be a contradiction to the desirable operation of BIST in the system due to system temperature and power consumption requirements. In the case that the BIST is also targeted for system-level use, two modes of BIST operation can be designed: a continuous BIST mode of operation for burn-in testing and a single cycle BIST sequence mode for system-level operation where the TPG will stop generating test patterns at the end of the BIST sequence. Alternatively, the burn-in test fixture must continue to cycle the BIST initiation sequence to continue to stress the device if a single cycle of the BIST sequence is designed for system-level use.

6.1.4 Other Issues

A common approach to reduce the area overhead of the BIST circuitry is the re-use of existing system function logic; in other words, using existing system functions such as counters or registers to operate as BIST circuitry during the test mode of operation. If BIST is only targeted for manufacturing testing and will not be accessible during system-level testing, the BIST implementation can take advantage of the complete control of the device I/O by the test machine to further reduce area overhead of the BIST circuitry. For example, there is no need to incorporate input isolation and a test controller in the BIST implementation if it is only used for manufacturing testing.

Asynchronous interfaces within a device can require special attention in the BIST design when used for system-level testing as will be discussed in the next section. During manufacturing testing, the various clocks to the asynchronous timing regions can be controlled by the test machine to emulate synchronous operation of the entire device. As a result, there is no need for special circuitry to isolate BIST across asynchronous timing regions. This has the potential for increasing the fault coverage that can be obtained during manufacturing testing compared to system-level testing. Additional simulations of the BIST sequence in the manufacturing mode of testing will be required since the resultant BIST signatures for manufacturing testing would be different from those that would result during system-level testing. Logic simulation can be used to obtain the fault-free circuit signatures, but fault simulations in both manufacturing and system-level testing modes may be required to determine the actual fault coverage that would be obtained. These issues will become more evident after a discussion of system-level use of BIST.

6.2 System-Level Use of BIST

The accessibility of BIST at the system-level can lead to significant reductions in system-level testing time and costs by providing higher quality tests in much less time than can be typically obtained by system diagnostic software. This is because system diagnostic software must go through system interfaces to access the hardware to be tested. These interfaces usually operate at a much lower frequency due to high loading on processor busses. Many devices in a system may not contain a processor interface to facilitate diagnostic access such that these devices must be exercised indirectly to determine if they are fault-free and operational. Improving diagnostic run time also improves Mean Time To Repair (MTTR) and system availability [251]. The indirect access of devices within the system makes non-BIST-based diagnostic fault isolation and identification more difficult and less certain. This means that components must be replaced (usually PCBs are the replaceable units) and retested a larger number of times due to misdiagnosis before the system is again operational, potentially increasing the MTTR and reducing system availability, not to mention increasing field service time and costs.

BIST provides diagnostic resolution to the CUT it is testing. While diagnostic resolution at the device-level is usually much higher than necessary for field repairs, the diagnostic information can be sent with the faulty PCB back to the repair facility to indicate which device needs to be replaced in order to repair the PCB. This helps to reduce PCB repair costs; otherwise, board-level testing must be performed in order to determine the device(s) that should be replaced. Sometimes board-level testing cannot reproduce the same conditions that lead to the failure in the field and the PCB passes all board-level tests.[1] With diagnostic information from the field, the faulty components reported from field testing can be replaced with a reasonable probability that the failure will not occur again in the field in the event that PCB-level testing could not reproduce the failure that was observed during system-level testing. The accessibility of BIST at the system-level comes with additional design requirements that must be satisfied to make the BIST effective during system-level testing.

6.2.1 Requirements for System-Level Use of BIST

There are six specific requirements that need to be satisfied when designing a BIST approach in order to ensure that BIST can be effectively used during system-level testing:[2]

1. the ability to activate and deactivate the BIST sequence by system software,
2. the ability to read the BIST results by system software for determination of the fault-free/faulty status of the CUT,

1. This is affectionately referred to as the "bone pile" by many PCB test engineers.

2. While these may seem obvious, I observed cases (including one of my own designs) where lack of attention to one or more of these requirements prevented system-level use of BIST or required design changes and re-fabrication of an ASIC in order to use BIST at the system-level.

3. the isolation of system inputs to the CUT and the isolation of data between asynchronous timing regions within the CUT,
4. the complete initialization of the BIST circuitry as well as the CUT at the beginning of the BIST sequence and prior to enabling output response compaction,
5. the ability to control a fixed-length output response compaction sequence (for non-comparison-based ORAs) and then to hold the BIST results until they can be read by system software, and
6. the ability to prevent the patterns produced at the primary outputs of the CUT during the course of the BIST sequence from causing undesirable effects in the system.

The first two requirements address the accessibility of the BIST circuitry by diagnostic software during system-level testing. The ability to initiate the BIST sequence is most easily provided by a register in the processor interface that can be written to initiate the BIST sequence. This usually only requires a single bit (for *BIST Start*) that can be included in some other larger multiple-bit register to save address space. While not necessary, it is highly desirable to include a status bit that can be read by the diagnostic software to indicate when the BIST sequence is completed (a *BIST Done* bit). This allows the software to perform other tasks while polling the device periodically to see when the BIST sequence has finished. Otherwise, the diagnostic software may need to implement a timer function to determine when sufficient time has elapsed for the BIST sequence to complete and the BIST results are ready to read. The ability to poll a *BIST Done* indication leads to additional reductions in diagnostic run time since processing (presumably other testing) can be done in parallel while the BIST is executing (including initiating and polling other BIST sequences running in parallel). The ability to deactivate the BIST sequence is also not necessary but is a good precaution for those cases where prolonged operation of the BIST sequence would result in exceeding power consumption requirements of the system or package temperature ratings.

Diagnostic software must retrieve the BIST results to determine the pass/fail status of the CUT once the BIST sequence has completed. In the case of a comparator-based ORA, only a single bit is needed. The other types of ORAs require the reading of a complete multiple-bit signature. The register containing the BIST results (counter, SAR, MISR, MICA, or accumulator) should be directly accessible for reading on a processor interface or via the Boundary Scan interface whenever possible. An existing register (or group of registers when the signature size is greater than the processor data bus width) that is readable via the processor interface can double as an ORA while the *BIST Start* signal is active to reduce area overhead. It is also a good idea to make the register(s) writable for initialization of the ORA as well as for testing the processor interface. When a processor interface is not available on the component that contains the ORA for the BIST, some mechanism must be designed to get the BIST signature to an area of the system that does have a processor interface.[1] Devices incorporating Boundary Scan with INTEST or user-defined register capabilities can provide a solution in these cases for not only reading the BIST results but also controlling *BIST Start* and polling for *BIST Done* [100].

The last requirement is not of major concern in most systems since BIST is typically performed off-line while the system is out of service and can easily be re-initialized. However, this easily overlooked requirement can have surprising ramifications in some systems. For example, errors are commonly detected by concurrent fault detection circuits in other VLSI devices or on other PCBs in the system as a result of the execution of BIST. These errors can, in turn, cause interrupts to be generated that may invoke an unintended or unexpected response in another portion of the system. In this case, disabling fault recovery software for that portion of the system during BIST execution may be sufficient to avoid any problems. Another example is unintentional back driving of a bi-directional bus due to multiple active drivers in devices running BIST in parallel. In this case, the solution may be simply requiring that the BIST for those devices be executed in some sequence and not in parallel. Alternatively, blocking gates used for input isolation (discussed next) can also be used to isolate the effects of the BIST sequence in a device from the rest of the system by forcing the output to be held at a constant logic value. However, the use of this isolation technique should be minimized to avoid extra area overhead and performance penalties in the form of additional clock-to-output delays.[2]

6.2.2 Input Isolation

The third requirement for system-level use of BIST is isolation of system input data. This is necessary to ensure that BIST results are reproducible from one execution of the BIST sequence to the next for a fault-free CUT. The system data inputs are typically not within the control of diagnostic software and, therefore, the data must be considered to be unknown. If this system data enters the CUT during the BIST sequence, the output response of the CUT may change depending on the values of this

1. This can be tricky in some cases. In one application, at the end of the BIST sequence, we had to insert a 16-bit signature onto the data path in the system, route it through two other VLSI devices onto another PCB, and then load the signature into a RAM that could be read through a processor interface. This example of the difficulty in getting the BIST signature to the system software gives an indication of what the system diagnosticians are up against when BIST is not made available to them during system-level testing.

2. Clock-to-output delay is the maximum time before data becomes valid at the primary output after the active edge of the clock.

infiltrating system data. A difference of a single bit can cause a BIST failure indication in a fault-free device.

Input isolation can be obtained by the same techniques used for ad-hoc DFT controllability: multiplexers or blocking gates as illustrated in Figure 6.2. While a blocking gate imposes the least area overhead and performance penalty, only a fixed logic value is applied to the CUT during the BIST sequence. The multiplexer allows application of test patterns from the TPG to be applied to the CUT for higher fault coverage at the expense of additional area overhead and performance penalty compared to a blocking gate. Since input isolation is frequently done at the primary inputs to an ASIC, there is concern regarding the impact on set-up and hold time specifications of the device.[1] As a result, there may be some cases where blocking gates are sufficient and more desirable. An alternative to blocking gates is to use the INTEST feature in devices that incorporate Boundary Scan with the INTEST capability. This eliminates the area overhead and, more importantly, the performance penalty while providing the required input isolation capability. Normally test patterns can not be applied during the BIST sequence via the Boundary Scan registers, but the BIST sequence can be repeated with different logic values applied via the Boundary Scan cells in order to improve the fault coverage.

Another aspect of the third requirement is isolation of data between asynchronous timing regions within the CUT, since the asynchronous timing can produce an effect similar to unknown system data or un-initialized circuitry, causing a BIST failure indication in a fault-free CUT. For example, during different executions of the BIST sequence, the data may be transferred between the asynchronous interface on different clock cycles such that a different BIST signature is obtained for each execution of

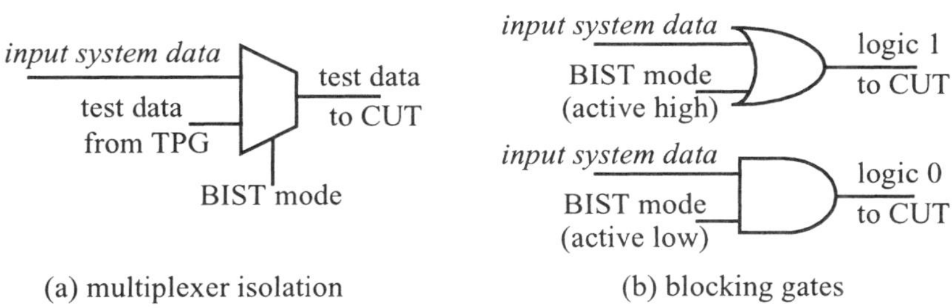

FIGURE 6.2 Examples of input isolation.

1. Set-up and hold times are the minimum times data must be held valid before and after the active edge of the input clock, respectively.

the BIST sequence, even though the CUT is fault-free. The same input isolation schemes (with the exception of Boundary Scan INTEST) can be used to partition the core of the device into separate CUTs. This may also imply that separate BIST circuits (TPG, ORA, and test controller) exist for each timing region since the BIST circuitry must also be clocked by the same clock as the CUT. Therefore, it is best to treat each asynchronous timing region as a separate CUT with separate BIST structures and apply isolation techniques at the interfaces. While single clock, synchronous designs do not require this consideration or modification, there are cases where asynchronous interfaces are unavoidable. One should also be careful of asynchronous resets/presets and tri-state busses to ensure they will not prevent the effective use of BIST at the system-level by causing non-reproducible BIST results.

6.2.3 Test Controller

The fourth and fifth requirements for system-level use of BIST also pertain to the reproducibility of the BIST results. First, the CUT as well as the BIST circuitry must be completely initialized to eliminate any unknown system data bits within the CUT from entering the ORA. Like isolation of the system data inputs, a single un-initialized bit left in the CUT can cause failing BIST indications in a fault-free circuit. This initialization is the responsibility of the last component of BIST, the test controller. When re-using existing system function circuitry (such as existing flip-flops) for constructing the BIST circuitry, the initialization is not always a simple task. While a global reset or preset sounds like a good solution, they tend to mask testability problems as well as create new ones. As a simple example, one must be careful not to reset the *BIST Start* register bit as part of the BIST initialization sequence.

The CUT can sometimes be initialized by enabling the TPG to begin applying test patterns to the CUT while the ORA is disabled. Once the CUT is fully initialized, the ORA can be enabled to begin output response compaction. The number of clock cycles required for initialization is usually the *sequential depth* of the CUT. The sequential depth of the CUT in this case is defined as the maximum number of flip-flops in any path from input to output. Feedback loops in the CUT can lead to longer initialization sequences and can sometimes prevent complete initialization. Therefore, potential initialization problems must be addressed. A method to determine if any initialization problems exist in a BIST implementation is described in the next section.

The fifth requirement is that the length of the BIST sequence must be constant from one execution to the next in terms of the time that the ORA is enabled and compacting output responses as well as in terms of the points in the test pattern sequence at which the ORA is enabled and disabled. The test controller must therefore apply a fixed length initialization sequence while the ORA is disabled, enable the ORA, and then disable the ORA at the same point in the test pattern sequence for every execution of the BIST sequence. Otherwise, one clock cycle difference in the initialization sequence or the output response compaction sequence can cause a BIST failure indication for non-comparator-based ORAs. This can easily be seen in the signature analysis sequence of Figure 5.4 where one clock cycle on either side of the correct point gives different signatures. This is not a difficult problem (easily implemented in the test controller design) but one that should not be ignored.

The test controller is generally not a complicated circuit but is essential for system-level use of BIST. A simple counter will usually suffice for the test controller and existing counters may be available in the design to double as a test controller during the BIST mode. The TPG itself can often serve for part, if not most, of the test controller function. Examples of this include the simple BIST design in Figure 1.5 and the March Y TPG in Figure 4.1. The remaining control to enable/disable the ORA can be accomplished with additional decoding logic as was illustrated in Figure 5.2 where the comparator-based ORA was disabled during write operations to the RAM. Once the ORA has been disabled at the end of the BIST sequence, the BIST results must be held until they can be read by the system diagnostic software, or transferred to a register that is accessible by system diagnostics, for comparison with the known results for the fault-free CUT.

Holding the BIST results in the ORA after completion of the BIST sequence implies additional logic that was not shown in the ORA implementation diagrams in Chapter 5. The clock-enabled flip-flop of Figure 2.11 provides an example of the logic that could be used in each flip-flop of the ORAs from Chapter 5 to facilitate enabling/disabling output response compaction during the BIST sequence as well as holding the final BIST results until read for the pass/fail determination. Therefore, the clock enable would be activated (high for the example in Figure 2.11) only during the portion of the BIST sequence for which output response compaction is desired, otherwise the ORA will hold its data. This adds one 2-to-1 multiplexer to each flip-flop of the ORAs in Table 5.1 for system-level use of the BIST approach.

6.2.4 Design Verification of BIST

Design verification of the requirements for system-level use and reproducibility of BIST results can be obtained through the following logic simulation procedure that makes use of the undefined logic value (2, 3, U, or X) to represent un-initialized logic values in most logic simulators [285]. The idea is to use the undefined logic value (denoted by '*U*' from this point) to indicate any unknown system data infiltrating from the system inputs or left over from insufficient or incomplete initialization. The simulation procedure (which we will call the *U's simulation*) is as follows:

1. Begin the logic simulation with an un-initialized circuit so that as much of the CUT and BIST circuitry as possible contains *U*s.
2. Through the logic simulation vector set, provide control for the absolute minimum number of critical input signals such as clock, synchronization inputs, and those required to initiate BIST.
3. Apply *U*s to all other system inputs during the entire logic simulation. Most simulators will allow a *U* as a valid input value, although this feature is rarely used.
4. Initiate the BIST sequence as early in the simulation as possible to avoid inadvertent initialization of the CUT by means other than the test controller. Initiate the BIST sequence in an as "system-like" manner as possible.
5. Monitor the ORA data bits throughout the logic simulation to observe any *U*s that might appear in the ORA.
6. Allow the logic simulation to continue well beyond the normal completion of the BIST sequence to look for any changes in the BIST results held in the ORA. Read the BIST results in an as "system-like" manner as possible.[1]

The procedure is usually fast and easy to run since this is a logic simulation and not a fault simulation. If a *U* does appear in the ORA it can be traced back to its source by repeating the simulation while monitoring different internal nodes in the circuit (a *U* is usually easy to spot in all the 1s and 0s of the rest of the logic). Once the source of the *U* is found, corrective measures can be taken and the *U*'s simulation repeated. It is important to monitor the ORA throughout the simulation and focus on the point that the first *U* appears, because once a *U* gets into an SAR, MISR, MICA, counter, or accumulator it rapidly fills the ORA making determination of its point of entry more difficult.

1. To illustrate the usefulness and importance of this simple *U's simulation* procedure, I hesitantly relate the following "war" story. I began using this procedure early in my career and was able to uncover a number of potential problems well before the design was fabricated. But, I inadvertently overlooked this procedure on one chip. There were no problems with the BIST at wafer, package, and board-level testing. One day, many months after the board containing this chip went into production, I received a phone call from the manufacturing facility to report a large number of PCBs that were failing system-level test but passing board-level test. The test engineer told me the system diagnostic failure was an incorrect BIST signature in this particular chip and asked what I thought could cause the problem. After some thought, I realized that I had forgotten to run the *U*'s simulation on that design. Within the next hour or so, I performed the *U*'s simulation and found (to my embarrassment) I had neglected to isolate one system input. Fortunately, it was a data input that system diagnostics had indirect control of via a register in another part of the system. As a result, system diagnostic software had to set that register bit to the appropriate logic value prior to executing the BIST sequence, a 'relatively painless' fix to the problem.

One can be reasonably confident that the BIST will work properly at the system-level once valid BIST results are obtained via this simulation method. The procedure does a good job of finding problems with requirements 3 through 5 and a decent job of finding problems with the first two requirements. An ASIC may be partitioned into separate CUTs and BIST structures due to asynchronous timing regions or different types of BIST approaches applied to different types of CUTs. Each CUT and its associated BIST circuitry should undergo this *U*'s simulation separately while the other CUTs remain un-initialized and their BIST sequences inactive. This will help ensure that the interfaces between the CUTs have been properly isolated from each other during the BIST mode.

6.2.5 Effective Fault Coverage

One disadvantage of system-level use of BIST is that the entire chip cannot be part of the CUT in most cases. This is usually due to access of BIST through a processor interface for initiating the BIST sequence, polling for *BIST Done,* and retrieving the BIST results at the end of the sequence. This means that portions of the processor interface in the device executing BIST must remain operational for system access before, during, and after the BIST sequence execution. As a result, the fault coverage obtained by BIST will be lowered for the overall device since the BIST circuitry is prevented from applying test patterns to those portions that must remain system accessible. Fortunately, a processor interface to a VLSI device is usually more controllable and observable than the rest of the device so that the portions not tested by the BIST architecture are easily testable by functional vectors at manufacturing test and by diagnostic software during system-level testing. Evaluating a BIST approach based on the lowered fault coverage of the overall chip, including the portions that must remain operational, may give a pessimistic impression of the effectiveness of the candidate BIST technique. Therefore, it is important to consider the fault coverage of the portion of the device that forms the CUT during the BIST sequence, including the BIST circuitry, but not including those portions reserved for system accessibility. The *effective fault coverage* provides information in addition to the overall fault coverage that facilitates more accurate evaluation and decision making concerning the effectiveness of a given BIST approach. The effective fault coverage, FC_{eff}, is given by:

$$FC_{eff} = \frac{D + xP}{T_C - U_C} \qquad \textbf{(6.1)}$$

where T_C and U_C are the total number of faults and undetectable faults, respectively, in the CUT, meaning only that portion of the chip that is being tested by the BIST circuitry as well as the BIST circuitry itself.

6.2.6 Diagnostic Evaluation

One problem that system diagnosticians face is determination of the fault detection and location capability of their diagnostic software. Aside from diagnostic run time, the ability to determine which PCB in a system is faulty during one pass of the diagnostic test is of critical importance to system down time. If it takes multiple replacements of suspect PCBs before the system is operational, the system down time increases. Physical fault insertion has been used to evaluate the effectiveness of system diagnostics in detecting and locating faults in systems that are too large for fault simulation. Physical fault insertion is a lengthy and expensive process which prolongs the product development cycle and increases the development costs. Since BIST gives diagnostic resolution to the CUT (whether that be a chip, a subcircuit within a chip, or an entire PCB), diagnosticians do not need to develop diagnostic software for that CUT. The need for physical fault insertion is reduced since the fault simulation results obtained from evaluating the BIST approach at the device-level give a quantitative measure of the fault detection capability that will be obtained with system diagnostic access of the BIST approach.[1]

6.2.7 Multiple BIST Sessions

BIST is rarely implemented as a single architecture for an entire VLSI device. This is usually due to different types of structures within a given chip, such as RAMs and general sequential logic, or multiple clocking regions of the general sequential logic of a given chip, such as high-speed and low-speed sections of the logic. As a result, the VLSI device may contain multiple BIST approaches. In some applications the multiple BIST approaches can be executed in parallel. However, due to the high data activity usually associated with BIST, power consumption/dissipation is often of concern in large chips [243]. The various BIST sequences may need to be scheduled to prevent exceeding power and/or temperature limitations. This can have an impact on the partitioning of the chip into smaller CUTs in order to control the power dissipation. Multiple BIST sessions offer a good solution to these problems but their execution in series results in an increase in total testing time. Therefore, we would like to execute as many BIST sequences in parallel as possible to reduce testing time while being careful to avoid exceeding power and temperature restrictions.

1. One diagnostic supervisor estimated a savings of three technical head-count months in diagnostic software development for every chip that incorporated BIST. The diagnosticians only needed to know how to access BIST and the resultant good circuit signature. About 50% of all the BIST implementations in the projects I worked on at Bell Labs were requested by system diagnosticians. That is a good indication of their endorsement of the value added by BIST to system-level testing.

Multiple BIST sessions can have implications on the test controller design as well. At the completion of a given BIST sequence, the test controller may need to communicate with other BIST controllers. This is particularly true if a sequence of BIST executions is needed to avoid excessive power dissipation, either at the device-level or at the board-level. This can be accomplished by connecting the *BIST Done* from one test controller to the *BIST Start* of another controller if the test controllers are designed such that the *BIST Start* and *BIST Done* are both active for the same logic value. As a result, BIST sequences are executed in a domino-like sequence. Once the final BIST sequence has been executed, as indicated by its *BIST Done*, the results from each BIST circuit can be read to determine the pass/fail status of the various CUTs.

In some BIST applications, area overhead and performance penalties in control paths can be eliminated by not incorporating control registers (that reside on a processor interface) in the CUT being tested by BIST. Instead, values written into the control registers prior to the execution of BIST and held static during the BIST sequence will determine the resultant signature at the end of the BIST sequence. The asynchronous operation of most processor interfaces prevents changing the control register values during the execution of the BIST sequence without risking failing BIST indications in a fault-free circuit. Higher fault coverage can be achieved by repeatedly executing the BIST while writing new logic values into the control registers between each execution of the BIST sequence. This can be an effective approach to avoiding area overhead and performance penalties in many applications, but this comes at the expense of multiple BIST sessions to achieve higher fault coverage. Note that during manufacturing testing the per clock cycle control of the test machine allows writing different control register values during a single BIST sequence without jeopardizing the integrity of the final signature. As illustrated in Figure 6.3, the TPG and ORA functions can be incorporated in existing registers on a processor interface (such as control registers and status registers, respectively) to facilitate easy access to BIST during

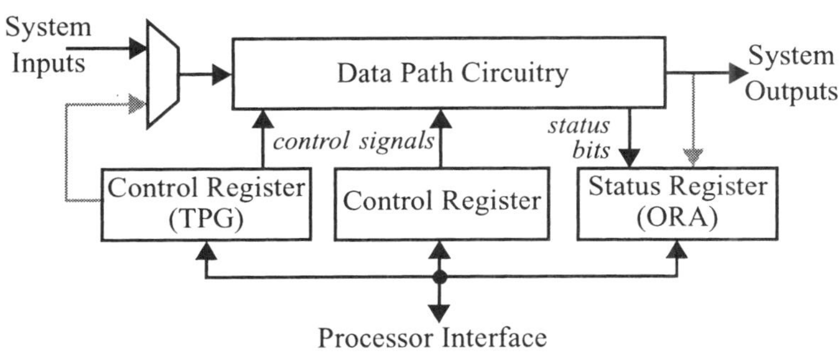

FIGURE 6.3 Application of BIST to chip with a processor interface.

system-level testing. Note that the TPG and ORA connections during the BIST mode of operation are shown by the gray lines and arrows. The TPG can be incorporated in control registers which drive control signals that can be changed on a per clock cycle basis. Normally static control signals (as indicated in the middle register in Figure 6.3) would be set to different values at the beginning of each BIST session. Any error outputs from concurrent fault detection circuits in the CUT should be fed into the ORA for compaction, illustrated by the "status bits" in the figure. This will be discussed in more detail in Chapter 15.

Another approach that involves multiple executions of BIST sequences is to partition the application into smaller CUTs in order to completely test a device while minimizing area overhead. A single TPG and single ORA can then be used to test each CUT in turn. Multiplexers are typically used to select which CUT is under test during a given BIST sequence. This brings us to the topic of generalized BIST architectures.

6.3 General BIST Architectures

There are several general classifications of BIST architectures that are sometimes used to describe BIST approaches [4]. BIST circuitry is often referred to as being either *centralized* or *distributed*. These two basic BIST architectures are illustrated in Figure 6.4. The centralized type of architecture can be effective in reducing area overhead and is sometimes called a *shared* BIST architecture. However, the centralized BIST architecture requires multiple executions of the BIST sequence since each CUT must be routed to the ORA independently through additional circuitry such as the multiplexer shown in Figure 6.4a. In the distributed architecture, each CUT has its own TPG and ORA such that BIST of the individual CUTs can be executed in paral-

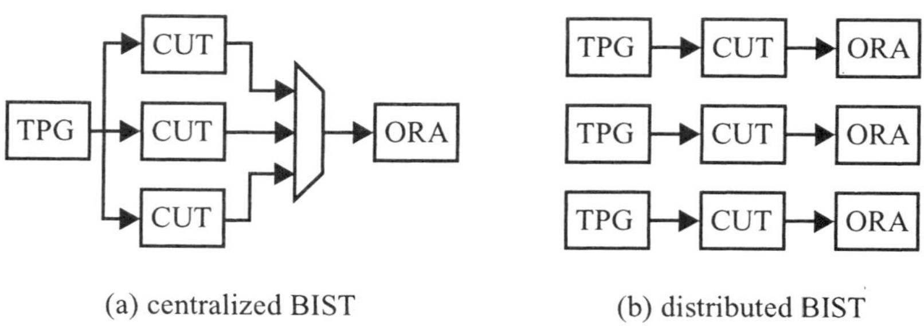

FIGURE 6.4 General BIST architectures.

lel. Therefore, there is a trade-off between test time and area overhead while diagnostic resolution to the faulty CUT is the same for both approaches.

When the TPG and ORA functions are implemented using existing flip-flops and registers of the CUT, the BIST architecture is generally referred to as *embedded*. Independent TPG and ORA functions added to the CUT are typically called a *separate* BIST architecture. Embedded BIST approaches are also referred to as *intrusive* or *invasive* BIST architectures since the BIST circuitry added to the existing flip-flops may also incur performance penalties in the critical timing paths of the CUT [241]. Intrusive BIST approaches often provide higher fault coverage in most applications than *non-intrusive* or *non-invasive* BIST architectures since the flip-flops inside the CUT offer better controllability and observability.

Another classification of BIST architectures is *on-line* and *off-line* which refers to the state of the system during the execution of BIST [5]. On-line BIST implies that the system is in-service and performing in its normal mode of operation. In off-line BIST, the system is out-of-service due to detection of a fault via concurrent fault detection circuits or for periodic testing to determine if the system is fault-free. Most BIST approaches can only be used for off-line testing since the BIST circuitry overwrites the normal system data passing through the CUT with test patterns. However, there are a few on-line BIST architectures that facilitate testing during normal system operation.

One final classification of BIST architectures is based on the application of test patterns to the CUT [50]. In some BIST applications, a new test pattern is applied with each clock cycle and these architectures are referred to as *test-per-clock* BIST systems. On the other hand, some BIST architectures use a scan chain as the basic mechanism for delivering test patterns to the CUT as well as for retrieving the output responses from the CUT. In these applications, as will be discussed in Chapter 10, the BIST is referred to as a *test-per-scan* system. The test-per-clock BIST systems result in shorter testing times in most cases.

The remainder of this book consists of chapters devoted to specific types of BIST approaches. Various BIST architectures for general sequential logic are discussed in Chapter 7 through Chapter 11. BIST approaches for regular structures such as RAMs, ROMs, and PLAs are discussed in Chapter 12. While FPGAs and CPLDs can be considered regular structures, their re-programmability and architectural features lead to more specialized BIST approaches that will be discussed in Chapter 13. The application of traditional digital BIST techniques to analog and mixed-signal systems requires special considerations for the BIST implementations as will be discussed in Chapter 14. Making use of existing concurrent fault detection circuits to help reduce the area overhead in BIST implementations is discussed in Chapter 15.

6.4 Overview of BIST Design Process

The best time to consider BIST for a given project is at the very beginning. Once the determination has been made to incorporate BIST, the next question is usually how to go about implementing BIST. When incorporating BIST into a design (a VLSI device for example), the following steps provide a general approach to most applications [1][3][135]:

1. Determine whether the BIST approach is to be used only for manufacturing, or if it should be accessible for system-level use as well. Almost all subsequent BIST design decisions will be a function of this determination.
2. Identify any unique aspects of the architecture that will be useful or difficult in the implementation and application of BIST. Examples include: a processor interface that can be used for system-level access, asynchronous timing regions that need to be partitioned and isolated, or critical timing paths that must avoid direct insertion of additional BIST logic.
3. Partition the design into subcircuits that break down according to general sequential logic (further partitioned along timing boundaries) and regular structures like RAMs, ROMs, FIFOs, etc. The reason for this partitioning is that these various types of structures are best serviced by different BIST approaches targeted for those specific structures.
4. Select a BIST technique for each subcircuit, including the BIST architecture: embedded or separate, centralized or distributed. This would also include selecting TPG type (LFSR, CAR, counter, FSM, etc.) and selecting ORA type (comparison or compaction-based). The selection of the BIST technique should take area overhead, performance penalties, and diagnostic resolution into consideration to ensure that manufacturing and/or system-level testing requirements are met.
5. Identify any CAD tools that exist (or need to be obtained and installed) that will assist in the implementation and evaluation of BIST. In fact, this particular step could come earlier in the process since the availability of CAD tools may assist in deciding which BIST approach to use for a given subcircuit.
6. Implement the BIST TPG and ORA circuitry and, if used for system-level testing, incorporate input isolation at boundaries of the subcircuits.
7. Run fault simulations to determine if the fault coverage obtained with the selected BIST approach meets requirements or expectations.
8. Consider power dissipation to determine if all BIST sequences can be executed in parallel to minimize test time, or if some, or all, of the BIST sequences must be executed in series in order to avoid excessive power dissipation.

9. Design the test controller(s) and consider interfacing the test controller(s) and ORA(s) with Boundary Scan to provide BIST accessibility from the Boundary Scan TAP or to a processor interface for system-level accessibility.
10. Run the *U*'s simulation to verify reproducibility of BIST results during system-level operation and to identify any initialization, input isolation, or test controller problems.
11. Run fault simulations for final determination of the fault coverage with various fault models of interest and/or *N*-detectability of gate-level fault models using elementary logic gates.

6.5 The Clock Enable Problem

There is one problem when applying BIST to general sequential logic that should be kept in mind for both manufacturing and system-level testing. The problem occurs when clock-enabled flip-flops are used in a circuit. This is a common design practice in single-clock, synchronous designs; for example, conversion from serial data to parallel data often uses this type of circuitry. To illustrate the problem, consider the circuit in Figure 6.5. While all flip-flops in this example circuit are clocked by the same clock on every rising clock edge, flip-flops A and B receive new data on every clock cycle but flip-flops C and D receive new data every *N* clock cycles based on the clock enable signal, *Enable*. As a result, the maximum delay that can be tolerated in the combinational logic between flip-flops A, B, and C must be less than the period of one clock cycle (denoted "delay<1cc") while the maximum delay for the combinational logic between flip-flops C and D must be less than the period of *N* clock cycles (denoted "delay<*N*cc"). If pseudo-random test patterns are applied to the *Enable* signal, flip-flops C and D could be enabled to receive new data more often than once every *N* clock cycles. This does not present a problem in flip-flop C since the maximum delay in its preceding combinational logic is less than one clock cycle. The problem results in flip-flop D since the data passing through its proceeding combina-

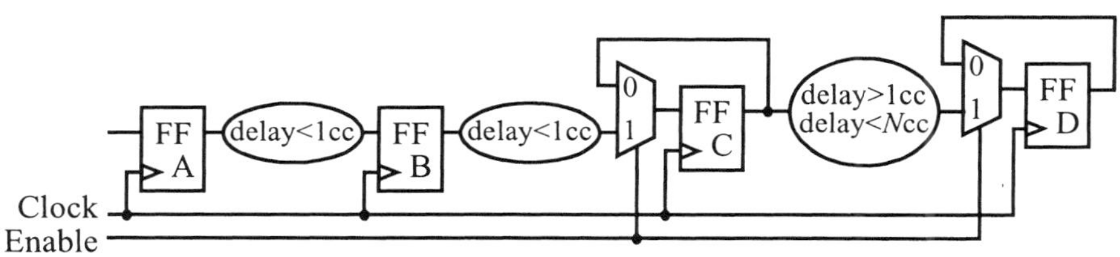

FIGURE 6.5 Example circuit with clock-enabled flip-flops.

tion logic has not had sufficient time to pass through and set-up at the input to flip-flop D. Depending on the propagation delay variations due to processing, temperature, and voltage, non-reproducible BIST results may be obtained such that a fault-free circuit will produce faulty BIST signatures.

This problem can be avoided by clocking the CUT at a reduced clock rate, say at least N times slower than the normal clock rate. This will guarantee reproducible BIST results regardless of how often the *Enable* signal is activated by the pseudo-random test patterns. However, it also means that the combinational logic between flip-flops A, B, and C will not be tested at-speed. As a result, delay faults in those combinational logic blocks will not be detected by the BIST. If delay faults exist and go undetected by the BIST, the circuit will fail during system operation and produce errors due to excessive delay in either the combinational logic network between flip-flops A and B or between flip-flops B and C. This is obviously a less than desirable scenario. While testing with the slower clock frequency does not lead to failing BIST results for a fault-free circuit, it requires implementation of a slower clock in the system for the BIST mode.

This example of testing with a slower clock frequency also points out another potential disadvantage of BIST. If BIST is used exclusively during all manufacturing testing and for system-level testing, any faults not detected by BIST may remain undetected but may cause failures of the system. It is a good idea to use other tests, such as functional vectors in conjunction with BIST during manufacturing testing to minimize the probability of defects escaping detection early in the testing process. This is especially true given the generally accepted "rule of ten" which says that the cost of finding a fault increases by an order of magnitude with each step in the testing process from device-level testing, to board-level testing, to system-level testing, and then to field testing [74]. While the vertical testability is an advantage of BIST, it can also be a potential problem if there is no testing orthogonality in the manufacturing testing process.

Another solution to the clock enable problem is to avoid applying pseudo-random test patterns to the *Enable* signal. Instead, the *Enable* signal would continue to be produced as if in the normal system mode of operation, even during the BIST mode. This clock enable problem is important to keep in mind when implementing BIST to avoid creating timing problems that will result in BIST failures while ensuring that BIST is testing what it is intended to test in the circuit. We will revisit this clock enable problem in Chapter 11 after we have investigated the various BIST approaches for general sequential logic.

CHAPTER 7

Built-In Logic Block Observer

We now begin looking at specific BIST architectures that have been implemented in a variety of practical applications. The Built-In Logic-Block Observer (BILBO) was the first BIST approach to be proposed and undergo wide spread use. Therefore, as a result of this historical distinction, it will be the first of the BIST architectures we will discuss. The architecture and operation of the BILBO along with the benefits and limitations of the approach are presented in this chapter.

7.1 BILBO Architecture

The BILBO is an embedded and off-line BIST architecture because it uses existing flip-flops from the CUT to construct the TPG and ORA functions [139]. The basic idea is to partition the CUT into groups of flip-flops and groups of combinational logic. Each group of flip-flops is augmented with additional logic to provide multiple modes of operation. When the BILBO functions as a TPG, it provides pseudo-random test patterns by operating as an LFSR. When the BILBO functions as an ORA, it performs multiple-input signature analysis by operating as a MISR [138]. The application of BILBO to a CUT in its simplest form is illustrated in Figure 7.1, where the flip-flops at the primary inputs and outputs are used to construct two BILBOs. The BILBO at the primary inputs generates pseudo-random test patterns as a TPG and the BILBO at the primary outputs operates as a MISR-based ORA. As a result, the

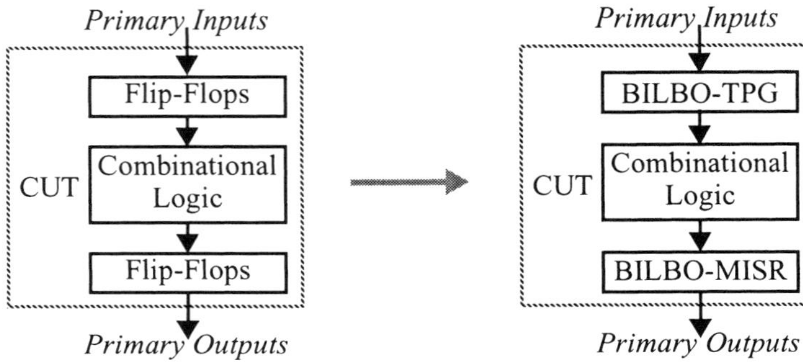

FIGURE 7.1 Simple BILBO application to a CUT.

BILBO is a test-per-clock BIST approach since a new test pattern is applied to the CUT and a new output response is compacted during each clock cycle of the BIST sequence.

The additional gates added to the existing flip-flops to create the BILBO (as it was originally proposed) are illustrated in Figure 7.2 for a 4-bit BILBO implementation. For each flip-flop, one exclusive-OR gate, one AND gate, and one NOR gate are added [138]. For each BILBO, one multiplexer, one inverter, and one or more exclusive-OR gates (needed to construct a primitive characteristic polynomial) are added. Note that in Figure 7.2, the LFSR mode of operation uses an external feedback implementation for the characteristic polynomial, $P(x) = x^4 + x + 1$.

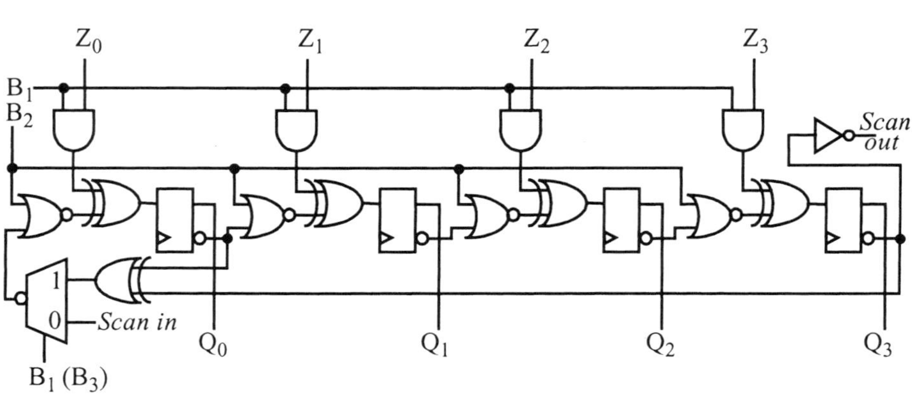

FIGURE 7.2 Original BILBO implementation.

There are two control leads (B_1 and B_2) used to control the modes of operation of the BILBO as summarized in Table 7.1. During BIST operation, the MISR mode is used for output data compaction. In order to provide a TPG during BIST, the *Z* inputs to the BILBO must be held at a constant logic 0. This forces the MISR to now operate as a LFSR generating a maximum length sequence of pseudo-random test patterns. This requires additional logic and/or additional control inputs to the BILBO not shown in Figure 7.2. For example, additional AND gates can be added to the *Z* inputs to force all *Z* inputs to logic 0 under the control of a third control input, B_3, when the BILBO is to operate as an LFSR-based TPG [4]. Alternatively, each of the 2-input AND gates at the *Z* inputs, shown in Figure 7.2, can be expanded to 3-input AND gates with B_3 driving the third input. An alternate approach is to provide independent control of the multiplexer using an additional B_3 control input (as shown in parentheses in Figure 7.2) [111]. This latter modification to the BILBO provides the modes of operation summarized in Table 7.2. Therefore, the BILBO as originally proposed was not directly capable of operating as a TPG.

TABLE 7.1 Original BILBO modes of operation.

B_2	B_1	Mode of Operation
0	0	Shift (scan) mode ($Q_i <= Q_{i-1}$)
0	1	MISR (BIST) mode ($Q_i <= Z_i \oplus Q_{i-1}$)
1	0	Initialization mode ($Q_i <= 0$)
1	1	System mode $Q_i <= Z_i$)

TABLE 7.2 Expanded BILBO modes of operation.

B_3	B_2	B_1	Mode of Operation
0	0	0	Shift (scan) mode ($Q_i <= Q_{i-1}$)
1	0	1	MISR (BIST) mode ($Q_i <= Z_i \oplus Q_{i-1}$)
1	0	0	TPG (BIST) mode ($Q_i <= 0 \oplus Q_{i-1}$)
X	1	0	Initialization mode ($Q_i <= 0$)
X	1	1	System mode ($Q_i <= Z_i$)

Later BILBO implementations used the basic flip-flop illustrated in Figure 7.3 along with its accompanying table of modes of operation [111]. The primary difference between this modified BILBO flip-flop and the flip-flop used in the design proposed for the original BILBO implementation is a re-ordering of the modes of operation with respect to B_1 and B_2. In addition, the scan chain uses the *Q* output of the flip-flop as opposed to the $\overline{Q}$ output. A "MOS friendly" alternative to the two AND gates in

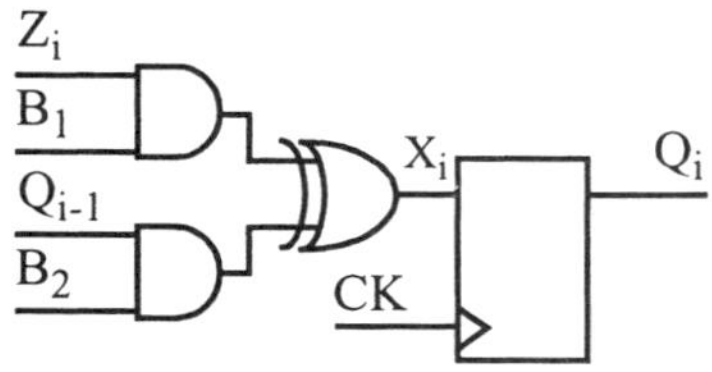

B_2	B_1	X_i	Mode of Operation
0	0	0	Initialization ($Q_i <= 0$)
0	1	Z_i	System ($Q_i <= Z_i$)
1	0	Q_{i-1}	Scan ($Q_i <= Q_{i-1}$)
1	1	$Z_i \oplus Q_{i-1}$	BIST ($Q_i <= Z_i \oplus Q_{i-1}$)

FIGURE 7.3 Modified BILBO flip-flop.

this flip-flop is to use NAND gates; the functional modes of operation are the same regardless of the use of AND or NAND gates.

A final modification to the BILBO overcomes the TPG mode of operation problem discussed above with minimum area overhead and without the need for the third control input, B_3 [50]. This modification is illustrated in Figure 7.4 for a 4-bit implementation where the modified BILBO flip-flop of Figure 7.3 is used (with NAND gates instead of AND gates). The modes of operation for this version of the BILBO are given in the associated table as a function of the two control inputs, B_1 and B_2. The BILBO is initialized using the scan mode of operation. This version of the BILBO has been more frequently used in applications since the mid-1980's than the original version of the BILBO.

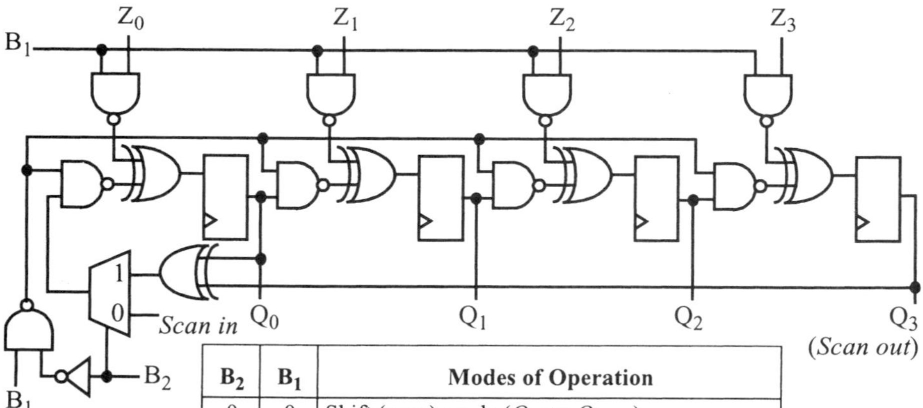

B_2	B_1	Modes of Operation
0	0	Shift (scan) mode ($Q_i <= Q_{i-1}$)
0	1	System mode $Q_i <= Z_i$)
1	0	LFSR (BIST TPG) mode ($Q_i <= Q_{i-1}$)
1	1	MISR (BIST ORA) mode ($Q_i <= Z_i \oplus Q_{i-1}$)

FIGURE 7.4 Modified BILBO and modes of operation.

7.2 BILBO Operation

Multiple test sessions are required to completely test the CUT due to the two BIST modes of operation for the BILBO as a TPG and as a MISR. Consider, for example, the CUT illustrated in Figure 7.5 that has had the BILBO approach applied. Two test sessions are required in order to test all of the groups of combinational logic. In the first test session, combinational logic blocks *A* and *C* are under test since their driving BILBOs are in the TPG mode of operation and the BILBOs that are driven by the outputs of these combinational logic blocks are in the MISR mode. Combinational logic block *B* is not under test during this test session and must wait until the second test session when its driving BILBO can be configured as a TPG and the BILBO it drives is configured as a MISR [140].

The test sessions start by putting the BILBOs into the scan mode (or shift mode) of operation to shift in the BILBO initialization values. Non-zero initialization values are required for the TPGs to produce their pseudo-random test pattern sequences. While the MISRs can be initialized to all 0s, the TPGs cannot and, as a result, the reset mode incorporated in the original BILBO implementation is of no use for BIST operation as a TPG.

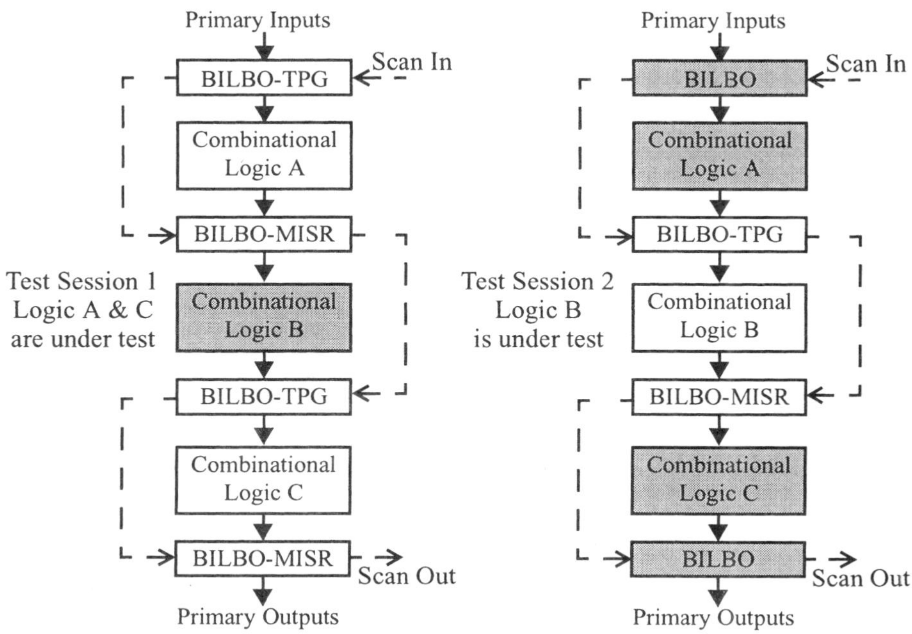

FIGURE 7.5 BILBO application and operation.

After initialization, the BILBOs are put into their appointed mode of operation for that test session (either TPG or MISR) and the BIST sequence is executed. At the completion of the BIST sequence, the scan mode is used to retrieve the BIST results in the BILBOs that performed the MISR operation to determine the pass/fail status of the combinational logic blocks that were under test during that test session. Note that the flip-flops of the CUT should be completely tested during any test sessions that they are actively used in either the TPG mode or the MISR mode. The additional BILBO circuitry will be tested as the various modes of operation are activated. While the results of one test session are being shifted out in the scan mode, the initialization values can be shifted into the BILBOs for the next test session. More than two test sessions are often required in realistic applications of the BILBO since it is rare to find the kind of pipe-lined structure illustrated in Figure 7.5. Therefore, determining the minimum number of test sessions, as well as which BILBOs should be TPGs and ORAs during a given test session, is an important part of implementing BILBO and can be difficult in some applications.

7.3 Test Session Scheduling

The goal of minimizing testing time and cost requires determining the minimum number of test sessions that must be applied to a CUT that has been implemented with multiple BILBOs. This is referred to as *test session scheduling* [2]. One technique for test session scheduling is to represent the CUT as a directed graph with each node representing a group of flip-flops that will form a BILBO while the edges represent the combinational logic between each group of flip-flops. This graph is referred to as a *register adjacency graph*. Graph coloring algorithms or clique partitioning algorithms can be used to obtain the minimum number of test sessions [68].

An example CUT and its associated register adjacency graph are illustrated in Figure 7.6. In the register adjacency graph, registers are represented by vertices, while the combinational logic blocks between two sets of registers are represented by the directed edges of the graph, with the direction indicating data flow [260]. Note that the edges of the graph in Figure 7.6b are labeled to more easily see the combinational logic blocks represented by the edges, but this is not commonly done in most algorithms for scheduling BILBO test sessions [263]. The basic idea of a test scheduling algorithm is to find the minimum number of groups of registers that have no edges between vertices within a given group as illustrated in the two steps of Figure 7.6b. This is because the register(s) driving the inputs to a combinational logic block under test must be TPGs while the register(s) at the output of the combinational logic block must be MISR-based ORAs during that test session.

After step 1 in Figure 7.6b, we can see that there will be no problem with registers R1 and R3 performing as TPGs in the test session with registers R2, R4, and R5 performing as MISR-based ORAs. This will test combinational logic blocks A and C. However, register R6 cannot be a TPG at the same time as register R3 since there is an edge between registers R6 and R3 representing combinational logic block B. Proceeding to step 2, we can see that during another test session registers R2 and R6 can perform as TPGs with register R3 performing as a MISR-based ORA to test combinational logic block B. In order to test the remaining combinational logic block D, a third test session will consist of register R5 operating as a TPG while register R6 operates as a MISR-based ORA. As a result, this example will require a minimum of three test sessions as illustrated in Figure 7.6c. To completely test combinational logic block B, we concatenate registers R2 and R6 as a single TPG with a primitive

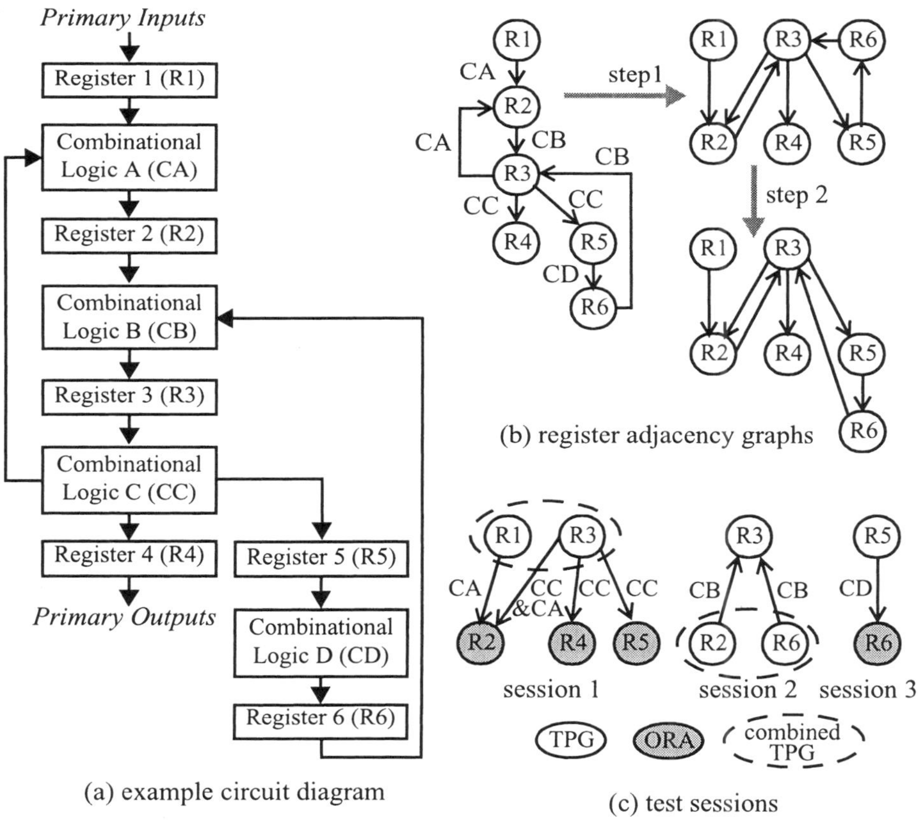

FIGURE 7.6 Example of BILBO test session scheduling.

characteristic polynomial constructed for the combined size of the two registers [24]. Otherwise, if registers R2 and R6 were separate TPGs with identical characteristic polynomials, then they would apply identical test patterns to both sets of inputs to combinational logic block B. This could significantly lower the fault coverage that would be achieved for this logic block. Similarly, registers R1 and R3 must be concatenated with a new primitive polynomial in order to completely test combinational logic block A. This concatenation of registers leads to additional area overhead for the BILBO implementation due to extra exclusive-OR gates and multiplexers required to create the alternate characteristic polynomials for the concatenated TPGs. Note that if the concatenated registers always perform together as TPGs and ORAs, they can be considered as a single BILBO register with a single characteristic polynomial. However, this is often not the case and multiple characteristic polynomials may be required for concatenated registers as illustrated in Figure 7.6c.

7.4 BIST Controller

Multiple test sessions require that a BIST controller be designed to sequence through the test sessions for system-level use of the BILBO approach. In the worst case, the BIST controller must produce a separate set of control leads (B_1 and B_2) to each individual BILBO resulting in a maximum of $2N$ control leads for N BILBOs. Obtaining the minimum number of test sessions not only minimizes the total testing time but also minimizes the number of separate control leads that must be supplied to the various BILBOs by the BIST controller. This in turn helps to minimize the logic and design complexity of the BIST controller [263]. The number of control signals and complexity of the BIST controller is also increased by those cases where we need to concatenate registers to form larger TPGs for complete testing of a given combinational logic block in order to control which characteristic polynomial is to be used for a given test session. Therefore, test session scheduling must be determined before the design of the BIST controller can begin.

The resultant design of the BIST controller consists of a basic FSM design that produces and distributes the control signals to the various BILBO implementations in the CUT. In addition, the BIST controller may include one or more counters to determine when the BIST sequence has completed for each test session. The size of the counter(s) will be a function of the TPG forming the largest degree LFSR for that test session [43]. Alternatively, by starting the BILBOs in the all 1s state, the return to the all 1s state can easily be decoded for the largest BILBO in a given test session to provide an indication of the end of the test session; with this technique minimal additional logic is required for the BIST controller. The FSM portion of the BIST controller can be designed using any FSM design methodology, but one-hot encoding techniques are good candidates for BILBO implementations [187].

Since the number of test control leads has a direct impact on the design of the BIST controller, minimizing the number of these control leads is also important [21]. Given the control leads (B_1 and B_2) of the modified BILBO in Figure 7.4, the total number of control leads can be reduced to N+1 for N BILBOs. This is because all BILBOs operate in the same mode when performing either the system or shift modes of operation [127]. As a result, a single B_2 control lead can be shared between all BILBOs since it switches the BILBOs between the system/shift and the TPG/ORA modes of operation. The B_1 control lead must be dedicated to each BILBO since it will control whether a BILBO is a TPG or an ORA for a given test session. The total number of B_1 control leads can be further reduced in the cases where groups of BILBOs always operate as TPGs or ORAs during the same test session, in which case a single B_1 control lead can be shared within a group. When a BILBO is not being used during a given test session, its control value for B_1 can be considered as a don't care for that test session; this facilitates minimization of the number of control leads. For example, the BILBO implementation illustrated in Figure 7.6 would require a total of four control leads for the three test sessions: a single B_2 shared by all BILBOs and three B_1 control leads supplied to R1/R3, R2/R4/R6, and R5, respectively, as illustrated in Figure 7.7. During shift and system modes of operation (when B_2=0), all of the B_1 control leads must provide the same logic value to all BILBOs.

	Test Session		
Register	**1**	**2**	**3**
R1	TPG	X	X
R2	ORA	TPG	X
R3	TPG	ORA	X
R4	ORA	X	X
R5	ORA	X	TPG
R6	X	TPG	ORA

→

	Test Session		
Register	**1**	**2**	**3**
R1/R3	TPG	ORA	X
R2/R4/R6	ORA	TPG	ORA
R5	ORA	X	TPG

FIGURE 7.7 Test control leads for BILBO implementation in Figure 7.6.

7.5 Concurrent BILBO

A problem that occurs in many practical applications of BILBO is referred to as *register self-adjacency*. This happens when the output of a register feeds back to the input of a combinational logic block that, in turn, drives that same register, as illustrated in Figure 7.8a [111]. As a result of this feedback, the BILBO (R2 in Figure 7.8b) must act in both TPG and MISR modes at the same time in order to test the combinational logic block. More specifically, the register operates as a MISR while the resultant signatures being compacted are simultaneously used to provide input test pattern bits. This could lead to a reduction in fault coverage since a maximum length pseudo-random sequence cannot be guaranteed [198].

A solution to this problem of register self-adjacency is the Concurrent BILBO (CBILBO) [333]. The CBILBO facilitates simultaneous TPG and ORA operation by creating an additional register. A 4-bit CBILBO implementation is illustrated in Figure 7.9 where the addition register is at the top of the figure and operates as a MISR. The lower register in Figure 7.9 performs as a LFSR-based TPG during BIST mode and as the normal system flip-flops during the scan and system modes of operation. The modes of operation of the CBILBO as controlled by B_1 and B_2 are given in the associated table. Since the MISR and TPG are now separate registers, they can operate independently such that the TPG portion of the CBILBO can produce a maximum length pseudo-random sequence while the MISR simultaneously compacts the output responses from the combinational logic as shown in Figure 7.8c. The CBILBO facilitates complete testing of the CUT in only one test session since the additional register for MISR-based ORA allows simultaneous TPG and ORA operation. Therefore, the CBILBO eliminates the problem of register self-adjacency as well as the need for test session scheduling. The price that must be paid is doubling the number of flip-flops in the CUT which can lead to significant area overhead in most practical applications, particularly since the extra register for the MISR is not used during normal system operation.

The MISR is initialized to the all 0s state whenever B_2=0 and n clock cycles (where n is the number of bits in the CBILBO) are applied in either the system or scan mode of

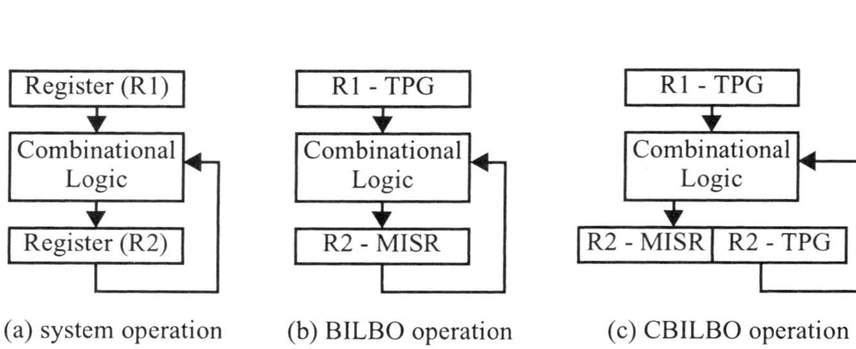

FIGURE 7.8 Register self-adjacency in BILBO applications.

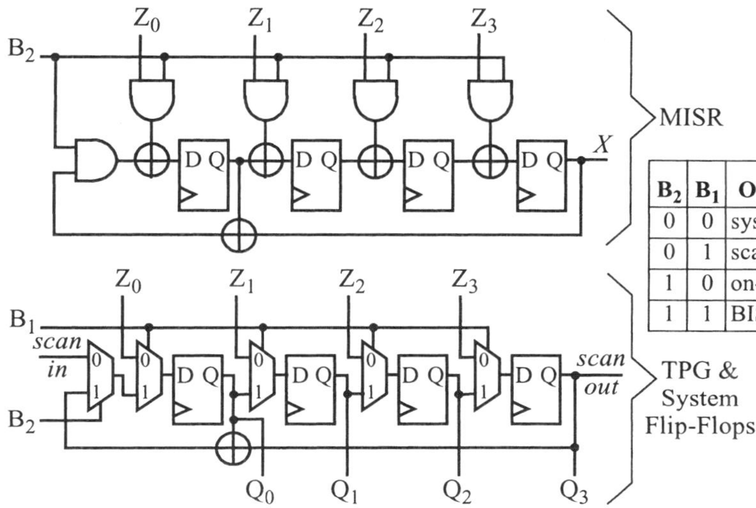

B_2	B_1	Operation Mode
0	0	system mode
0	1	scan mode
1	0	on-line checking
1	1	BIST mode

FIGURE 7.9 Concurrent BILBO structure.

operation. The original CBILBO, as proposed in [333], includes additional combinational logic in the TPG register to implement an LFSR that will produce all 2^n possible test patterns, including the all 0s test pattern. Therefore, the TPG could also be initialized to the all 0s state. The TPG shown in Figure 7.9 does not contain this additional logic and would be initialized to any state other than the all-0s state via the scan mode of operation.

An "on-line checking" mode of operation is provided by the MISR portion of the CBILBO during system operation to perform the equivalent of signature analysis on the data passing through the system. For general sequential logic, this would require additional hardware to control the boundaries of the data word, as well as the ability to compare the resultant signatures for on-line fault detection. Therefore, the usefulness of this mode of operation for general sequential logic is somewhat doubtful. However, this on-line checking can be effectively used for FSMs with fixed state sequences with little additional hardware [333].

Regardless of whether the MISR portion of the CBILBO is used for on-line checking or off-line BIST, the resultant signatures from the MISR portion of each CBILBO must be retrieved in order to determine the faulty/fault-free status of the CUT. When B_1=1 and B_2=0, the data in the MISR portion of the CBILBO shown in Figure 7.9 will shift data from left to right [333]. However, the left-most AND gate in the MISR portion of the CBILBO also serves to initialize the MISR to the all 0s state. As a result, there is a separate scan path for each CBILBO to retrieve multiple signatures. Alternatively, a continuous scan chain can be formed through all of the MISRs by including a 2-to-1 multiplexer at the input to the first flip-flop in the MISR portion of each CBILBO. This also requires an additional control input B_3 as well as an additional scan out primary output of the CUT since this MISR scan chain is separate from the normal TPG scan chain.

In summary, the CBILBO eliminates the need for test session scheduling since the entire CUT can be tested in a single test session. Furthermore, the CBILBO overcomes the problem of register self-adjacency. However, a major area overhead associated with CBILBO implementations for the entire CUT results from the doubling of all flip-flops in the CUT. An area efficient alternative to using all CBILBOs in the CUT is to only incorporate CBILBOs in cases where register self-adjacency occurs while using normal BILBOs of all other registers [24].

7.6 Benefits and Limitations

The maximum amount of logic added to the CUT for implementing BILBO and CBILBO are given in Table 7.3 in terms of the number of flip-flops, exclusive-OR gates, 2-to-1 multiplexers, and elementary logic gates (denoted 'Gates' in the table). This excludes additional logic required to implement the BIST controller. The expressions in the table can be used together to accurately predict the maximum amount of additional logic required to implement any combined BILBO/CBILBO approach to overcome register self-adjacency without incurring the larger area overhead of a complete CBILBO implementation. The performance penalty associated with a BILBO implementation is a 2-input NAND gate delay and a 2-input exclusive-OR gate delay in every path between flip-flops since all flip-flops are modified to create BILBOs. This means that these two delays have been inserted into every critical timing path in

TABLE 7.3 Area overhead of BILBO and CBILBO.

BIST Approach	Area Overhead				Performance Penalty
	Flip-Flops	Exclusive-ORs	Multiplexers	Gates	
BILBO	0	$N_{FF} + 3N_B$	N_B	$2N_{FF} + 3N_B$	1 XOR + 1 gate
CBILBO	N_{FF}	$N_{FF} + 6N_C$	$N_{FF} + N_C$	$N_{FF} + N_C$	1 MUX

where: N_{FF} = total number of flip-flops in CUT
N_B = number of BILBOs in CUT
N_C = number of CBILBOs in CUT

the CUT. The performance penalty associated with a CBILBO implementation is one 2-to-1 multiplexer delay in every path between flip-flops since existing flip-flops are modified to create the TPG portion of the CBILBOs. This means that the CBILBO incurs less performance penalty than the regular BILBO implementation. However, it is important to remember that since each CBILBO doubles the number of flip-flops in those registers that are replaced by CBILBOs, the larger area overhead will tend to spread out the physical layout, adding to the performance penalty as a result of additional routing delays.

By using primitive polynomials, maximum length pseudo-random sequences are generated by each BILBO or CBILBO operating as a TPG and applied to the combinational logic circuits at the output of those flip-flops. If the all 0s test pattern is not generated by the TPG, the combinational logic is not exhaustively tested. The scan mode can be used to apply the all 0s test pattern and to shift out the results. However, in most CUTs the combinational logic is pseudo-exhaustively tested since an m-input combinational logic circuit usually consists of many single output combinational logic functions with fewer than m inputs. These single output combinational logic functions are referred to as *logic cones* [32]. A logic cone has a vertex consisting of a primary output or the input to a flip-flop. Tracing back through the circuit, the cone of logic consists of all logic gates found in the paths between the vertex of the logic cone and its inputs, which consist of primary inputs and/or outputs of flip-flops. The inputs to a combinational logic cone are sometimes referred to as the set of *dependencies* for that logic cone [286]. An n-bit subset of the m-bit patterns ($n < m$) produced by an m-bit LFSR implementing a primitive characteristic polynomial contains all possible 2^n possible patterns including the all 0s pattern. Therefore, each combinational logic cone that has fewer inputs than the number of bits in the BILBO driving the logic cone is exhaustively tested, achieving detection of all detectable single stuck-at gate-level faults. This high fault coverage is the primary advantage of BILBO.

Since a new test pattern is generated and applied to the combinational logic each clock-cycle, the BILBO is referred to as a *test-per-clock* BIST approach [50]. The size of the BILBOs determines the test time associated with each test session. The test length of a given test session is determined by the largest BILBO that is configured as a TPG in that test session. For an n-bit BILBO, the test sequence will require 2^n-1 clock cycles. Therefore, BILBOs must be kept to some reasonable maximum size, less than 22 for example, to avoid extremely long test sessions.[1] At the same time,

1. We once investigated the possibility of implementing the BILBO approach in a 256-point digital FFT circuit [209]. The largest logic cone in the circuit had 1,806 dependencies. This made a BILBO implementation impractical, even though the architecture was nicely pipe-lined, since there was no way to run the maximum length pseudo-random sequence, not to mention the problem of finding a primitive polynomial for n=1,806.

BILBOs must be kept to some reasonable minimum size, greater than 10 for example, in order to minimize the probability of signature aliasing when the BILBO is operating as a MISR; recall that the classical bound on signature aliasing is 2^{-n} for an n-bit MISR. Therefore, grouping the existing flip-flops of the CUT into appropriately sized BILBO registers can be difficult for many applications, particularly when we must concatenate registers to construct larger TPGs.

Register Transfer Level (RTL) VHDL and Verilog descriptions are typically partitioned into registers through the definition of related flip-flops that would naturally form registers. For example, a VHDL *bit_vector* or *std_logic_vector* [255] would imply a natural register that could correspond to a BILBO implementation. As a result, description of the system function in VHDL or Verilog offers a good opportunity to begin consideration of a possible BIST implementation using the BILBO or CBILBO approach [134]. However, there are typically any number of single flip-flops (defined as a *bit* or a *std_logic* in VHDL, for example) that must be considered for concatenation with other registers in order to obtain optimal sized BILBOs. In addition, the register self-adjacency problem can be common in synthesized designs when clock-enabled circuits are incorporated as a result of incompletely specified signals.

As in the cases of the TPGs discussed in Chapter 4 and the ORAs discussed in Chapter 5, an item of interest is how well do the BILBO and CBILBO test themselves during the BIST sequence. Table 7.4 gives the single stuck-at gate-level fault coverage of an 8-bit BILBO implementation using the modified design illustrated in Figure 7.4. The fault simulations proceed through the modes of operation in the order given in Table 7.4 to emulate the sequence of operations that would be used during off-line testing. It is important to note that only 92.4% fault coverage is obtained during the normal BIST sequence including scan mode initialization, BILBO operation as a TPG and as an ORA, and using the scan mode to retrieve the resultant signature at the end of the BIST sequence for pass/fail determination. In order to detect the remaining faults, the BILBO must be operated in the system mode. This implies that 100% fault

TABLE 7.4 Modified BILBO fault coverage during operation.

% Faults Detected	Cumulative Fault Coverage	Mode of Operation
57.4%	57.4%	Scan mode (*for initialization*)
14.2%	71.6%	TPG (BIST) mode
20.3%	94.9%	MISR (BIST) mode
0%	94.9%	Scan mode (*to retrieve signature*)
5.1%	100%	System mode

coverage can be obtained for the BILBO implementation during manufacturing testing since the CUT can also be tested in its system mode of operation. However, only 92.4% fault coverage can be obtained for the BILBO circuitry itself during system-level use of BIST since the system mode of operation will not be under test.

Table 7.5 gives the single stuck-at gate-level fault coverage of an 8-bit CBILBO implementation using the design illustrated in Figure 7.9. The fault simulations proceed through the modes of operation in the order given in Table 7.5 to emulate the sequence of operations that would be used during off-line testing. Note that only 87% fault coverage is obtained during the normal BIST sequence including initialization using the scan mode, the single test session BIST mode, and the scan mode to retrieve the resultant signature at the end of the BIST sequence for pass/fail determination. In order to detect the remaining faults the CBILBO must operate in the system mode and in the on-line test mode for retrieval of the on-line signature. This implies that 100% fault coverage can be obtained for the CBILBO implementation during manufacturing testing but only 87% fault coverage can be obtained for the CBILBO circuitry during system-level use of BIST since the system mode of operation will not be under test.

TABLE 7.5 Concurrent BILBO fault coverage during operation.

% Faults Detected	Cumulative Fault Coverage	Mode of Operation
25.6%	25.6%	Scan mode (*for initialization*)
11.1%	36.7%	BIST mode
50.3%	87.0%	Scan mode (*to retrieve signature*)
8.2%	95.2%	System mode
4.8%	100%	On-line test mode (*with signature retrieval*)

It is important to note that the BILBO and CBILBO implementations (as illustrated in Figure 7.4 and Figure 7.9, respectively) do not include the ability to hold the signatures in the MISR until read by the system diagnostic software. Instead, the scan mode must be entered immediately after the BIST sequence has completed and the signatures shifted out from each BILBO in the MISR mode and/or the MISR portion of each CBILBO. This type of signature retrieval is not a problem during manufacturing testing. However, system-level testing can be much more difficult if the processor used to compare the expected signatures operates asynchronously to the CUT. Additional holding registers, external to the CUT, may be needed for transfer of the resultant BIST signatures to the processor for pass/fail determination. Alternatively, an additional 2-to-1 multiplexer can be added to each flip-flop to facilitate a clock enable function as shown in Figure 2.11, which can be used to hold the final signature until it can be retrieved for pass/fail determination for that test session. This adds to the area overhead and performance penalty of the BILBO and CBILBO approaches. For BILBO, the area overhead is an additional 2-to-1 multiplexer for every flip-flop and the performance penalty is an additional 2-to-1 multiplexer delay in every path between flip-flops. For the CBILBO, the 2-to-1 multiplexer is only added to the flip-flops used for the MISR. As a result, the normal system function does not incur any additional performance penalty other than an increase in routing delays due to the additional logic increasing the area of the chip and associated routing lengths.

The area overhead can be reduced by not constructing BILBOs from all the registers in the CUT. This approach can work well in some pipe-lined architectures. In other applications, the fault coverage may be significantly reduced when not all registers are replaced with BILBOs [4]. While BILBO encounters implementation problems in many applications, it does work well for pipe-lined circuits where the circuit is already partitioned into registers and combinational logic blocks. The pseudo-exhaustive testing of the combinational logic means that fault simulation may be eliminated since high fault coverage is guaranteed. This is the major advantage of BILBO and makes it worth keeping in mind as a candidate for such applications.

CHAPTER 8 Pseudo-Exhaustive BIST

The high fault coverage provided by BILBO is due to its ability to perform pseudo-exhaustive testing of the CUT. This ensures that every combinational logic cone in the circuit is tested with an exhaustive set of test patterns. In other words, each n-input combinational logic cone is tested with all 2^n possible test patterns. Collectively, the complete combinational logic network is pseudo-exhaustively tested. This guarantees detection of all detectable gate-level faults in the CUT without the need for fault simulation. While BILBO has the problems of register self-adjacency and test session scheduling, other pseudo-exhaustive BIST approaches have been developed in which these problems are non-existent. In this chapter, we investigate these pseudo-exhaustive BIST approaches.

8.1 Autonomous Test

The first pseudo-exhaustive BIST approach to be proposed after BILBO was called *Autonomous Test* [174]. In its most basic form, Autonomous Test adds an LFSR, that includes logic to generate the all 0s state, for the TPG and a MISR for the ORA. A 2-to-1 multiplexer is added to the input of each flip-flop in the sequential circuit to facilitate application of the exhaustive test patterns to the combinational logic cones as well as to retrieve the output responses for transfer to the MISR. While the multiplexer at the input to the CUT flip-flops seems somewhat similar to scan design-based

DFT, there is a major difference in the way the flip-flops of the CUT are loaded with the test patterns. Instead of a serial shift, Autonomous Test uses the 2-to-1 multiplexers to perform a parallel load of the flip-flops. This is illustrated in Figure 8.1 for an N-input, K-output sequential CUT with M flip-flops. To implement the BIST approach, an $N+M$-bit LFSR is added along with a $K+M$-bit MISR. The N-bits of the LFSR supply test patterns to the primary inputs via the input isolation multiplexers while the remaining M-bits of the LFSR supply test patterns to the M flip-flops of the CUT via the 2-to-1 multiplexers at the inputs to the flip-flops. The MISR then compacts the output responses from the K primary outputs and from the outputs of the M flip-flops.

Two clock cycles are required to apply a test pattern and retrieve the results. In the first clock cycle, the test pattern produced by the TPG is transferred into the flip-flops of the CUT via the multiplexers; this occurs when the BIST controller output labeled *TVenable* is a logic 1. During the next clock cycle, *TVenable* is set to a logic 0 such that the flip-flops of the CUT apply the test pattern while clocking in the output response from the combinational logic. To avoid skipping any test patterns produced by the TPG and ensure pseudo-exhaustive testing, the LFSR must be disabled from proceeding to the next state whenever *TVenable* = 0. Therefore, the LFSR is either clocked with an independent clock from that of the flip-flops or else the TPG is constructed from active high clock enabled flip-flops. The BIST controller consists of a toggle flip-flop plus decoding logic for the $N+M$ outputs of the TPG to determine when the LFSR has cycled through all 2^{N+M} possible states. While it does not matter

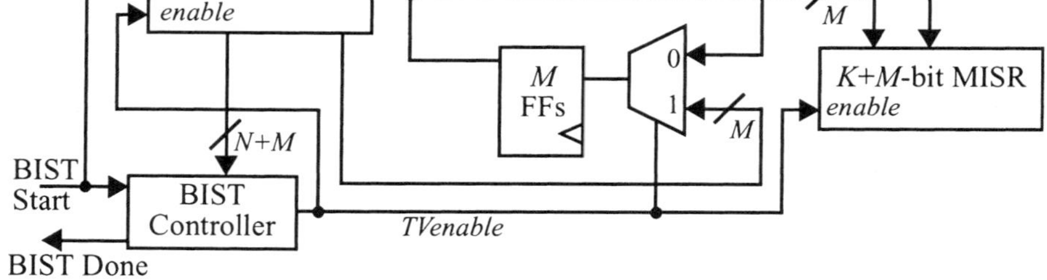

FIGURE 8.1 Autonomous Test basic architecture.

if the MISR compacts during every clock cycle, including the test patterns as they enter the flip-flops or the CUT, it can also be constructed from active high clock enabled flip-flops such that only the output responses of the CUT are compacted for the final signature as shown in the figure. In addition, the size of the MISR can be reduced to less the $K+M$-bits by using concentrators at its inputs.

The significance of the two clock cycle test vector application and response retrieval sequence is that the test patterns are applied in the normal system mode of operation. As a result, the actual system mode logic is tested at-speed to detect any faults that may affect system operation [174]. The ability to completely test the system mode logic at-speed is an important attribute of Autonomous Test. Otherwise, the MISR could compact the outputs directly from the combinational logic while the 2-to-1 multiplexers could be moved to the output of the M flip-flops to facilitate application of test patterns directly from the TPG to the combinational logic. In this case, however, the M flip-flops would not be tested, nor would the paths from the combinational logic to the flip-flops or from the flip-flops to the combinational logic.

The Autonomous Test architecture, as illustrated in Figure 8.1, provides exhaustive testing since all possible patterns and states are applied to the CUT, including combinational logic and flip-flops. This guarantees 100% fault coverage for all detectable gate-level and bridging faults. Therefore, no fault simulation is required for this approach; only a logic simulation is required in order to obtain the final signature. The area overhead consists of a 2-to-1 multiplexer at the input to every flip-flop and every primary input to the CUT. Therefore, the performance penalty is comparable to that of scan design-based DFT, one 2-to-1 multiplexer delay in every path between flip-flops. In paths from primary inputs to flip-flops, a performance penalty of two 2-to-1 multiplexers is incurred. The additional logic and flip-flops required to construct the TPG and MISR is the major area overhead since we triple the number of flip-flops in the CUT, with one set of flip-flops for the LFSR and another set for the MISR. In addition, the TPG has extra flip-flops corresponding to the number of primary inputs while the MISR may have extra flip-flops to account for the primary outputs. An economical alternative for the MISR is to use concentration to reduce the number of bits in the MISR.

8.2 Hardware Partitioning

One problem with this basic application of Autonomous Test, as illustrated in Figure 8.1, is that when the number of primary inputs plus the number of flip-flops becomes large, say $N+M > 25$, the test time becomes excessive. This means most practical applications cannot incorporate Autonomous Test as described thus far. One solution is to partition the CUT into smaller sequential circuits with multiplexers and execute the Autonomous Test for each partitioned CUT in turn during multiple BIST sessions [175]. This is illustrated in Figure 8.2 where it is assumed that $N + M \geq X + Y$ such that the N+M-bit LFSR can supply exhaustive test patterns[1] to both subcircuits (M and X are the number of flip-flops in subcircuits A and B while N and Y are the number of inputs to subcircuits A and B, respectively). The BIST controller is not shown in the figure but would be operated in a similar manner to that of Figure 8.1. An additional control lead would be used to select the subcircuit under test for a given test session if multiplexers are used to select which set of combinational logic and flip-flops will have its output responses compacted by the MISR. Alternatively, a concentrator could be used to linearly combine the output responses and, as a result,

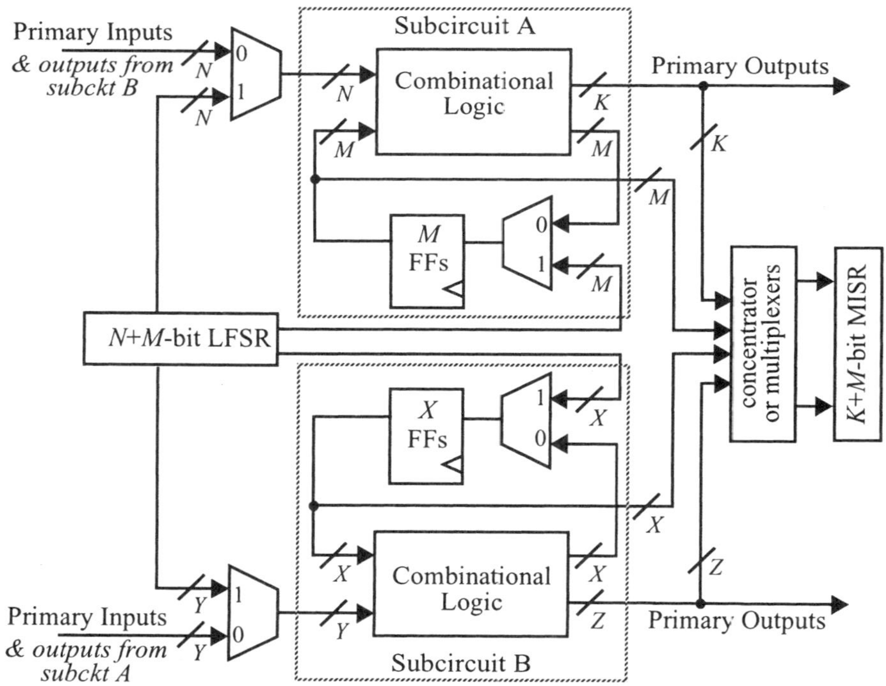

FIGURE 8.2 Partitioned Autonomous Test architecture.

1. This assumes the LFSR is constructed to generate the all 0s test pattern.

both subcircuits can be tested simultaneously. While the subcircuits of partitioned architecture shown in Figure 8.2 appear to be independent or disjoint, in fact the subcircuits could be interconnected where some of the inputs to *Subcircuit A* could be outputs from *Subcircuit B*, and vice versa. In addition, there can be more than two partitioned subcircuits in this architecture and flip-flops in the CUT may fan-out to multiple subcircuits. This allows Autonomous Test to be applied to many more practical applications. The only restriction that remains with this partitioned architecture is that the size of the LFSR be equal to the maximum number of inputs (including flip-flop outputs) to the combinational logic network in any of the subcircuits. This approach is often referred to as *hardware partitioning* or *physical segmentation* [175]. The advantage of this partitioned Autonomous Test approach is that the size of the TPG LFSR will be smaller and more practical, reducing both area overhead and test time.

8.3 Pseudo-Exhaustive Self-Test

Another solution to the problem of a CUT with a large number of primary inputs and flip-flops is to partition the CUT for pseudo-exhaustive testing. In this case, the combinational logic does not receive all possible 2^{N+M} test patterns but every combinational logic cone does receive exhaustive test patterns. In other words, all possible 2^n test patterns are applied to every n-input logic cone [175]. This technique is sometimes referred to as *cone segmentation*. The basic idea is to backtrace from the logic cone vertex (a primary output or a flip-flop input) to find the complete set of logic cone dependencies (primary inputs and/or flip-flop outputs). The next step is to assign an output of an N-bit LFSR to each logic cone input via a multiplexer. The result is that the logic cone is exhaustively tested in the test mode. The flip-flops at the vertex of every logic cone can be combined to form a MISR for compaction of the output responses of the logic cones as illustrated in Figure 8.3 [345]. This eliminates the additional MISR that was incorporated in the Autonomous Test approach while completely testing the flip-flops of the CUT. This approach is called Pseudo-Exhaustive Self-Test (PEST) [345].

There are five logic cones in the PEST example of Figure 8.3, some of which share logic as indicated by the overlapping triangles. The key step in obtaining exhaustive testing of the individual logic cones is the assignment of the N-bits of the LFSR to the logic cone inputs so that every logic cone gets a unique bit of the LFSR at each of its inputs while the total number of unique bits required, N, is a minimum. This keeps the size of the LFSR to a minimum and the resultant test time to a practical value. Typical values of N are on the order of 20 to 24 [345]. Primary outputs that are directly driven

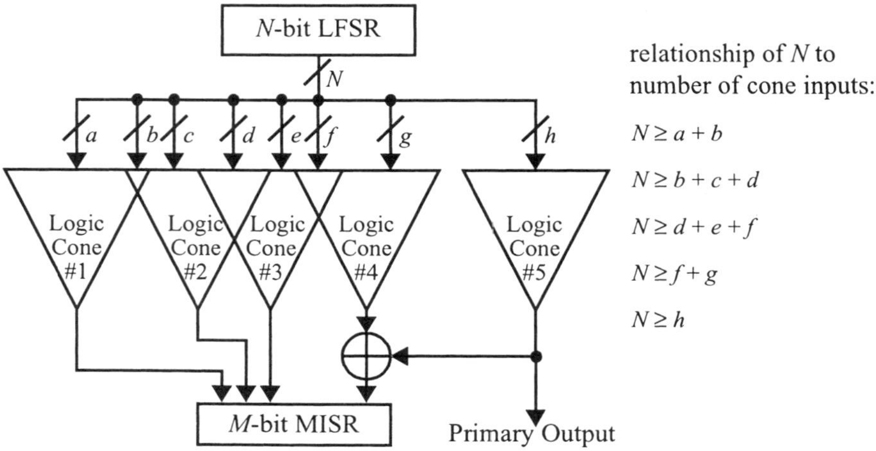

FIGURE 8.3 Cone segmented pseudo-exhaustive self-test.

by the vertex of a logic cone and not by a flip-flop can be linearly combined, via a concentrator, with other logic cone outputs entering the MISR during the BIST mode as illustrated in the right most logic cone of the example in Figure 8.3.

In the PEST approach, 2-to-1 multiplexers are incorporated at the inputs to the combinational logic cones of the CUT to facilitate insertion of the test patterns. The basic PEST cell that is used to replace the existing flip-flops of the CUT is illustrated in Figure 8.4 along with a table showing the functional modes of operation. The BIST circuitry at the input of the flip-flop is the same as the BILBO flip-flop shown in

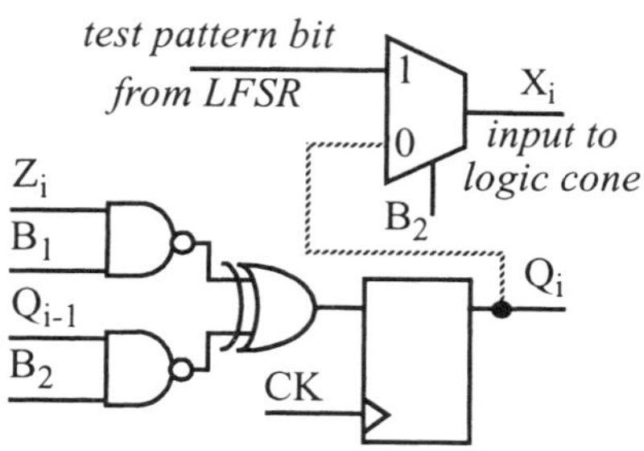

B_2	B_1	Flip-Flop Modes of Operation	MUX Output X_i
0	0	Initialization ($Q_i <= 0$)	don't care
0	1	System ($Q_i <= Z_i$)	Q_i
1	0	Scan ($Q_i <= Q_{i-1}$)	don't care
1	1	BIST ($Q_i <= Z_i \oplus Q_{i-1}$)	LSFR output

FIGURE 8.4 PEST flip-flop and modes of operation.

Figure 7.3. The shift mode allows the resultant signature to be shifted out of the MISR at the end of the BIST sequence but there is no mechanism to hold the signature until it can be read by system software for the pass/fail determination. This is a problem similar to that discussed for BILBO implementations and must be addressed for system-level use of the PEST approach. The 2-to-1 multiplexer at the output of the flip-flop is used to insert the test patterns from the LFSR.

One of the main differences with PEST compared to Autonomous Test is that the connections between the flip-flops of the CUT and the inputs to the logic cones are not tested during the PEST sequence (indicated by the dashed line in Figure 8.4). The area overhead and performance penalty per CUT flip-flop are larger for this approach than that for Autonomous Test. For every flip-flop in the PEST approach there are two 2-input NAND gates plus a 2-input exclusive-OR gate in addition to the 2-to-1 multiplexer that is incorporated in the Autonomous Test approach. In addition to the 2-to-1 multiplexer delay inserted in every path between flip-flops in the Autonomous Test approach, the PEST approach has an addition performance penalty of a 2-input NAND gate delay and a 2-input exclusive-OR gate delay.[1] However, the PEST approach does not require the two clock cycle test vector loading and application sequence, making the BIST controller a simpler design and the test time half as long. In addition, the total area overhead is not as large for PEST since the TPG LFSR is fixed with an upper limit of around 20 to 24-bits while the MISR is constructed from the flip-flops of the CUT, as compared to more than tripling the number of flip-flops in the CUT to implement the TPG and MISR for Autonomous Test [345]. The main problem encountered in PEST is large logic cones with many dependencies that require an LFSR of a size that prevents a practical test time.

8.4 Sensitized Partitioning

Large logic cones can be dealt with in pseudo-exhaustive BIST approaches using partitioning techniques. One technique is referred to as *sensitized partitioning,* sometimes called *sensitized path segmentation* [317]. In this technique, the logic cone is partitioned through path sensitization techniques similar to the Path Sensitization Algorithm described in Chapter 2. The result of that algorithm was a sensitized path from the fault site to a primary output for observation of the fault behavior. Just as

1. While the area overhead and performance penalties seem high, this PEST approach saw widespread use in ASICs at Bell Labs during the late 1980's and early 1990's, primarily due to the fact that synthesis of PEST was completely automated and high fault coverage was guaranteed without the need for fault simulation.

importantly, paths are sensitized from the primary inputs to the fault site for controlling the logic value at the fault site. The objective of sensitized partitioning is to decompose the large logic cone into smaller logic cones that are interconnected such that the vertex of one logic cone is a dependency to another logic cone. Paths are sensitized through these smaller logic cones to provide controllability and/or observability of the other logic cones. By maintaining the logic values that sensitized the path, the smaller logic cones can be exhaustively tested.

The sensitized partitioning technique has been used to test Triple Modular Redundancy (TMR) and *N*-tuple Modular Redundancy (NMR) fault tolerant circuits by sensitizing a path through the majority voting circuit to test each replicated system module independently and then to test the majority voter itself [268]. An example of this technique is illustrated in Figure 8.5 for a TMR implementation. The TMR circuit is completely testable as long as the triplicated sets of primary inputs to the replicated circuit modules can be manipulated independently [269]. By applying a test pattern that will produce a logic 1 at the output of one of the circuit modules (denoted *Test Vector X* input to the second circuit module in the figure) while applying a test pattern that will produce a logic 0 at the output of another of the circuit modules (denoted *Test Vector Y* input to the third circuit module in the figure), a path is sensitized through the majority voting circuit (as indicated by the gray line). As a result, the first circuit module can be exhaustively tested while the inputs to the second and third circuit modules are held constant. In this manner, each of the three circuit modules can be exhaustively tested in turn. Finally, by applying various combinations of *Test Vector X* and *Test Vector Y* to the primary inputs of the replicated circuit modules, the majority voting circuit can be exhaustively tested. This technique facilitates detection of all single and multiple gate-level faults as well as all delay faults [76]. Assuming

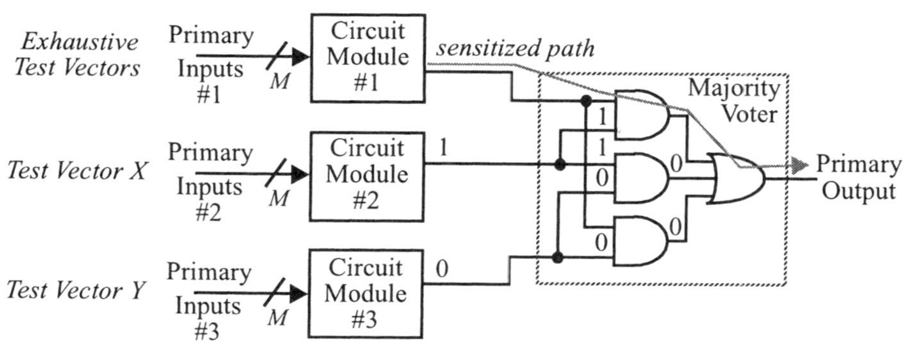

FIGURE 8.5 Sensitized partitioned pseudo-exhaustive testing.

each circuit module has M inputs, we have reduced the size of the logic cone from $3M$ to M using sensitized partitioning. In the general case of NMR implementations, the size of the logic cone is reduced from NM to M, where N is the degree of redundancy (the N-tuple in NMR). As a result, the number of test patterns that need to be applied for pseudo-exhaustive testing is $N2^M+2^N$ instead of 2^{NM}, which corresponds to a significant reduction in test time. Note that the 2^N test patterns are used for exhaustive testing of the majority voting circuit.

In some circuits, pseudo-exhaustive testing is achieved through the regularity of the structure. Examples include array multipliers, Iterative Logic Arrays (ILAs), parity trees, etc., which are composed of identical cells in a one or multi-dimensional array. The structure can be tested with a constant number of test patterns regardless of the size of the circuit; these structures are said to be *C-testable* [4]. To illustrate this, let us look at a particularly interesting case of the parity tree or concentrator circuit discussed in Chapter 5 and illustrated here in Figure 8.6b. This entire circuit (in fact, any parity tree of any size) can be pseudo-exhaustively tested with only four test vectors [185]. The principle behind the technique is illustrated in Figure 8.6a where exhaustive testing of an exclusive-OR gate produces a sequence of patterns which when combined with either of the two input sequences will exhaustively test another exclusive-OR gate. As a result, we obtain three sequences of bits (denoted **S**, **R**, and **T** in the figure), any two of which can be exclusive-ORed to produce the third. Therefore, starting at the output of the parity tree, we can assign any of the three sequences (**S** in this example) and then assign the other two sequences to the two inputs to the exclusive-OR gate that drives the output of the parity tree. Working our way backward through the parity tree, we assign input sequences until we have reached the primary inputs to the parity tree. The final result is a set of four test vectors which will exhaus-

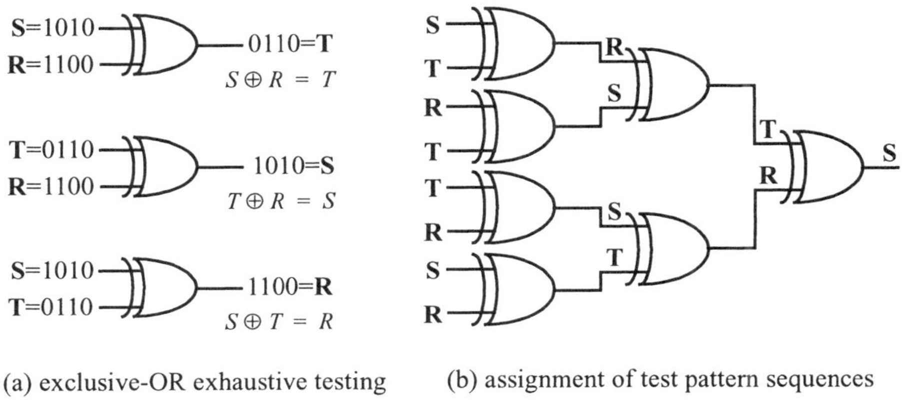

(a) exclusive-OR exhaustive testing (b) assignment of test pattern sequences

FIGURE 8.6 Pseudo-exhaustive testing of parity trees [185].

tively test every exclusive-OR gate in the parity tree simultaneously such that the entire circuit is pseudo-exhaustively tested, regardless of the size of the parity tree.

8.5 Test Point Insertion

Another technique for dealing with large logic cones is to insert test point multiplexers in the same manner as ad-hoc DFT techniques to facilitate physical partitioning of the logic cones to make smaller ones. A simple example of test point insertion is illustrated in Figure 8.7, where a test point multiplexer is inserted at the output of the logic common to both unpartitioned logic cones. As a result of the test point, three logic cones have been created, but the maximum number of inputs to any partitioned logic cone is three, instead of four as in the unpartitioned case. Furthermore, the set of inputs to a given logic cone is independent (disjoint) from the sets of inputs to the other logic cones. The output of the logic cone that represents the shared logic in the original circuit is fed to the MISR (perhaps via a concentrator) for output response compaction. The *Test Pattern* input to the multiplexer is used to insert test patterns from the TPG to test the other partitioned logic cones normally driven by the shared logic.

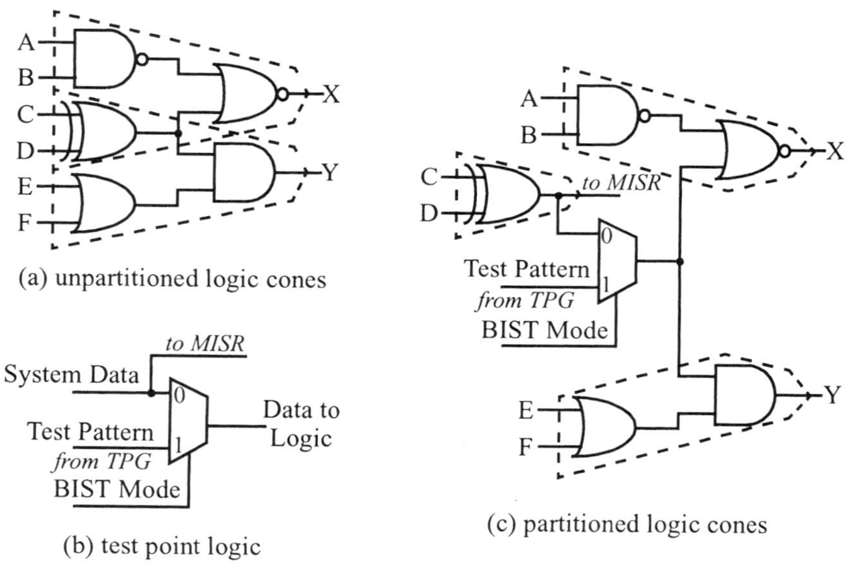

FIGURE 8.7 Test point insertion to partition logic cones.

This technique can also be applied to large logic cones where there is no shared logic between logic cones. The result is that large logic cones can be decomposed into much smaller logic cones to make pseudo-exhaustive BIST techniques practical for any general sequential logic circuit. The area overhead and performance penalty is one 2-to-1 multiplexer per test point inserted. While the area overhead may not be a problem, the performance penalty could be a problem in critical timing paths since it is typically the very large logic cones that create the critical timing paths in the design.

There are a number of techniques that can be used to select the best sites for insertion of test points. Aside from decomposing large logic cones into smaller logic cones for practical test time consideration, test points are selected to provide the maximum improvement to fault coverage. This is particularly important in the case of faults that are difficult to detect with pseudo-random test patterns when exhaustive and pseudo-exhaustive testing are not applied due to the test time considerations. Test point insertion is used by other BIST approaches that will be discussed in subsequent chapters.

8.6 Benefits and Limitations

The primary benefit of pseudo-exhaustive BIST approaches is guaranteed high fault coverage without the need for fault simulation. Only logic simulation is needed to determine the final signature for the fault-free CUT. In its basic form, without contending with large logic cones, design automation is reasonably easy [328]. CAD tools have been developed for automatic insertion of the PEST approach including test point insertion [345]. The main problem with these pseudo-exhaustive BIST approaches is encountered in large, complex CUTs where the logic cones have a large number of dependencies. Partitioning techniques must be used to partition those large logic cones to make the approach useful for practical applications [318].

While PEST has a larger performance penalty compared to Autonomous Test, the total area overhead is lower with PEST. The area overhead and performance penalties associated with the three approaches described in this chapter are summarized in Table 8.1 in terms of the number of flip-flops that must be added to the CUT along with the number of exclusive-OR gates (XOR), 2-to-1 multiplexers (MUX), and elementary logic gates (denoted 'Gates' in the table). It should be noted that the area overhead and performance penalties given in the table do not include test point insertion. In addition, the area overhead and performance penalty given in the table does not include logic required to hold the resultant signature in the MISR at the completion of the BIST sequence during system-level until it can be read for the pass/fail determination. The BIST controller is also not considered in the area overhead. Finally, it should also be noted that the area overhead for the Autonomous Test approaches assume a MISR size equal to the number of flip-flops in the CUT, N_{FF}, plus the number of primary outputs, N_{PO}. This can be reduced by using concentrators in conjunction with smaller MISRs.

TABLE 8.1 Area overhead and performance penalty of pseudo-exhaustive BIST.

BIST Approach	Area Overhead				Performance Penalty
	Flip-Flops	Exclusive-ORs	Multiplexers	Gates	
Autonomous Test	$N_{PI}+2N_{FF}$ $+ N_{PO}$	$N_{FF} + N_{PO} + 6$	$N_{PI} + N_{FF}$	$N_{PI}+2N_{FF}$ $+ N_{PO}$	1MUX (FF to FF) 2MUX (PI to FF)
Partitioned Autonomous Test	N_{PImax} + $2N_{FFmax}$ + N_{POmax}	$6 + N_{FFtot}$ + N_{POtot}	$N_{PItot} + N_{FF}$	$N_{PI}+2N_{FF}$ $+ N_{PO}$	1MUX (FF to FF) 2MUX (PI to FF)
PEST	N	$N_{FF} + 3$	N_{FF}	$2N_{FF} + N$	1MUX+1Gate+1XOR

where: N = number of bits in TPG
N_{FF} = total number of flip-flops in CUT
N_{PI} = number of primary inputs to CUT
N_{PO} = number of primary outputs of CUT
N_{FFmax} = maximum number of flip-flops in a CUT
N_{PImax} = maximum number of primary inputs to a CUT
N_{POmax} = maximum number of primary outputs from a CUT
N_{PItot} = total number of primary inputs to all CUTs
N_{POtot} = total number of primary outputs from all CUTs
N_{FFtot} = total number of flip-flops in all CUTs

An additional consideration in area and performance penalties of pseudo-exhaustive BIST approaches is the routing from the TPG to the flip-flops of the CUT. As opposed to a serial scan chain routing through the VLSI device in the BILBO approach or in scan design-based DFT, a test pattern bus, consisting of a number of bits equal to the size of the TPG, must be routed through the chip in order to deliver the test patterns to the logic cone inputs. This could be a 20 to 24-bit bus in practical applications and could account for considerable increase in area which, in turn, will add performance penalties since the normal system logic of the device will be spread out with longer routing segments and larger routing delays.

CHAPTER 9

Circular BIST

The problems associated with BILBO (register self-adjacency and multiple test sessions) and pseudo-exhaustive BIST (partitioning and large logic cones) applications are non-existent in the next three BIST approaches: Circular BIST [265], Circular Self-Test Path (CSTP) [142], and Simultaneous Self-Test (SST) [32]. These three BIST approaches are very similar and are frequently referred to simply as Circular BIST. While design automation of the BILBO and pseudo-exhaustive BIST approaches was complicated by the above problems, Circular BIST was developed with the intent of easy design automation for synthesis through CAD tools.[1] This chapter describes the architecture and operation of these three approaches, their similarities and differences, as well as their benefits and limitations.

9.1 Circular BIST Architecture

Like the BILBO and pseudo-exhaustive BIST, Circular BIST (including CSTP and SST) is an embedded, off-line, and test-per-clock BIST architecture because it uses existing flip-flops from the CUT to construct the TPG and ORA functions [264]. The

1. The primary motivation for the development of Circular BIST was a direct result of the introduction of a behavioral model synthesis system [287] since the designer was removed from intimate knowledge of the gate-level implementation of the synthesized design.

basic idea of the Circular BIST, SST, and CSTP approaches is to partition the CUT into flip-flops and combinational logic. The flip-flops are augmented with additional logic to operate in a BIST mode to test the combinational logic as well as the flip-flops themselves. The basic application of Circular BIST to a CUT is illustrated in Figure 9.1 (note the similarity to the scan design-based DFT implementation of Figure 3.4). The "circular" nature comes from the fact that the output of the last flip-flop in the chain is connected to the Q_{i-1} input of the first flip-flop in the chain to form a circular shift register type of arrangement.

The Circular BIST approach forms a large MISR-like structure from the existing flip-flops in the CUT. In the BIST mode of operation, the output responses of the CUT are linearly combined (via exclusive-OR gates) with the contents of the circular BIST chain while the resulting contents of the chain are used as the next test pattern to the CUT. Though the circular BIST chain performs output response compaction in a manner similar to a MISR used for signature analysis, there is no primitive characteristic polynomial associated with the circular BIST chain. The circular feedback path is equivalent to a characteristic polynomial $P(x)=x^n+1$, where n is the number of flip-flops in the circular BIST chain. As a result, there is no need to worry about finding primitive polynomials for circular BIST chains of varying lengths. An alternative way of looking at the BIST approach is to construct one large BILBO from all flip-flops in the CUT using $P(x)=x^n+1$ as the characteristic polynomial and then to operate the

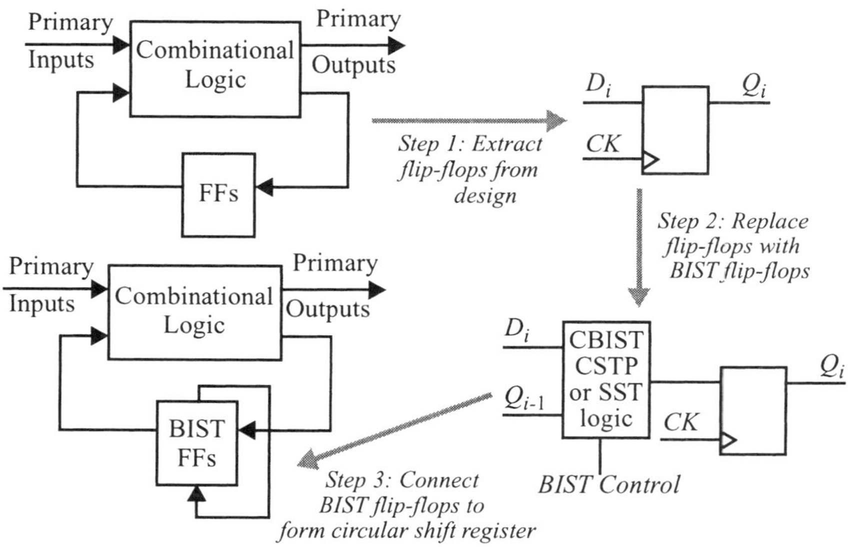

FIGURE 9.1 Basic implementation of Circular BIST, SST, and CSTP.

BILBO only in the MISR mode with the compacted signatures serving as the test patterns to the combinational logic. This eliminates the need for test session scheduling since there is only one test session. In addition, there is no concern about register self-adjacency since the MISR is simultaneously serving as the TPG. The test patterns produced by the compaction of the output responses of the CUT with the contents of the circular BIST chain have characteristics similar to those of random patterns [133]. However, the test patterns are not guaranteed to provide pseudo-exhaustive testing like the BIST approaches in the previous two chapters.

One of the main differences in Circular BIST, SST and CSTP is the logic that is added to the flip-flops and the resultant modes of operation that it facilitates, as illustrated in Figure 9.2. The CSTP flip-flop has two modes of operation, system mode and BIST mode, and requires an exclusive-OR gate and a 2-to-1 multiplexer for a total of six elementary logic gates (assuming a 2-to-1 multiplexer and a 2-input exclusive-OR gate consist of three elementary logic gates each). The SST flip-flop pro-

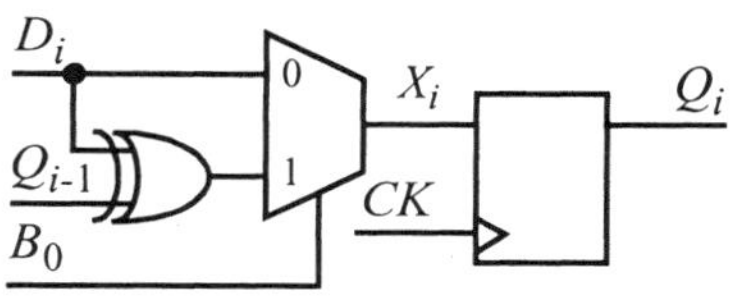

B_0	X_i	Mode
0	D_i	System
1	$D_i \oplus Q_{i-1}$	BIST

(a) CSTP flip-flop and modes of operation

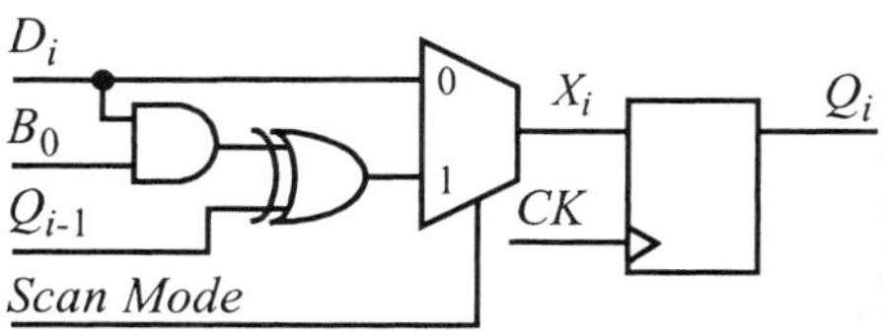

Scan Mode	B_0	X_i	Mode
0	X	D_i	System
1	0	$Q_{i\text{-}1}$	Shift (Scan)
1	1	$D_i \oplus Q_{i-1}$	BIST

(b) SST flip-flop and modes of operation

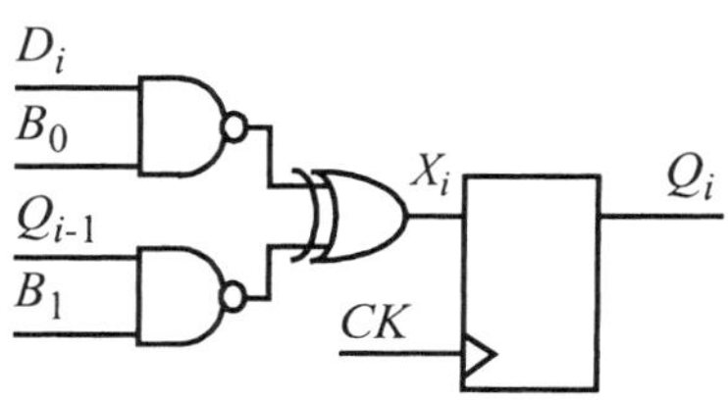

B_1	B_0	X_i	Mode
0	0	0	Initialization
0	1	D_i	System
1	0	$Q_{i\text{-}1}$	Shift (Scan)
1	1	$D_i \oplus Q_{i-1}$	BIST

(c) Circular BIST flip-flop and modes of operation

FIGURE 9.2 Comparison of Circular BIST, SST, and CSTP flip-flops.

vides three modes of operation at the expense of an additional AND gate and test mode control signal, when compared to CSTP. The Circular BIST flip-flop has four modes of operation and requires an exclusive-OR gate and two NAND gates for a total of five elementary logic gates. Therefore, the Circular BIST flip-flop is slightly more area efficient than the CSTP and SST flip-flops. The CSTP flip-flop does not provide for initialization of the CUT. As originally proposed, the CSTP approach assumed that all flip-flops in the CUT contained a reset or preset capability controlled by a global signal [141]. Since global initialization is not necessarily common in most practical applications, an additional gate is required between the multiplexer and the input to the flip-flop in order to provide for complete initialization of the CUT. The SST flip-flop provides a shift mode for scan design testing which facilitates both initialization and retrieval of the BIST results. Therefore, the Circular BIST flip-flop provides a superset of the modes of operation of both the CSTP and SST flip-flops and has become the most frequently used flip-flop in BIST applications [177]. Therefore, the remainder of this chapter will focus on the use of this flip-flop.

The major goal of Circular BIST (as opposed to SST and CSTP) is to provide system-level testing access with minimal area overhead. As a result, Circular BIST uses a slightly more complicated implementation than CSTP or SST. During the flip-flop replacement process shown in Figure 9.1, not all flip-flops are replaced such that combinational and sequential logic are left to be tested by the circular BIST chain, as opposed to only combinational logic when all flip-flops are replaced. This *selective replacement* of flip-flops not only reduces area overhead but also facilitates avoiding the insertion of BIST logic at the inputs to flip-flops that reside in critical timing paths in the CUT. Hence, performance penalties can be avoided.

Input isolation circuitry is incorporated in the form of either multiplexers or blocking gates (as illustrated in Figure 9.3) in order to use the Circular BIST approach during system-level testing. Using multiplexers for input isolation has the advantage over blocking gates in allowing application of the test patterns produced by the circular BIST chain to the primary inputs of the CUT for higher fault coverage. Finally, a SAR or MISR can be incorporated into the circular BIST chain as illustrated in Figure 9.3. This SAR or MISR is constructed with a primitive polynomial and facilitates further compaction of the data in the circular BIST chain into a smaller and more manageable signature that can be more easily accessed via a processor interface. In addition, the SAR or MISR can supply test patterns to the BIST mode inputs of the input isolation multiplexers and can compact the output responses of primary outputs of the device that are not driven directly by flip-flops incorporated in the circular BIST chain, as indicated by the dashed line in Figure 9.3. For those primary outputs that are driven directly by flip-flops in the circular BIST chain, there is no need for further data compaction of these primary outputs since those data values are already being compacted by the circular BIST chain and then by the SAR or MISR.

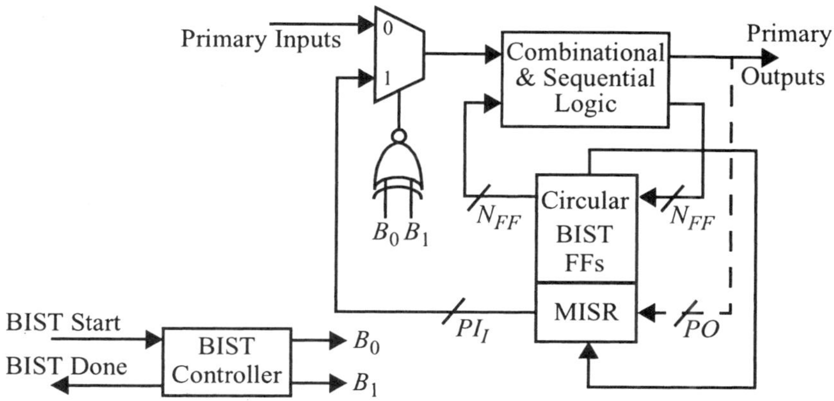

FIGURE 9.3 Complete Circular BIST architecture.

9.2 Circular BIST Operation

The first phase of the BIST sequence is the initialization phase in which all Circular BIST flip-flops are put into the initialization mode (B_0=B_1=0) and are reset to logic 0. The BIST controller must maintain the CUT in the initialization mode until all other flip-flops in the CUT have been initialized, when selective replacement of existing flip-flops with Circular BIST flip-flops has been used. This requires a number of clock cycles equal to the sequential depth of the CUT, possibly more clock cycles if there are feedback loops that had not been cut by the Circular BIST flip-flops. In this case, the sequential depth of the circuit is the maximum number of non-BIST flip-flops in any path between flip-flops in the circular BIST chain or between a primary input and a Circular BIST flip-flop. Feedback in the sequential CUT (when selective replacement has been used) can require the initialization sequence to be longer than the sequential depth of the CUT and, in some cases, can even prevent complete initialization. However, proper initialization of the CUT, including the length of the initialization sequence, and the associated design of the BIST controller for the initialization sequence can be easily verified using the *U*'s simulation described in Chapter 6. If all flip-flops in the CUT have been replaced with Circular BIST flip-flops, the sequential depth is 0 and the initialization mode requires only one clock cycle.

The second phase in the BIST sequence puts the Circular BIST flip-flops in the BIST mode of operation (B_0=B_1=1) for simultaneous test pattern generation and output response compaction by the circular BIST chain. During the first clock cycle of the BIST mode of operation, the output response of the CUT to the logic 0s (the initialization value for the BIST flip-flops) in the circular BIST chain is compacted by being exclusive-ORed with the contents of the circular BIST chain and shifted in the circular BIST chain. This compacted result is then used as the test pattern for the next clock cycle in the BIST sequence. This process of output response compaction and using the compacted data as the test pattern for the next clock cycle in the BIST sequence continues for the duration of the BIST sequence. The compacted output responses continue to cycle around the circular BIST chain and are further compacted in the SAR or MISR by its characteristic polynomial circuitry in the case of the Circular BIST architecture illustrated in Figure 9.3 where an SAR or MISR is incorporated.

At the completion of the BIST sequence, the resultant BIST signature must be read for comparison with the expected signature for the pass/fail determination. In SST and CSTP applications, the solution is to shift out the resultant signature. While SST contains a scan mode, CSTP (as originally proposed) does not. One solution for CSTP is to continue the BIST sequence for an additional N_{FF} clock cycles, where N_{FF} is the number of CSTP flip-flops, while monitoring a primary output that is directly driven by a CSTP flip-flop [141]. This facilitates observing the contents of the chain of flip-flops during manufacturing testing under the control of an external test machine, but this is not a feasible solution in most applications for system-level testing due to asynchronous processor interfaces running at a slower clock frequency. An alternative approach for both CSTP and SST is to connect the circular chain of flip-flops to the Boundary Scan TAP for retrieval of the BIST results (the BIST chain contents) at the end of the BIST sequence. This implies that the clock be changed from the system clock to TCK (test clock input to the Boundary Scan interface) without altering the contents of the BIST chain.

For the Circular BIST architecture illustrated in Figure 9.3, only the contents of the SAR or MISR need to be read at the completion of the BIST sequence. As a result, only the contents of the SAR or MISR need to be held after the completion of the BIST sequence until read by the system via a processor or the Boundary Scan interface for the pass/fail determination. Once the BIST results have been retrieved by the system, the CUT can be returned to the system mode of operation. An alternative is that the BIST controller automatically returns the CUT to the system mode of operation via the BIST control signals (B_0=1 and B_1=0) while holding the contents of the BIST results in the SAR or MISR for retrieval by the system. This latter approach can significantly reduce power consumption after the BIST sequence has been completed and the system is waiting for retrieval of the final BIST signature.

Additional fault coverage may be obtained by the application of deterministic test vectors using the scan mode of operation for the Circular BIST (or SST) flip-flop during manufacturing testing. These additional faults that might be detected are a result of states not visited during the length of the BIST sequence or due to *limit cycling* (which will be discussed later in this chapter). A scan mode can be incorporated in the CSTP approach at the expense of an additional 2-to-1 multiplexer or by expanding the existing multiplexer to a 3-to-1 multiplexer. In order to make use of the shift mode for partial scan testing (full scan testing if all flip-flops in the CUT are incorporated in the circular BIST chain), an additional 2-to-1 multiplexer must be inserted in the feedback that forms the circular BIST chain as illustrated in Figure 9.4. In this way the circular chain can be broken during scan mode testing. A similar modification can be made to SST and CSTP (if the CSTP flip-flops include a scan mode of operation).

The *Scan In* signal can use any primary input to the CUT excluding those required to execute the BIST sequence such as system clock or synchronization signals. The *Scan Out* signal can be a primary output that is driven by a flip-flop incorporated in the circular BIST chain, but the scan mode multiplexer must be inserted at that point in the circular chain. ATPG software for scan design or partial scan design-based DFT can be used to obtain scan mode test vectors for additional fault coverage for those faults not detected during the BIST sequence. In addition, it has been observed that for manufacturing testing the input isolation can be disabled during the BIST mode with random test patterns applied to the primary inputs to obtain an additional 1% to 5% increase in fault coverage [296].

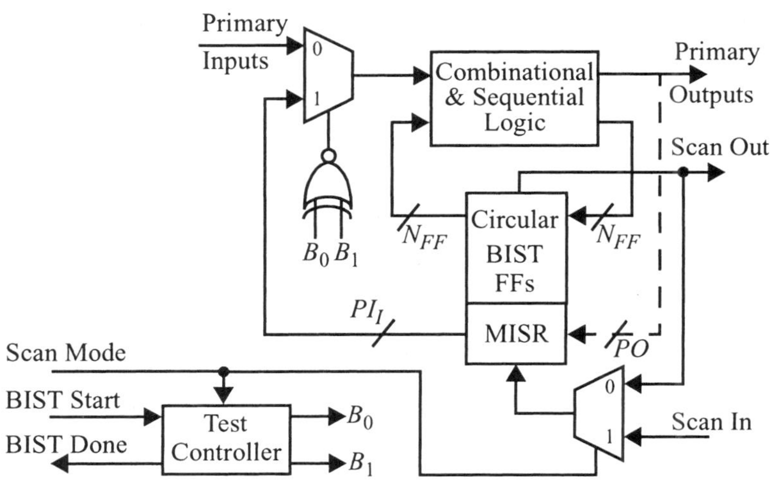

FIGURE 9.4 Access to Circular BIST scan mode.

9.3 BIST Controller

The BILBO and other BIST approaches, that use LFSRs or counters for the TPG, provide an advantage over Circular BIST when it comes to the design of the test controller. Counters and LFSRs repeat in a predictable manner and this repetition of the TPG can be used to make the design of the test controller simpler with lower area overhead by simply decoding states. As a result, the test controller may not be a completely separate entity, as was the case in the March Y algorithm TPG of Figure 4.1. In Circular BIST, including SST and CSTP, the TPG function is a by-product of the ORA and does not repeat in the same predictable manner as a counter, LFSR, or FSM. Therefore, the test controller must be a separate entity designed to control the length of the initialization sequence and BIST sequence. This is essential for the operation of Circular BIST during system-level testing.

The length of the BIST sequence is always the main concern in designing the BIST controller. The longer the BIST sequence, the better chance for higher fault coverage but this also requires a larger BIST controller. In addition, a longer BIST sequence increases testing time and cost during manufacturing testing, from wafer-level testing through board-level testing. While a shorter BIST sequence will reduce testing time as well as the area overhead of the BIST controller, it may also reduce fault coverage that can be obtained with Circular BIST. Assuming N_{FF} flip-flops in a circular BIST chain and that all the flip-flops in the CUT are incorporated in the circular BIST chain, the maximum length, L, of the BIST sequence would be $L \le 2^{N_{FF}}$. Since N_{FF} can easily be in the hundreds or even thousands for many practical applications, it is not possible to run the BIST sequence for the maximum possible length. Fortunately, most applications of Circular BIST achieve reasonably high fault coverage (greater than 90%) with BIST sequence lengths from tens to hundreds of thousands of clock cycles [178]. This would correspond to a counter-based or LFSR-based BIST controller on the order of 13 to 18 bits.

An example controller for Circular BIST is illustrated in Figure 9.5, and consists of an N-bit counter and a synchronous Set/Reset flip-flop for the *BIST Done* indication. The *BIST Start* signal is used to clear the counter and reset the Set/Reset flip-flop. When the *BIST Start* signal is activated (to a logic 1), the counter begins to count. For the first 2^{N-M} clock cycles, the BIST controller puts the Circular BIST flip-flops into the initialization mode (B_0=B_1=0). From that point in time until the *BIST Done* is activated (to a logic 1), the controller puts the Circular BIST flip-flops in the BIST mode of operation (B_0=B_1=1) for 2^N-2^{N-M} clock cycles. During this time, the SAR or MISR is enabled for output response compaction by using the B_1 control signal as an active high enable input to the SAR or MISR. When the counter "rolls over", the carry-out sets the *BIST Done* to a logic 1 and the SAR or MISR is disabled from further com-

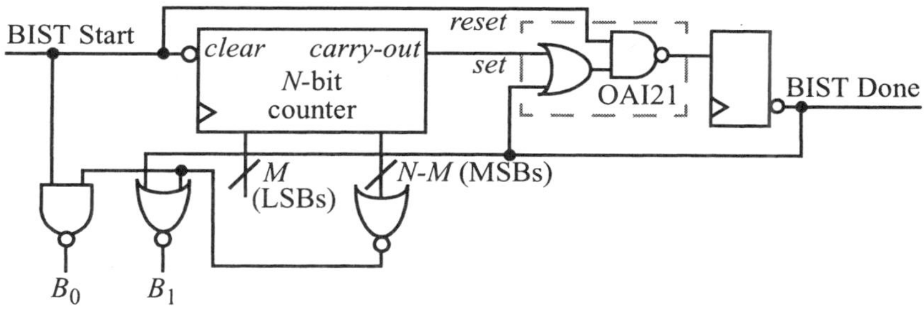

FIGURE 9.5 Example Circular BIST controller.

paction of the data circulating through the circular BIST chain. Whenever the *BIST Start* signal is returned to a logic 0, the BIST controller returns the Circular BIST flip-flops to the system mode of operation.

9.4 Selective Replacement of Flip-Flops

Selective replacement of flip-flops in the CUT with Circular BIST flip-flops is important since it allows the designer to control the area overhead and performance penalty. The main issue is selecting which flip-flops to replace. Avoiding performance penalties is usually the first priority. Therefore, any critical timing paths in the CUT should be identified to determine if replacing the destination flip-flop in a given path will cause the delay to exceed the limit for the required system clock operating frequency. Flip-flops that do not lie in critical timing paths are candidates for replacement. Area overhead becomes the second priority with the object being to maintain high fault coverage and initializability while minimizing the number of flip-flops replaced. In order to maintain initializability, it is important to insure that all feedback loops have at least one Circular BIST flip-flop in the loop. This includes clock-enabled flip-flops, like the one illustrated in Figure 2.11.

Every flip-flop lies at the vertex of a combinational logic cone. The set inputs to each combinational logic cone (the set of dependencies of the logic cone) consists of primary inputs to the CUT and/or the outputs of other flip-flops in the CUT. The number of dependencies to any given logic cone will range from 1 to C_{in}, where C_{in} is the number of inputs to the largest logic cone in the CUT. One way to ensure that all feedback loops have a Circular BIST flip-flop for initialization is to replace any flip-flop at the vertex of a logic cone with three or more dependencies. Logic cones with less than three dependencies are easily testable and will not have feedback loops [264]. Therefore, this is a safe and conservative selective replacement strategy to reduce area overhead. In some circuits, this selection criteria replaces every flip-flop, so the area overhead penalty is not always reduced. This selective replacement algorithm was applied to seven production ASICs where it replaced about 75% of the flip-flops on average to obtain fault coverages ranging from 90% to 94% [265]. However, studies in selective replacement have indicated that only about 60% of the flip-flops need to be replaced to achieve high fault coverage for most applications [296].

The prior work done on selection of partial scan flip-flops provides fertile grounds for the selective replacement of Circular BIST flip-flops [13]. For example, including every loop-cutting flip-flop in the selection set of flip-flops for replacement has been shown to ensure that every flip-flop in the circuit can be initialized. The circuit can be represented as a directed graph (also called a dependency graph) where every vertex is a flip-flop and every edge indicates that combinational logic joins the two flip-flops. Replacing every loop-cutting flip-flop means that every loop in the directed graph contains at least one BIST flip-flop. However, not all loop cutting flip-flops need be replaced to ensure initializability and high fault coverage [296]. Conversely, only including loop-cutting flip-flops is not sufficient to guarantee high fault coverage. In general, the fewer the flip-flops replaced, the lower the fault coverage [265].

One approach to selective replacement breaks the process into two stages [214]. The first stage involves replacing loop-cutting flip-flops, referred to as the primary selection set. The second stage involves replacing flip-flops that significantly increase fault coverage, referred to as the secondary selection set. This secondary selection set can be chosen by four different methods [214]. The first method is a selection based on the number of flip-flops feeding into the logic cones of each flip-flop; this is similar to the approach described above where flip-flops at the vertices of logic cones having 3 or more dependencies were replaced. The second method calculates the number of signal paths between each pair of flip-flops, where the path is not broken by another flip-flop; this is reconvergent fan-out when multiple paths exist. Controllability and observability measurements of each flip-flop comprise the third and fourth selection methods, respectively. The controllability measurement is based on the number of combinational and sequential node assignments needed to set a flip-flop to a logic 1 or a logic 0. Similarly, the observability measurement is based on the number of combinational and sequential node assignments needed to propagate a logic 1 or a logic 0 to a primary output. These four methods can be used individually or as a weighted sum where the candidates with the largest associated number would represent the best candidate flip-flops for replacement since a large number of any of these attributes indicates a potential testability problem. When the fault coverage of each method individually was compared in a study, the reconvergent fanout method gave the best secondary selection set [214]. When combinations of the four methods were compared, the reconvergent fanout method combined with the observability measurement method resulted in the highest fault coverage. A final consideration is sequential depth, the largest number of flip-flops in a signal path from a point of controllability to a point of observability. In this case, BIST flip-flops are considered to be points of controllability and observability along with primary inputs and primary outputs. In general, the smaller the sequential depth, the better the fault coverage [214].

Another selective replacement study for replacement of flip-flops using the scan flip-flops of partial scan DFT is also applicable to Circular BIST [92]. In this case, an algorithm was devised that produces a selection set given a target fault coverage. The algorithm starts with all flip-flops included in the selection set. Each flip-flop is temporarily removed, and a profit function is computed. The flip-flop that decreases the profit the least is permanently removed from the set. The fault coverage for the new selection set is computed and stored. The process continues to examine the effect of removing each flip-flop from the selection set and permanently removing the worst candidates for replacement until no flip-flops are left in the set; this results in an ordered list of flip-flops. At this point, the minimal set of flip-flops needed to get the desired fault coverage can be chosen from the resultant ordered list by replacing those flip-flops which had the highest profit function (the largest impact on the fault coverage). To verify that this selection actually gives the desired fault coverage, fault simulation must be performed. If the actual fault coverage is unacceptable, the next flip-flop in the list is added, and the circuit is re-simulated. This continues until the target fault coverage is achieved. In this partial scan flip-flop replacement study, the algorithm was applied to several circuits to find the number of partial scan flip-flops needed to be replaced to obtain fault coverages of 99% and 100% [92]. It was observed that the percentage of flip-flops needed to obtain 100% fault coverage can be as much as four times higher than the percentage of flip-flops needed to obtain 99% fault coverage.

An important item in selective replacement is to include flip-flops at the primary outputs for observability during system-level testing [296]. Designers concerned with area overhead can replace as few as one-third of the flip-flops to obtain high fault coverage normalized with respect to area overhead. Even though fault coverage is lower with one-third replacement, the fault coverage may be higher than that obtained by diagnostic software, which applies functional tests. In addition, the part is tested at speed with very little performance penalty. This results in significant reductions in diagnostic run time since the BIST sequence can be executed at system clock frequencies in multiple chips in parallel while diagnostic software would normally apply the functional tests to one circuit at a time via low speed communication interfaces. Diagnostic resolution at the device-level is achieved as opposed to system diagnostics interpreting the testing results and attempting to determine the faulty device.

9.5 Register Adjacency

Register adjacency is one problem that can occur in Circular BIST (including CSTP and SST) implementations whenever $Q_i=f(Q_{i-1})$. The worst-case scenario of register adjacency is illustrated in Figure 9.6 (using CSTP flip-flops in this example) where for the normal system operation $D_i=Q_{i-1}$ such that the output of the exclusive-OR gate is always logic 0 ($X_i=0$) during the BIST mode of operation. This blocks the propagation of fault detection information through the circular BIST chain and prevents the generation of a logic 1 test pattern to the CUT at the point of the register adjacency. This worst-case example of register adjacency will not occur with most selective replacement algorithms since only the flip-flops at the vertex of large combinational logic cones are replaced. However, the possible effects of register adjacency still exist to a lesser degree in larger combinational logic cones whenever $D_i=f(Q_{i-1})$. When a path is sensitized from Q_{i-1} to D_i, $X_i=0$ (or else $X_i=1$, depending on whether the logic function of the CUT is inverting) for that clock cycle, regardless of the value of Q_{i-1} [265]. These examples of register adjacency are referred to as *unit-distance* or *distance*-1 register adjacency. Examples of distance-2 register adjacency have also been suggested for implementations where selective replacement of flip-flops is used [52]. In distance-2 register adjacency, a path from a Circular BIST flip-flop travels through a non-BIST flip-flop and reconverges at the exclusive-OR gate of a Circular BIST flip-flop located two flip-flops along the circular BIST chain. In fact, there can exist distance-*d* register adjacency if the distance between two flip-flops in the circular BIST chain is *d* and there also exists a path between these two flip-flops through the CUT which contains *d* non-BIST flip-flops [52].

Register adjacency not only destroys the fault detection data being accumulated in the circular BIST chain but also reduces the "randomness" of the test patterns being produced and applied to the CUT. As a result, the existence of register adjacency can reduce the fault coverage that can be obtained for a given application. Ordering the flip-flops in the circular BIST chain to avoid functional dependencies can minimize the effects of register adjacency [265]. An alternate approach is to introduce additional flip-flops in the circular BIST chain to avoid register adjacency but this approach is less desirable due to the increase in area overhead [25].

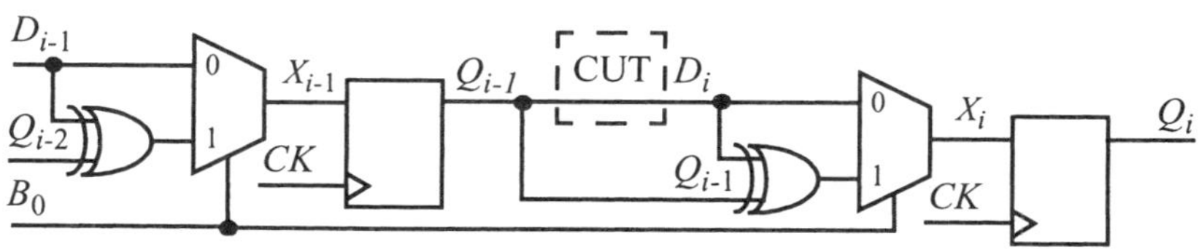

FIGURE 9.6 A worst-case example of register adjacency.

The theoretical estimate of the fault coverage that can be obtained with CSTP was shown to be near 100% [208]. However, this estimate was based on assumptions that did not prove to be valid in practical implementations including the absence of register adjacency and the lack of input isolation. In application to production ASICs, the typical fault coverage that has been observed is much closer to the 90% to 95% range. In some Circular BIST applications, fault coverage has been observed to be as low as 30% [65]. Initially, the reason for low fault coverage was attributed to register adjacency in the circular BIST chain [265]. But the existence of register adjacency did not fully account for the reduced fault coverage that can occur with Circular BIST.

9.6 Limit Cycling

More recently, the low fault coverage obtained in some Circular BIST applications has been correlated with cycles in the *state transition graph* of the CUT in the BIST mode of operation [47]. These cycles result from the CUT and the circular BIST chain forming a closed system since the primary inputs must be isolated from unknown system data that would prevent the successful reproduction of the correct BIST signature for a fault-free circuit from one execution of the BIST sequence to the next. As a result, only a specific sequence of states will be visited during the BIST depending on the initial state of the circular BIST chain as well as the initial state of non-BIST flip-flops in the CUT and the state transition graph for the CUT in the BIST mode of operation. The state transition graph and the resultant state transition sequence during Circular BIST is a function of the initialization value of the circular BIST chain and CUT, the ordering of the BIST chain, the combinational logic between the BIST flip-flops, and the manner in which the CUT is isolated from the rest of the system. Isolation of the primary inputs to the circuitry under test can be accomplished with blocking gates, multiplexers, or flip-flops. Input isolation multiplexers and flip-flops facilitate the application of test patterns from the circular BIST chain to the primary inputs but also add additional potential fault sites and logic area overhead.

As a result of the closed loop nature of the Circular BIST approach, the states generated by the Circular BIST sequence will repeat after a certain number of clock cycles. Every state in the cycle provides a distinct test vector to the CUT under test. In order to have good fault coverage, one needs to apply an adequate number of distinct test patterns. The number of clock cycles after which the states begin to repeat is referred to as the *cycle length*. Maximizing the cycle length, in terms of the number of states visited in the state transition graph from the initial state of the BIST sequence to the completion of the first cycle, will provide the best potential for maximizing the fault coverage obtained with Circular BIST. For example, in the state transition graph given in Figure 9.7 we see that initializing the CUT to state *a* will provide the maximum cycle length (9 states: *a-b-d-e-f-h-i-j-k*) since we will visit the largest number of states before the sequence of states begins to repeat (repetition states: *e-f-h-i-j-k*). An approach to finding this optimal initial state (or *head state* as it is sometimes referred to) for the circular BIST chain based on Binary Decision Diagrams (BDDs) was proposed and shown to significantly enhance the fault coverage that can be obtained for some circuits [65]. However, there are two problems that can be encountered with this solution to the limit cycling problem: 1) the computational complexity of finding the optimal initial state can be significant for large circuits and 2) there is no alternative if the fault coverage resulting from the optimal initial state is not sufficient. Due to the complexity of most circuits, it is difficult to determine the state transition graph for the circuit in the BIST mode of operation and whether a given circuit will experience limit cycling resulting in low fault coverage. The designer would like to be able to determine whether a given circuit suffers from the effects of limit cycling by some simple method other than fault simulation of the complete CUT in the Circular BIST mode or derivation of the state transition graph.

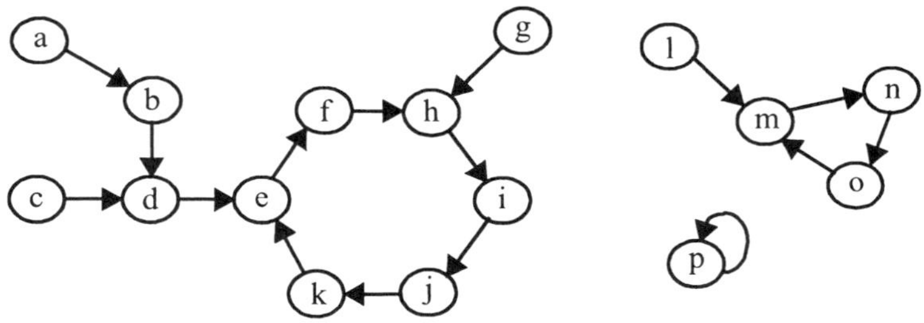

FIGURE 9.7 Example state transition graph for a CUT in Circular BIST mode.

9.6.1 Design Guidelines for Limit Cycling

Design guidelines have been proposed based on simple parameters that facilitate predetermination of whether limit cycling will result in low fault coverage for a given circuit with Circular BIST [278]. These design guidelines were developed and verified through fault simulations of numerous example circuits including the ISCAS'89 sequential benchmark circuits [45]. The parameters found to be most useful in determining the likelihood of encountering low fault coverage due to limit cycling are N_{FF}, the number of flip-flops in the circular BIST chain, and C_{in}, the number of inputs to the largest combinational logic cone in the CUT; more specifically, the ratio of N_{FF} to

C_{in}. The results of some of these fault simulations are shown in Figure 9.8 as a function of the ratio of N_{FF} to C_{in}. From the fault simulation data, it was found that sequential circuits incorporating Circular BIST generally achieve good fault coverage and do not suffer from limit cycling when the following relationship is satisfied:

$$\frac{N_{FF}}{C_{in}} > 2 \tag{9.1}$$

This condition is referred to as 'case 2' circuits since the ratio of N_{FF} to C_{in} is greater than 2. It was also found that a given circuit will almost always suffer from limit cycling and low fault coverage when the following relationship holds:

$$\frac{N_{FF}}{C_{in}} < 1 \tag{9.2}$$

This is referred to as 'case 0' circuits since the ration of N_{FF} to C_{in} is less than 1. Finally, it was found that a given circuit may suffer from limit cycling and low fault coverage in the intermediate case (referred to as 'case 1' circuits) given by:

$$1 \le \frac{N_{FF}}{C_{in}} \le 2 \tag{9.3}$$

Therefore, by extracting the values of N_{FF} and C_{in} from a circuit to be implemented with Circular BIST, the designer can determine if limit cycling will be a potential

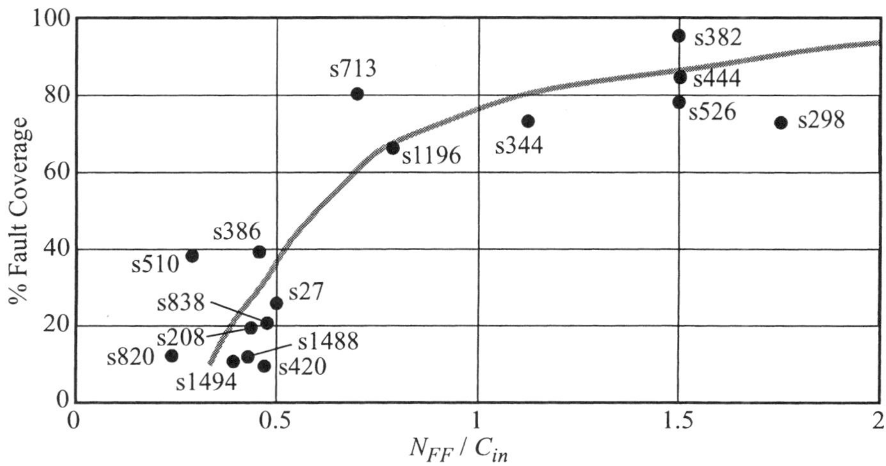

FIGURE 9.8 Limit cycling in ISCAS'89 benchmark circuits.

problem prior to the actual implementation of Circular BIST. Given the conditions which lead to low fault coverage, as predicted by Equation (9.2) and Equation (9.3), an obvious question is whether these are realistic conditions encountered in practical applications since most ASICs contain hundreds to thousands of flip-flops which may be incorporated in the circular BIST chain. In fact, there is often the need to partition ASICs into multiple circular BIST chains with input isolation between the chains. These situations are common in ASICs with asynchronous clocking boundaries which must be isolated from each other and clocked by their individual clocks in order to obtain reproducible BIST results during system-level testing. Otherwise, the asynchronous system clocks can lead to incorrect final signatures, which would indicate the ASIC is faulty when it is actually fault-free. Similarly, synchronous circuits often require partitioning at clocking region boundaries when multiple frequency clocks are used within the same ASIC. This can even occur in single-clock, single-edge, synchronous designs where some of the flip-flops are enabled in order to clock in new data at a much lower data rate than the actual clocking frequency (the clock enable problem). In these applications, the lower data rate portions of the ASIC are likely to have critical timing path delays that are longer than the main clock period. If the entire ASIC is implemented as one large circular BIST chain, one of two possible situations is encountered. When the higher frequency clock is used to clock the BIST chain, the delays in the critical paths for the lower frequency clocking boundaries will cause incorrect (and non-reproducible) BIST results. When the lower clocking frequency is used, significant portions of the ASIC will not be tested at-speed. In either case, additional clock multiplexing circuitry is required. Therefore, these clocking issues for at-speed system-level testing lead to partitioning of the ASIC into multiple circular BIST chains with input isolation between the clocking regions. The smaller, partitioned circuits can easily fall within the limit cycling problem conditions of Equation (9.2) or Equation (9.3). Conversely, some sequential circuits can have surprisingly large combinational logic cones which can also lead to these problem conditions.

9.6.2 Hardware Solutions to Limit Cycling

If any set of m flip-flops in the circular BIST chain produces all possible 2^m patterns, then the cycle length is at least 2^m. Therefore, hardware solutions to limit cycling seek to construct an m-bit register with additional logic to provide a cycle length in the circular BIST chain of greater than or equal to 2^m. The use of flip-flops for input isolation (as illustrated in Figure 9.9) has the effect of increasing the number of potential states that can be visited during the BIST sequence. In the following discussion of three hardware solutions to limit cycling, we will assume that the additional register will also be used to provide the input isolation function. However, designers can locate the register at any point in the circular BIST chain.

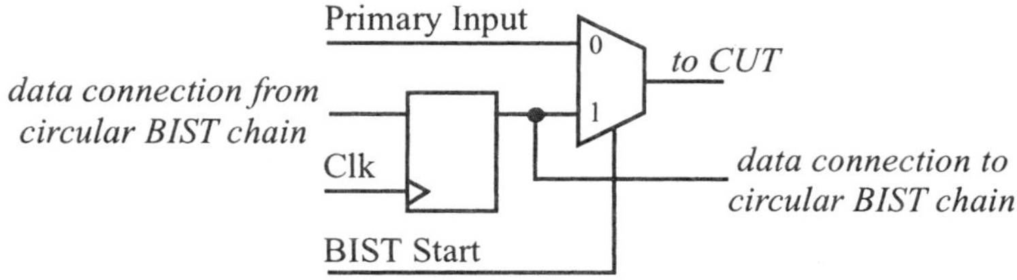

FIGURE 9.9 Input isolation shift register.

The first approach is to add a shift register to the circular BIST chain where each flip-flop added doubles the number of possible states in the state transition graph that can be visited during the BIST sequence [72]. Of course doubling the number of possible states does not imply that the number of states visited will also double, but doubling the number of possible states does imply that the number of states visited may increase. This approach is the simplest of the three with the lowest area overhead. Experimentation with this approach revealed significant improvements to fault coverage but also revealed cases where the cycle length and fault coverage decreased with increases in the number of flip-flops in the shift register, only to increase again when additional flip-flops were included in the shift register. As a result, the effectiveness of this approach is unpredictable and requires fault simulation.

The insertion of additional flip-flops in the circular BIST chain has been used as a remedy to register adjacency [25]. This not only changes the state transition graph for the CUT in BIST mode of operation but also increases the number of possible states in the state transition graph. As a result, it is difficult to distinguish between the removal of register adjacency and the increase in the number of states in the state transition graph when improvements in fault coverage are obtained. Similarly, additional flip-flops have been used for input isolation to the CUT [66], which will also have the same effect of increasing the number of states in the state transition graph. Therefore, when designers use flip-flops for input isolation, these additional flip-flops should be included in the calculation of N_{FF}. It should also be noted that the flip-flops used for input isolation along with the flip-flops in the original circuit may not be sufficient to prevent limit cycling if the resultant circuit falls into the 'case 0' or 'case 1' circuit classifications. Conversely, using flip-flops for input isolation can be unnecessary area overhead when the circuit falls into 'case 2' classification such that area overhead savings can be realized by using multiplexers for input isolation.

The second hardware approach to overcoming limit cycling inserts a shift register with additional combinational logic that will cause the shift register to produce a pseudo-random sequence as a function of the state of the shift register and the logic values coming from the output of the circular BIST chain [278]. The structure of this approach (referred to as the *target register*) is illustrated in Figure 9.10 where, in essence, a nonlinear circuit function (denoted as the target register logic) emulates the behavior of an LFSR with a primitive characteristic polynomial. As a result, the cycle length of the BIST sequence is guaranteed to be greater than 2^m where m is the number of flip-flops in the target register. This is because the target register begins with the all-zeros value and then cycles through all possible 2^m logic values before repeating any states. Since the cycle length obtained with this approach is predictable, the fault coverage obtained is also somewhat predictable in that increasing the size of the target register increases the cycle length and increases the resultant fault coverage that will be obtained with this modified Circular BIST implementation. The construction of the target register consists of a three step process as follows:

1. Implement a shift register of desired length m to be incorporated in the circular BIST chain.
2. Determine the additional logic which will produce all possible 2^m patterns in the m-bit target register. This determination is made by breaking the circular BIST chain at the serial input to the target register (denoted 'y' in Figure 9.10) and constructing a primitive polynomial feedback circuit for the target register such that the target register behaves as a SAR. Perform a logic simulation of the broken circular BIST chain with the target register acting as an LFSR while collecting the simulation results at the output of the circular BIST chain (at the point of the break, y) and the contents of the target register. Initialize the SAR with all-0s and then feed a single logic 1 into the target register (at point y in the figure) to begin generation of the pseudo-random sequence. If it can be determined when the first logic 1 will emerge from the circular BIST chain, the single logic 1 can be fed into the target register at that point to create a natural sequence of patterns. Otherwise, the single logic 1 can be fed in immediately which will ensure that, in the final implementation, the circular BIST chain will not "lock-up" in the all-0s state. The simulation should proceed until the target register has produced all 2^m possible patterns while collecting the results of each clock cycle at the outputs of the target register (x^1 through x^m) and output of the broken circular BIST chain (denoted y in Figure 9.10).

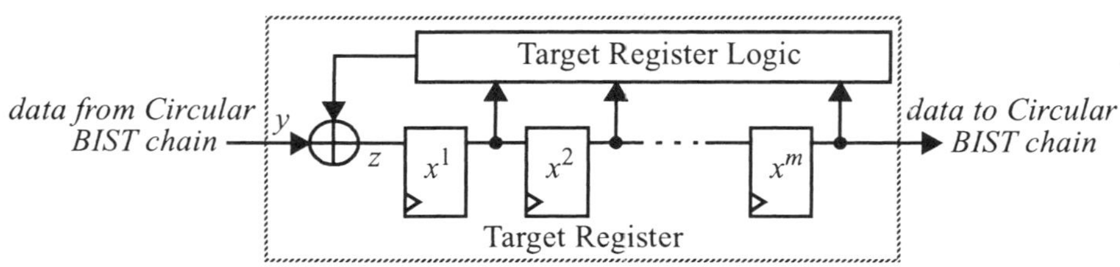

FIGURE 9.10 Target register implementation in Circular BIST.

3. Minimize the logic values obtained during the simulation to obtain a single output combinational logic function for the final target register implementation. This is done by minimizing the truth table defined by inputs x^1 through x^m with the output values defined by $y \oplus z$ such that the resultant logic expression is exclusive-ORed with the output of the circular BIST chain to produce the desired input value z to the target register. The resultant synthesized logic function is incorporated between the circular BIST chain and the target register flip-flop outputs (as illustrated in Figure 9.10). In the fault-free circuit, the shift register and feedback logic function, in conjunction with the output of the circular BIST chain will produce the same 2^m pseudo-random patterns as the LFSR simulation.

One disadvantage of this target register approach is that the complete implementation requires a multiple step process since the pseudo-random patterns produced by the target register are a function of the circular BIST chain output. Therefore, if the CUT changes or the ordering of the circular BIST chain, the target register must be resynthesized. In addition, the logic area overhead incurred by the target register implementation can be large (particularly for target registers of 12-bits or more) and tends to grow exponentially with the size of the target register. For example, a 12-bit target register requires an average of just over 100 gates for the feedback logic [278]. However, the target register approach gives consistently high fault coverage.

A final approach incorporates an LFSR external to the circular BIST chain with the serial output of the LFSR combined with the contents of the circular BIST chain at some point via an exclusive-OR gate. This injects a pseudo-random sequence into the circular BIST chain and results in a cycle length for the BIST sequence of greater than 2^m-1 where m is the number of flip-flops in the LFSR. The area overhead for this approach is only slightly higher than that of the shift register approach due to the exclusive-OR gates required for the primitive characteristic polynomial and to linearly combine the output of the LFSR with the contents of the circular BIST chain. This approach provides higher fault coverage than the shift register approach with lower area overhead than the target register approach. This LFSR-based approach is similar to the complete Circular BIST approach illustrated in Figure 9.3 where an SAR or MISR is added to the circular BIST chain to provide further compaction of the output response data in the circular BIST chain. During the initial clock cycles of the BIST sequence the constant logic values at the inputs to the SAR or MISR allow it to operate as an LFSR with a primitive characteristic polynomial to produce and apply pseudo-random sequences to the circular BIST chain, thus reducing the problems of limit cycling. Any of these hardware approaches can be combined with the technique for finding the optimal head state described in [65] to help maximize fault coverage in Circular BIST applications. The scan mode would be used to load the head state.

9.7 Benefits and Limitations

Circular BIST has been frequently used in the general sequential logic of ASICs since the mid-1980s. Circular BIST's success has been largely due to low area overhead and performance penalties as well as the ease with which the technique can be automated in CAD tools [314]. The total number of gates added when implementing one of the Circular BIST approaches in a CUT is given in Table 9.1 in terms of the number of flip-flops, exclusive-OR gates, 2-to-1 multiplexers, and elementary logic gates (Gates). An additional 2-input exclusive-NOR and 2-to-1 multiplexer are required for inclusion of the scan testing feature in the circular BIST chain. The area overhead can be controlled by selective replacement of flip-flops.

TABLE 9.1 Area overhead of Circular BIST.

BIST Approach	Area Overhead				Performance Penalty
	Flip-Flops	Exclusive-ORs	Multiplexers	Gates	
CSTP	0	N_{FF}	$N_{FF} + N_{PI}$	N_{FF}	1 MUX + 1 gate
SST	0	N_{FF}	$N_{FF} + N_{PI}$	N_{FF}	1 MUX
Circular BIST	0	N_{FF}	N_{PI}	$2N_{FF}$	1 XOR + 1 gate
where: N_{FF} = number of flip-flops in CUT replaced by BIST flip-flops N_{PI} = number of primary input to CUT requiring input isolation					

The worst case performance penalty associated with a Circular BIST implementation is a 2-input NAND gate delay and a 2-input exclusive-OR gate delay in every path between flip-flops when the destination flip-flop is replaced with a Circular BIST flip-flop. However, critical timing paths can be avoided during the selective replacement of flip-flops to avoid performance penalties. This ability to avoid performance penalties in critical timing paths is one of the main advantages of Circular BIST over the BILBO and other pseudo-exhaustive BIST approaches.

The Circular BIST approach can be easily automated as was the case in scan design-based DFT. The CAD tools to support Circular BIST implementations can also incorporate selective replacement of flip-flops as well as ordering of the circular BIST chain to minimize register adjacency. In addition, the design guidelines for limit cycling can be used to indicate when limit cycling might present a problem in terms of low fault coverage. In such cases, one of the hardware solutions and/or the technique for finding the optimal head state [65] can be used to improve fault coverage. The main limitation of the Circular BIST techniques is that high fault coverage is not guaranteed. However, most practical applications have obtained single stuck-at gate-level fault coverage in excess of 90% [178].

CHAPTER 10

Scan-Based BIST

The success of scan design-based DFT through the late 1970s and early 1980s, particularly the high degree of automatic synthesis via CAD tools, made the extension of scan designs to BIST an obvious candidate for implementation and experimentation. While some of the problems of BILBO and pseudo-exhaustive BIST were eliminated with Circular BIST approaches, the performance penalty of most of these approaches was greater than that of scan design-based DFT. BILBO and Circular BIST incur a 2-input NAND gate delay and an exclusive-OR gate delay while scan design incurs only a 2-to-1 multiplexer delay which is comparable to that of the 2-input exclusive-OR gate and, therefore, less than that of the BILBO and Circular BIST approaches. This chapter explores the scan-based BIST approaches that have evolved to become one of the most frequently used BIST approaches for general sequential logic at the present time. The architecture and operation of various scan-based BIST approaches will be discussed along with the problems that have been encountered and solutions that have been proposed.

10.1 Scan BIST Architectures and Operation

The obvious approach to creating a BIST architecture from a scan design is to incorporate a TPG in the form of an LFSR at the *Scan In* input to the scan chain and an ORA in the form of an SAR at the *Scan Out* output of the scan chain, as illustrated in

Figure 10.1a. This architecture works well for manufacturing use of the BIST when the primary inputs can be supplied with test patterns from the external test machine. However, for system-level use of the scan-based BIST, system input isolation is required along with the ability to apply test patterns to the primary inputs and output data compaction of the primary outputs (as shown in gray in Figure 10.1b). The BIST controller provides the necessary *Scan Mode* control to switch the flip-flops in the scan chain between system mode (to apply test patterns and retrieve output responses) and shift mode (to shift in test patterns from the TPG and shift out the resultant output responses to the SAR or MISR). In addition, the BIST controller must disable output response compaction by the MISR until valid output responses are available at the primary outputs and at the *Scan Out* output of the scan chain.

Operation of this scan BIST approach would begin with activation of the *BIST Start* input, at which time the BIST controller would initialize the LFSR and the MISR, isolate the primary inputs by selecting the alternate inputs to the input isolation multi-

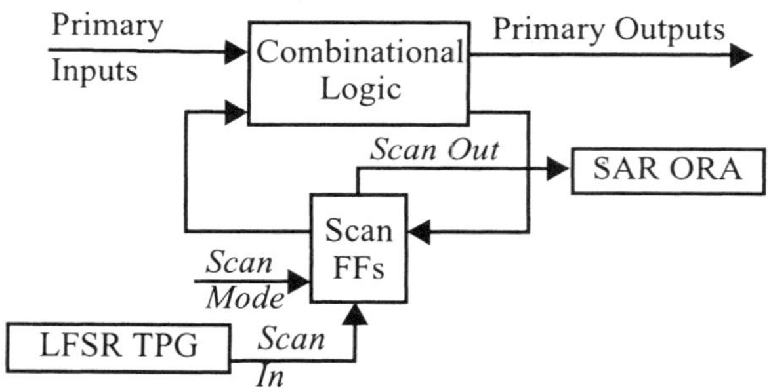

(a) manufacturing use only

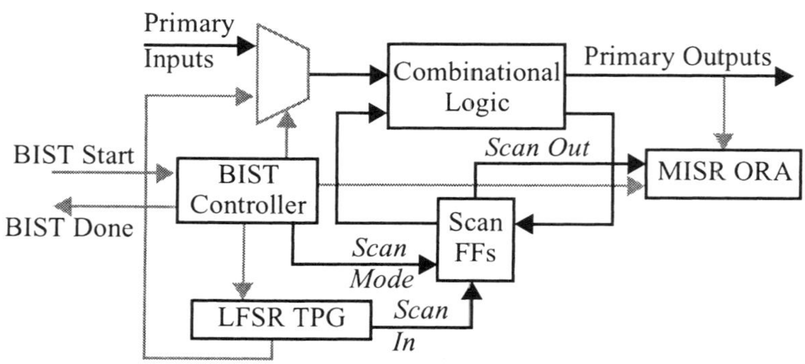

(b) system-level use

FIGURE 10.1 Basic scan BIST architecture.

plexers, and put the scan chain into the shift mode. After initialization, the LFSR begins to generate pseudo-random test patterns that are shifted into the scan chain. When sufficient clock cycles have elapsed to have the scan chain filled with a test pattern from the LFSR, the BIST controller puts the scan chain into the system mode of operation for one clock cycle to apply the test pattern to the CUT. During this clock cycle, the LFSR applies the test pattern to the primary inputs via the input isolation multiplexers. The CUT output responses to the test patterns are clocked into the scan chain during this clock cycle. The BIST controller enables the MISR to begin output response compaction from the primary outputs and puts the scan chain back into the shift mode for the next clock cycle. As the output responses to the first scan vector are shifted out, the MISR compacts the data exiting the scan chain while the LFSR produces the next scan vector shifting into the scan chain. The MISR continues to compact the output responses on every clock cycle until the end of the BIST sequence. Upon completion of the BIST sequence, the MISR is disabled until the resultant signature can be read for the pass/fail determination of the BIST.

The length of the BIST sequence is determined by the number of scan vectors needed to obtain the desired fault coverage. As a result the BIST controller can be constructed from a counter with $M+N$ bits where the M least significant bits control the shifting of test patterns into and out of the scan chain. Each time the M least significant bits are all 1s, the scan chain is put into the system mode for one clock cycle. As a result, $M = \lceil \log_2(N_{FF}) \rceil$ where N_{FF} is the number of flip-flops in the scan chain and $N = \lceil \log_2(N_{TV}) \rceil$ where N_{TV} is the number of scan vectors to be applied to obtain the desired fault coverage. This simple $M+N$-bit counter-based BIST controller design is illustrated in Figure 10.2a. Note that the *BIST Start* input signal is also used to preset the LFSR to the all 1s state, reset the $M+N$-bit counter, and reset the MISR (not shown in the figure) to the all 0s state whenever *BIST Start* = 0.

The count value for switching the scan chain to the system mode can be the exact count for the number of flip-flops in the scan chain, N_{FF}, plus one clock cycle to put the scan flip-flops in the system mode of operation for application of the test vectors and capture of the output responses. Similarly, the scan vector count can also be the exact count for the number of scan vectors to be applied, N_{TV}, to minimize the test time. The main problem is determining the number of scan vectors to be applied to obtain the desired fault coverage. This determination can be made via fault simulation of the BIST sequence. By separating the counters a more optimized BIST sequence is obtained in terms of testing time at the expense of a slightly more complicated BIST controller design as illustrated in Figure 10.2b. In this case, the Shift Control is a counter that counts to $N_{FF}+1$, then repeats. The active high carry-out of the Shift Control counter is used to put the scan flip-flops into the system mode for one clock cycle and also to enable the Pattern Control counter to increment the scan vector count. The Pattern Control counter can be a simple N-bit binary counter or it can be designed to

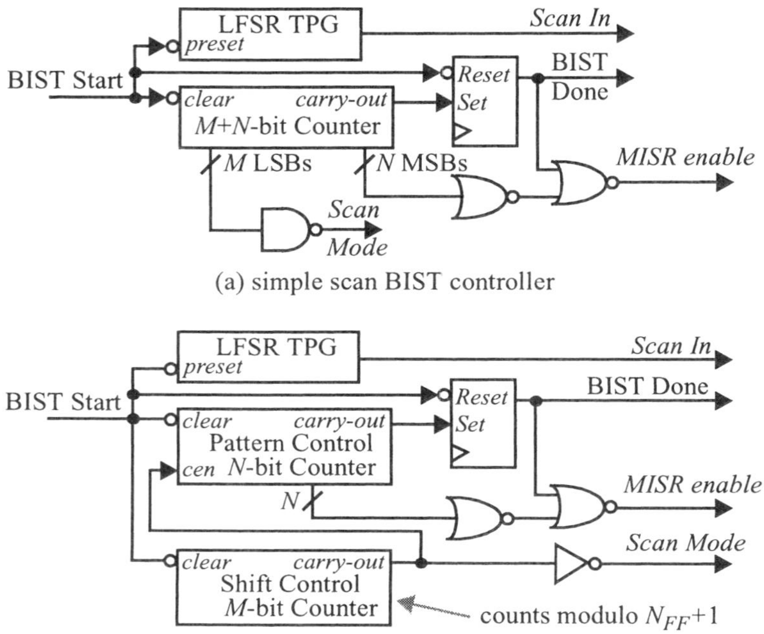

(a) simple scan BIST controller

(b) more optimized scan BIST controller

FIGURE 10.2 Basic scan BIST controller design.

count to a value of N_{TV} before setting the *BIST Done* indication. Note that the NOR gates in Figure 10.2 are used to generate an active high enable signal for the MISR such that no output response compaction takes place until after the first scan vector has been applied and the MSR is disabled when *BIST Done* goes active.

This basic scan BIST approach is a test-per-scan, embedded BIST architecture in that the scan chain is created from existing flip-flops. However, it can also be considered a separate BIST architecture for designs which already contain a scan chain since the flip-flops of the LFSR and MISR are not part of the existing system function such that the TPG and ORA functions and are separate from the CUT. The BIST circuitry (TPG and ORA) can also be considered to be centralized since the TPG and ORA functions can be used for multiple scan chains or multiple VLSI devices with individual scan chains. A number of variations of this basic scan BIST architecture have been proposed and implemented since the early 1980s. In the following subsections, we will explore some of these approaches. Of specific interest will be their applicability to system-level use as well as the similarities and differences to the basic scan BIST architecture discussed above.

10.1.1 The First Scan BIST?

One of the first scan BIST approaches proposed combined the TPG and ORA functions into a single SAR and used the compacted output response data as the scan vectors being shifted into the scan chain [247]. This reduced the area overhead of the overall approach since the TPG and ORA functions are external to the logic performing the normal system function as illustrated in Figure 10.3. In this approach, two counters are used for the BIST controller: one counter to count the number of flip-flops in the scan chain for shifting in scan vectors (referred to as the Shift Control) and this, in turn, drives the other counter to count the number of scan vectors applied to the CUT (referred to as the Pattern Control). During the initialization portion of the BIST sequence the SAR and the scan chain are loaded with all 1s; this allows the SAR to generate non-zero test patterns at the beginning of the BIST sequence.

In the results reported for an application of this approach, 250 scan vectors were applied to the CUT and 96% fault coverage was obtained [247]. The primary inputs were not isolated but were held at a constant logic value in this approach, nor were the primary output responses compacted in the SAR. Obviously, the approach was intended for manufacturing test use only. Some of the faults not detected by the BIST approach were reported to be in the BIST circuitry; the comparator and expected good circuit signature (denoted '*good ckt sig*' in Figure 10.3) are likely candidates for these faults as discussed in Chapter 5. Input isolation, similar to that illustrated in Figure 10.1b, can easily be added to this approach to facilitate system-level use. In

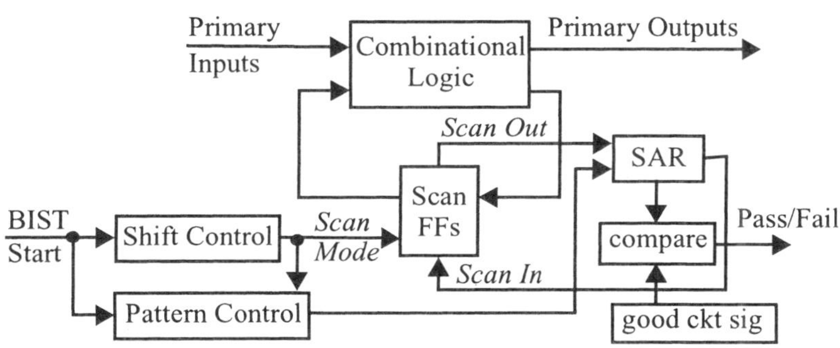

FIGURE 10.3 Architecture of what appears to be the first scan BIST approach.

addition, the signature can be read from the SAR to avoid the additional area overhead of on-chip storage and comparison with the known good circuit signature. Deterministic test vectors developed for the CUT using scan design-based ATPG CAD tools required only about 25 scan vectors. Therefore, the scan BIST approach required approximately ten times the number of scan vectors to achieve comparable fault coverage. To overcome the additional test time, it was suggested that multiple scan chains be used to reduce the time required for shifting in scan vectors and shifting out the resultant responses [247].

10.1.2 Random Test Socket

The Random Test Socket (RTS) approach was initially used for manufacturing testing of VLSI devices and PCBs that incorporated scan design-based DFT [31]. The basic architecture of the RTS approach is illustrated in Figure 10.4. The M-bit LFSR at the primary inputs makes it apparent that the approach was not initially intended for system-level use. As originally proposed and implemented, both LFSRs, the SAR and the N-bit MISR, along with the BIST controller were all external to the VLSI device or PCB under test. However, this BIST circuitry could easily be incorporated inside a VLSI device that includes scan design-based DFT or on a PCB to test multiple VLSI devices that incorporate scan chains. To complete the BIST implementation, input isolation multiplexers can be incorporated at the primary inputs of the chip or PCB along with the LFSRs, SAR, MISR and BIST controller to make the approach accessible at the system-level. The separate LFSRs for the scan chain and the primary inputs require additional area overhead compared to the basic scan BIST approach of Figure 10.1. Similarly, the separate SAR for the scan chain and MISR for the primary

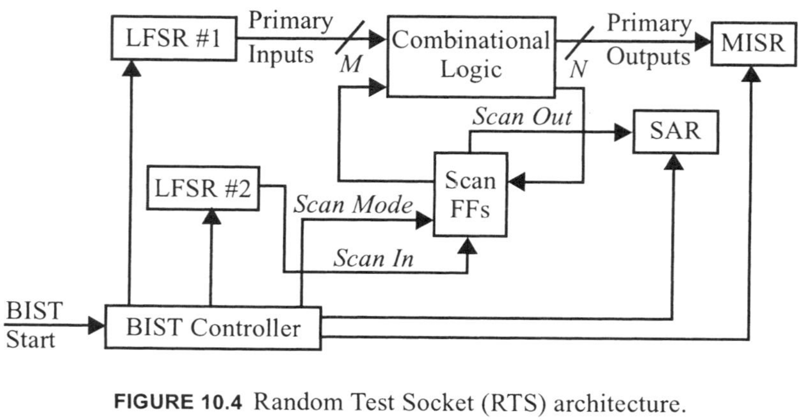

FIGURE 10.4 Random Test Socket (RTS) architecture.

outputs require additional area overhead that may be undesirable for implementing in a VLSI device. Alternatively, the LFSRs could be combined, as could the SAR and MISR.

The operation of RTS begins with the activation of the *BIST Start* signal, at which time the LFSRs, SAR and MISR are initialized, followed by a pseudo-random test pattern being shifted into the scan chain. Once the scan chain is full, the scan flip-flops are placed in the system mode of operation for one clock cycle, at which time the SAR and MISR are enabled for data compaction. The process of shifting in pseudo-random test vectors while the output responses from the previous vector are shifted out continues, interrupted only by the single clock cycles in the system mode of operation to apply the scan vector and capture the output responses. In the original description of the RTS approach, separate clocks were supplied to the LFSRs, SAR, MISR and scan chain for independent control of each function, for a total of five separate clocks [31]. The primary reason was to clock LFSR #1 and the MISR only once per scan vector. However, only a single clock needs to be applied, letting LSFR #1 continue to supply new test patterns each clock cycle while LFSR #2 is shifting in new scan vectors and while the SAR and MISR continue to compact the output responses. The only additional control signal that needs to be supplied (besides the *Scan Mode* input to the scan chain) is an enable to the SAR and MISR to enable output response compaction once the scan chain is initialized. This enable to the SAR and MISR can also be used to hold the resultant signatures at the end of the BIST sequence until read for the pass/fail determination.

The RTS approach was later implemented as a true BIST approach in VLSI devices with only minor modifications for input isolation [79]. A later implementation of a modified version of RTS was referred to as the LSSD On-Chip Self-Test (LOCST) approach [148]. The main difference between these two implementations and the RTS approach as shown in Figure 10.4 is the use of a single LFSR and SAR with a shift register at the primary inputs and outputs of the CUT connected in series with the LFSR, SAR and scan chain of the CUT. One limitation of the approach was the long test time due to the many clock cycles required to shift the scan vectors and responses versus only one clock cycle of test vector application time (test-per-scan). When LFSR #1 and the MISR continue to function during the time that the scan vector and output response are being shifted in and out, the combinational logic at the primary outputs will continue to be tested on a test-per-clock basis. This logic will be completely tested early in the BIST sequence such that the test time is still dominated by the test-per-scan nature of the approach. A solution to this problem of long testing time is to implement multiple scan chains to reduce the number of clock cycles required to shift in the test vectors and shift out the output responses. This leads to the next scan-based BIST architecture.

10.1.3 STUMPS

The Self-Test Using MISR/Parallel SRSG (STUMPS) approach significantly reduces test time of the previous scan BIST approaches by taking advantage of multiple scan chains to reduce the number of clock cycles required to shift in scan vectors and to shift out the output responses of the CUT. Note that the acronym SRSG stands for Shift Register Sequence Generator which is the same as an LFSR-based TPG [32]. The basic STUMPS architecture is illustrated in Figure 10.5. An N-bit LFSR-based TPG is incorporated to supply test patterns to multiple scan chains in the CUT with each of the *Scan In* inputs connected to a different stage of the LFSR. Similarly, the *Scan Out* outputs from each of the N scan chains drive the inputs of an N-bit MISR for output response compaction. As originally described, STUMPS was used to test multi-chip modules (MCMs) where each scan chain represented the scan chain in an individual chip. In this MCM application, the LFSR and MISR were implemented on a separate STUMPS test chip. However, STUMPS can easily be applied at the PCB-level for VLSI devices that incorporate scan design or in single chips with multiple scan chains.

Once the BIST sequence is initiated, the BIST controller initializes the LFSR and MISR and puts the scan chains in the scan mode of operation. The pseudo-random patterns produced by the LFSR are shifted into the scan chains. After N_{FF} clock cycles, where N_{FF} is the number of flip-flops in the longest of the N scan chains, the

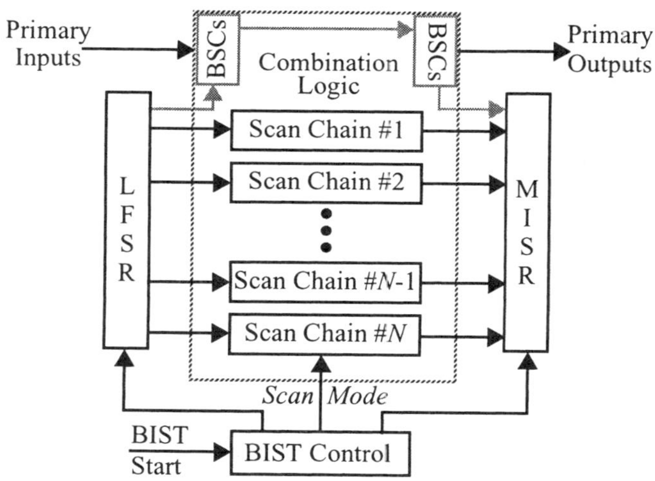

FIGURE 10.5 STUMPS architecture.

BIST controller puts the scan chains in the system mode of operation for one clock cycle in order to apply the test vectors to the combinational logic and capture the output responses in the scan chains. At this point in time the MISR is enabled to begin output response compaction. As the output responses to the scan vector are shifted out and compacted by the MISR, the patterns being generated by the LFSR are shifted in as the next scan vector to be applied to the combinational logic. This process continues until sufficient scan vectors have been applied to obtain the desired fault coverage. The BIST controller shown in Figure 10.2b could be used for STUMPS. The primary inputs and outputs were not addressed in the original STUMPS description, but the LFSR and MISR can be expanded based on the number of primary inputs and outputs, respectively, to facilitate complete system-level testing. In addition, input isolation multiplexers could be inserted at the primary inputs in a manner similar to those shown in Figure 10.1. Alternatively, Boundary Scan cells in conjunction with an INTEST capability could be used as an additional scan chain to provide test patterns to the primary inputs and to capture CUT responses at the primary outputs, with a subsequent shift into the MISR for compaction. This is indicated by the blocks in Figure 10.5 labeled BSCs (for Boundary Scan cells) along with the gray arrows to indicate the complete scan chain path. This implies that the Boundary Scan chain can be manipulated with the rest of the scan chains during the BIST mode.

10.2 Linear and Structural Dependencies

There are some inherent problems in the scan BIST approaches that have been observed in practical applications. These problems include *structural dependencies* and *linear dependencies* which arise from the way in which LFSRs and CA registers are constructed. While structural dependencies create problems in multiple scan chains, as in the STUMPS architecture, linear dependencies create problems in any scan BIST architecture. As a result, structural and linear dependencies can both occur and can interact in multiple scan chain architectures.

10.2.1 Structural Dependencies

An obvious problem in scan BIST arises from the shifting nature of the test patterns produced in adjacent stages of an external feedback LFSR (as was illustrated in Figure 4.7). This means that adjacent scan chains are receiving the same scan vectors but shifted by one flip-flop in the scan chain. As a result, there are combinations of patterns that cannot be produced in adjacent scan chains [32]. To illustrate this problem of structural dependencies, there are 16 possible combinations of four bits, but there are only 8 of these 16 possible combinations (or one-half) in any 2×2 array of bits produced by the external feedback LFSR. This is illustrated in Table 10.1 where the same 8 combinations are produced by any two adjacent bits in the external feedback LSFR while the other 8 possible combinations are never produced. Only 128 out of the 512 (or one-fourth) possible combinations of nine bits in any 3×3 array of bits are produced by an external feedback LFSR [32]. The situation gets worse with larger array sizes. As a result of these structural dependencies preventing certain combinations of test patterns in the scan chains, some faults in the combinational logic may not be detected, depending on how the combinational logic is connected to the scan chains. For example, if a 4-input NOR gate was connected to the four bits shown in the table for an external feedback LFSR, two of the inputs stuck-at-0 would not be detected. This reduces the fault coverage that can be obtained.

TABLE 10.1 Structural dependencies in LFSRs and CAR of Figure 4.7.

2×2 patterns	External FB LFSR	Internal FB LFSR	CA Register	2×2 patterns	External FB LFSR	Internal FB LFSR	CA Register
00 00	any two adj bits	bits 1&2 bits 4&5	bits 1&2 bits 7&8	00 01			bits 1&2
10 00			bits 1&2	10 01	any two adj bits	bits 1&2 bits 4&5	bits 1&2 bits 7&8
01 00	any two adj bits	bits 1&2 bits 4&5	bits 7&8	01 01			
11 00				11 01	any two adj bits	bits 1&2 bits 4&5	bits 7&8
00 10	any two adj bits	bits 4&5	bits 7&8	00 11		bits 1&2	
10 10		bits 1&2		10 11	any two adj bits	bits 4&5	bits 7&8
01 10	any two adj bits	bits 4&5	bits 1&2 bits 7&8	01 11		bits 1&2	bits 1&2
11 10		bits 1&2	bits 1&2	11 11	any two adj bits	bits 4&5	bits 1&2 bits 7&8

The internal feedback LFSR is not quite as bad as the external feedback LFSR, particularly at the flip-flops representing non-zero coefficients of the characteristic polynomial where more possible combinations are produced. For example, the characteristic polynomial used for the 8-bit internal feedback LFSR in Figure 4.7 was $P(x) = x^8 + x^6 + x^5 + x + 1$. For example, bits 5 and 6 as well as bits 6 and 7 produce all 16 possible combinations of bits for a 2×2 array of bits. However, bits 1 and 2 produce only 8 of the 16 possible combinations and 4 of those combinations are different from those produced by adjacent bits with zero coefficients. This can be seen in Table 10.1 by comparing the patterns produced by bits 1 and 2 with the patterns produced by bits 4 and 5 which have zero coefficients. Non-zero coefficients exist in no more than three flip-flops in the LFSR since there are, at most, only three non-zero coefficients in most primitive polynomials used to implement LFSRs for TPGs. This is because we are typically interested in primitive polynomials with the fewest non-zero coefficients in order to minimize the number of exclusive-OR gates required for the LFSR implementation. Therefore, most of the bits in a large internal LFSR produce test patterns that suffer from these structural dependencies.

Cellular Automata (CA) registers are better than either LFSR implementation as can be seen in Figure 4.7. In fact many of the adjacent bits produce all 16 possible combinations for a 2×2 array of bits. However, structural dependencies still exist in some of the test patterns produced by a CA register [30]. It is most obvious near the end of the register where the null boundary conditions produce shifting patterns similar to the LFSR. As shown in Table 10.1, bits 1 and 2 produce only 8 of the 16 possible combinations. Similarly, bits 7 and 8 produce only 8 combinations but 4 of these are different from those combinations produced by bits 1 and 2. While cyclic boundary conditions would help avoid this problem, it has been conjectured that there are no large CA registers with cyclic boundary conditions that produce maximum length pseudo-random sequences [30].

10.2.2 Linear Dependencies

Another problem encountered in scan BIST applications is linear dependencies in the test patterns produced by any single bit of an LFSR or CA register [29]. There are some problems associated with linear dependencies that are obvious but others are not so obvious. To illustrate the obvious problem, consider the 4-bit internal feedback LFSR of Figure 4.3b where the last bit in the LFSR is used to supply test patterns to the *Scan In* input of a 10-bit scan chain, as shown in Figure 10.6. The LFSR will produce a 15-bit sequence before it begins to repeat such that if we shift in exactly 10 bits to use as a scan vector, then another 10 bits, and so on, only three unique scan vectors

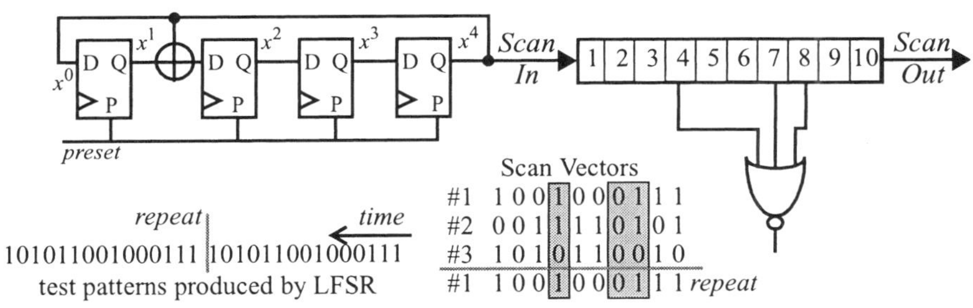

FIGURE 10.6 Linear dependencies example.

will be applied to the CUT before we begin to repeat the same scan vectors. In the example in Figure 10.6, two of the three test patterns to the 3-input NOR gate are repeated such that only two different scan vectors are applied (as shown in the shaded columns). As a result, we only detect the NOR gate output stuck-at-0 and stuck-at-1 but do not detect any of the inputs stuck-at-0.

Therefore, we want to avoid the case where N_{FF} and 2^n-1 have a common divisor (other than 1), where N_{FF} is the number of flip-flops in the scan chain and n is the number of bits in the LFSR [131]. The number of clock cycles used to shift the scan vectors into the scan chain must be greater than or equal to N_{FF}. We can choose a number larger than N_{FF} in the design of the BIST controller in order for N_{FF} and 2^n-1 not to have a common divisor. We can also choose the value of n for the design of the LSFR. In this case, it is recommended that the value of n be greater than the largest *span* of any logic cone driven by the scan chain. The span is the number of scan flip-flops between the two inputs to a logic cone that are farthest away from each other in the scan chain [32]. It is further recommended that, if we desire N_{TV} scan vectors to be applied to the CUT, we choose the value of n to satisfy $2^n - 1 \geq N_{FF} \times N_{TV}$ to help reduce linear dependency problems [131]. Note in our example the span is 5 but n=4.

Following these design guidelines does not eliminate all of the problems of linear dependencies. Some of the problems are very subtle and to illustrate these we return to the example in Figure 10.6 where we can assume that we will shift in the scan vector for 11 clock cycles (with the 11th clock cycle used to apply the scan vector to the CUT while the 11th bit coming from LFSR is ignored). In this way the pseudo-random patterns overflow the scan chain, which is no problem. Since the numbers 11 and 15 do not have a common divisor, we end up applying all 15 possible scan vectors produced by the LFSR to the scan chain and the CUT in 165 clock cycles. When we investigate the test patterns contained within these 15 scan vectors that will be applied to the 3-input NOR gate, we find that only four test patterns are applied to the NOR gate. These vectors include {000, 110, 101, 011} which only detect the NOR gate output stuck-at-0 and stuck-at-1 but do not detect any of the inputs stuck-at-0. This is an example of a linear dependency that is a function of the characteristic polynomial of the LFSR and the relationship of the locations of the scan flip-flops driving the logic cone [32].

There are methods to look for these linear dependencies in the scan chain with respect to the characteristic polynomial, $P(x)$, of the LFSR and what is referred to as the *sampling polynomial.* This is a polynomial that describes the logic cone connections to the scan chain in terms of their position in the scan chain. For example, the scan chain connections of the 3-input NOR gate can be represented by the sampling polynomial $S(x) = x^8 + x^7 + x^4 = x^4 + x^3 + 1$. If $S(x) \bmod P(x) = 0$, then linear dependencies will exist in the test patterns being produced by the LFSR [29]. If this method does not indicate the existence of linear dependencies, there is a second method to check for linear dependencies. This method requires breaking the sampling polynomial $S(x)$ into its sub-polynomials and checking those sub-polynomials to determine if linear dependencies exist [32]. While structural dependencies are not as prevalent in the test patterns produced by CA registers, other than near the edges of the register when null boundary conditions are used, linear dependencies still exist [30].

10.3 Linear and Structural Dependency Solutions

There are a number of ways to overcome the problem of low fault coverage associated with structural and linear dependencies in scan BIST applications. One approach is to externally apply deterministic test patterns during manufacturing testing using the scan chain, but this does not improve the fault coverage of the scan BIST approach during system-level testing [30]. Of course, the deterministic test patterns can be stored on-chip in a ROM but this increases area overhead, design time, and risk to the project in the case of late design changes to the CUT. Another solution is to reorder the scan chain such that different sampling polynomials are created for the logic cones, but new problems may be introduced in some logic cones as problems are avoided for other logic cones. In addition, reordering the scan chain defeats the goal of low routing overhead and performance penalty that can be achieved with post-layout scan chain ordering. In this section, we explore some of the more reasonable approaches to alleviating the problems of linear and structural dependencies.

10.3.1 Reseeding LFSRs and Multiple Polynomials

The main problem of linear dependencies stems from the characteristic polynomial of the LFSR, including the degree of the polynomial, versus the sampling polynomials of the logic cones, including their degrees or spans. One technique for reducing the problem of linear dependencies is to use large LFSRs compared to the spans of the logic cones. As was mentioned in the previous section, a recommended design guideline is to choose the value of n to satisfy $2^n - 1 \geq N_{FF} \times N_{TV}$, where n is the number of bits in the LFSR, N_{FF} is the number of flip-flops in the (longest) scan chain and N_{TV} is the number of scan vectors to be applied [131]. This can lead to large LFSRs with greater than 20 to 25 bits such that exercising the maximum length sequence of the LFSR will not be practical from a testing time and cost standpoint. Therefore, only a portion of the maximum length sequence of the LFSR can be applied during the BIST sequence. By incorporating the ability to begin the LFSR at different points in its sequence, different sets of test patterns can be applied [240]. This is a simple hardware solution and requires additional circuitry to facilitate initialization of the LFSR to different states, or *reseeding*, prior to each execution of the scan BIST sequence in the suite of test phases that make up the complete BIST session [105].

Another approach is to use an LFSR with the ability to change characteristic polynomials [105]. This can be done by using the generic LFSRs illustrated in Figure 4.5 such that the characteristic polynomial can be changed as desired. By changing the characteristic polynomial, the linear dependencies are changed due to the different relationships to the sampling polynomials of the logic cones connected to the scan chain, even though the sampling polynomials remain constant. The programmable polynomial can be coupled with reseeding of the LFSR to further enhance the effectiveness of the overall approach and minimize the effects of linear dependencies on the fault coverage obtained with scan BIST. By making the LFSR a register that can be written from a processor interface or via Boundary Scan, the seed values can easily be changed at both manufacturing and system-level testing with new seed values used well after the fabrication of the VLSI device and with minimal area overhead. Similarly, different characteristic polynomials can be established from a second register that is written from a processor interface or via Boundary Scan. Polynomials and seed values that provide the best improvements to fault coverage can be investigated by fault simulations at anytime without adding risk to the project by attempting to determine and hardwire these values on-chip prior to fabrication.

10.3.2 Bit Manipulation

Weighted pseudo-random test pattern generation has been proposed as a solution to the problems of linear dependencies [189]. Another method to overcoming the problem of linear dependencies is to include additional circuitry that manipulates the bits produced by the LFSR as they enter the scan chain. There are two approaches to bit manipulation that are illustrated in Figure 10.7; one method is to flip the bits as shown in Figure 10.7a [346] and the other approach is to fix the bits to logic 1s or 0s as shown in Figure 10.7b [306]. The logic blocks shown as SC and PC represent the Shift Control counter and the Pattern Control counter, respectively. These are parts of the necessary control for the scan BIST approach and not additional logic required for either the bit flipping or bit fixing approach. The bit flipping and bit fixing approaches can use these count values to control when they will flip a bit or fix a bit to a logic 0 or a logic 1 as it shifts into the scan chain. The bit flip logic or bit fix logic is basically a decoder function which takes as input the state of the LFSR along with the states of Shift Control and Pattern Control to produce active control signals to the exclusive-OR gate (in the case of the bit flipping approach) or to the NOR gates (in the case of the bit fixing approach) [131]. Note that the *fix-to*-1 and *fix-to*-0 control signals are active high in the figure.

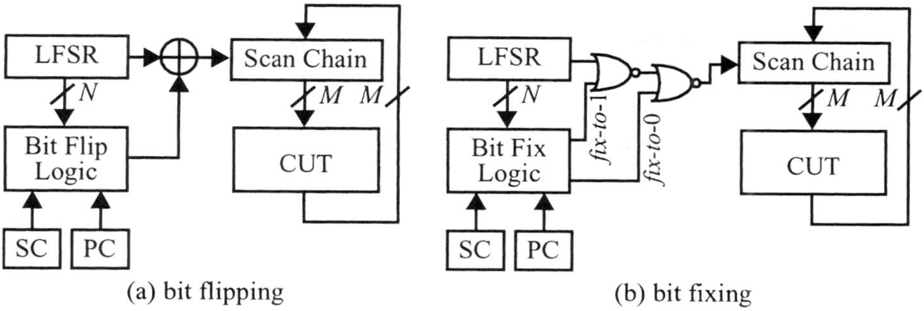

FIGURE 10.7 Hardware solutions to linear dependencies.

The basic procedure for designing the bit manipulation control logic is to first perform fault simulation using the scan BIST approach implemented without the bit manipulation control logic. The purpose of this fault simulation, aside from determining the fault coverage of the scan BIST approach without bit manipulation, is to establish which faults are not detected and which scan vectors produced by the LFSR do not detect faults during the BIST sequence. This latter point is important since we only want to modify those test patterns that are not detecting any faults, and not to modify those patterns that are detecting faults. Test patterns are then established for each of the undetected faults using ATPG software or through path sensitization and fault simulation. The main goal in this case is to find the test patterns that closely match those produced by the LFSR in order to minimize the number of bits that have to be manipulated which, in turn, should help to minimize the amount of logic required to manipulate the bits.

10.3.3 Test Point Insertion

The insertion of test points (points of controllability and observability using gates and/or multiplexers in a manner similar to ad-hoc DFT techniques) in the combinational logic of the CUT has also been proposed to improve the fault coverage for scan BIST approaches that encounter the problems of linear dependencies [301]. The procedure for inserting test points is similar to that of the process of designing the bit manipulation logic described above. Undetected faults from fault simulation of the scan BIST approach are used to determine the number, location, and type of test points [307]. The main problem with this approach as compared to the bit manipulation approaches is that the test point logic is inserted in the combinational logic of the CUT and, as a result, incurs performance penalties that may reside in critical timing paths. However, test point insertion can provide significant improvements to the fault coverage with only a few test points [310]. Furthermore, test point insertion can be used in conjunction with the other approaches described in this section [107].

10.3.4 Phase Shifters

Phase shifters are applied to multiple scan chain architectures like STUMPS to overcome structural dependencies in the scan vectors entering the scan chains. As a result, the phase shifters manipulate multiple bits in parallel as they leave the LFSR or CA register and enter the scan chains. The general structure of scan BIST architectures that incorporate phase shifters is illustrated in Figure 10.8. The initial phase shifter designs were composed of exclusive-OR networks (as illustrated in Figure 10.8) which removed structural dependencies but not necessarily linear dependencies [218]. The size of the LFSR, n, is chosen to satisfy $2^n - 1 \geq N_{FF} \times N_{SC}$ where N_{SC} is the number of scan chains and N_{FF} is the number of scan flip-flops in the longest scan chain. However, with a phase shifter, the size of the LFSR can actually be smaller than the number of scan chains; in other words n can be less than N_{SC}. This is because the phase shifter can produce additional outputs for the scan chains through its exclusive-OR network.

The underlying problem of structural dependencies in LFSRs, as was illustrated in Table 10.1 and Figure 4.7, particularly in the case of the external feedback LFSR, is the fact that the patterns coming from each stage of the LFSR are identical to those of the previous stage but delayed by one clock cycle. The basic idea of a phase shifter is

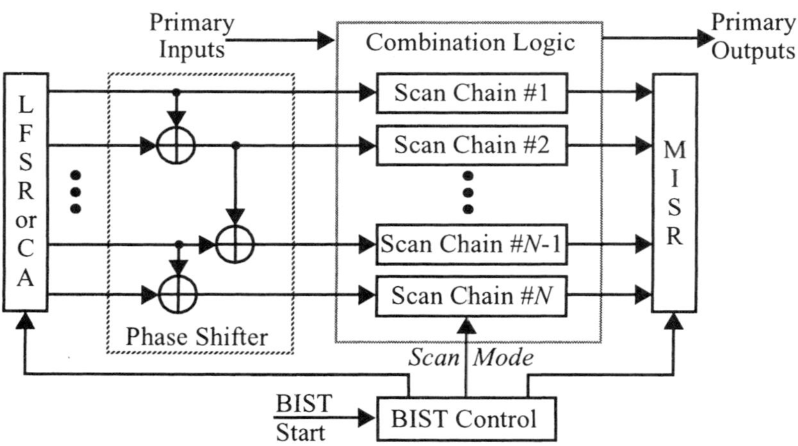

FIGURE 10.8 Scan BIST with phase shifter.

to produce patterns for each scan chain (or what is referred to as a *channel*) that are delayed by a controlled number of clock cycles with respect to the other channels. In this way the serial patterns can be aligned in a manner to remove the structural dependencies. One obvious approach is to incorporate shift registers of varying lengths to control this delay but of course that would incur unacceptable area overhead in cases where large delays are desired. Fortunately, a properly designed phase shifter exclusive-OR network facilitates that control without the need for shift registers or any additional flip-flops. While the mathematics behind the design of phase shifters can be complex [32], a simple algorithm based on simple logic simulation provides an easier design procedure [217].

Before presenting the design procedure, let us first look at the effects of the results of a phase shifter in terms of the patterns produced by an external feedback LFSR. The test patterns with and without a phase shifter are compared in Figure 10.9. The LFSR used in this case was the same 8-bit external feedback LFSR used to generate the patterns in Figure 4.7 with the patterns shown again here on the left for direct comparison with the outputs of the phase shifter shown on the right. The phase shifter was designed for a constant phase shift of nine clock cycles delay with respect to the previous channel; in other words, c_i is delayed from c_{i-1} by nine clock cycles, as denoted

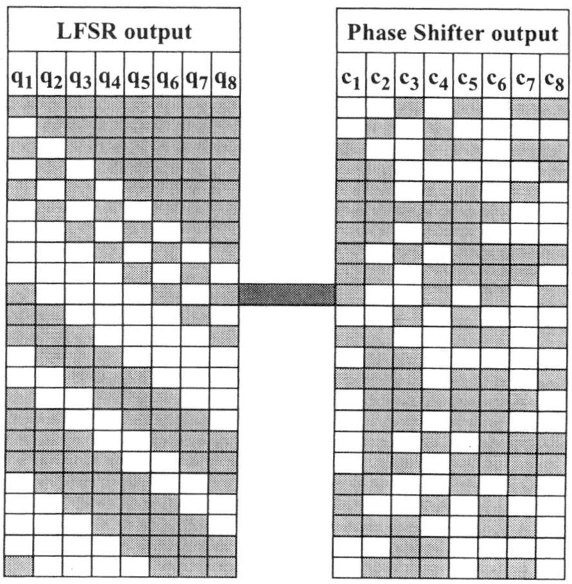

FIGURE 10.9 Pattern comparison of external feedback LFSR and phase shifter.

by the marker in between the tables. In addition, the phase shifter was designed to delay channel 1 (c_1) by nine clock cycles with respect to q_1 of the LFSR. While the LFSR produces only half of the 16 possible combinations of a 2×2 array of bits as shown in Table 10.1, the phase shifter produces all 16 combinations in all adjacent pairs of columns (for example, columns 4 and 5 produce 15 of the 16 possible combinations in the first 23 clock cycles of the sequence in Figure 10.9). Due to the constant phase shift of this particular design, a sequence of patterns can be seen to reappear as we move down and to the right in the table. However, a constant phase shift is not necessarily desirable in terms of minimizing the area overhead of the phase shifter design.

The design procedure for the synthesis of the phase shifter begins with the design of the LFSR that will be used in the scan BIST approach [218]. The LFSR can be either an internal feedback or an external feedback design but the size of the LFSR should follow the guidelines $2^n - 1 \geq N_{FF} \times N_{SC}$. The design of the phase shifter then proceeds as follows [217]:

Step 1: Construct the dual of the LFSR, where the dual of an external feedback LFSR is an internal feedback LFSR with the same polynomial, and vice versa.

Step 2: Initialize the dual LFSR to all 0s except for a single logic 1 in the ith stage of the LFSR and simulate for k clock cycles, where k is the maximum possible phase shift desired for any channel with respect to the ith stage of the LFSR.

Step 3: The logic 1s contained in the contents of the dual LFSR at the jth clock cycle, indicate the flip-flop outputs of the original LFSR to be exclusive-ORed together to create a channel with a phase delay equal to j clock cycles with respect to the ith stage of the LFSR.

A simple example of using this procedure is illustrated in Figure 10.10 for the 4-bit external feedback LFSR from Figure 4.4. Assume a 5-channel phase shifter is desired with the following phase shifts with respect to the x^1 output of the LFSR: c_1 phase

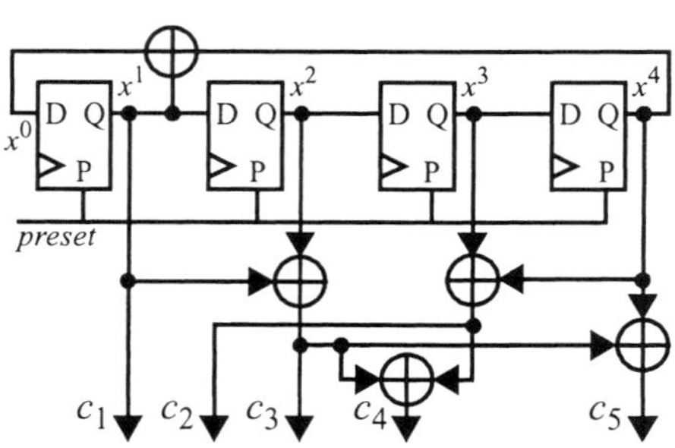

$c_1c_2c_3c_4c_5$	$c_1c_2c_3c_4c_5$
1 0 0 0 1	1 1 1 0 0
0 0 1 1 0	0 0 1 1 1
1 0 1 1 0	0 1 0 1 0
0 1 1 0 0	0 1 0 1 1
1 1 1 0 1	1 0 1 1 1
1 1 0 1 1	1 0 0 0 0
0 1 1 0 1	1 1 0 1 0
0 0 0 0 1	1 0 0 0 1 *repeat*

FIGURE 10.10 LFSR pattern sequence for with phase shifter.

shift = 0, c_2 phase shift = 6, c_3 phase shift = 4, c_4 phase shift = 12 = -3, and c_5 phase shift = 7. To obtain the exclusive-OR network we would create the dual internal feedback LFSR, initialize it to 1000 and simulate it for the number of clock cycles corresponding to the desired phase shift to get the contents of the dual LFSR. This information can be obtained from the state diagram for the internal feedback LFSR with the same characteristic polynomial that was given in Figure 4.3b. Starting in the 1000 state and counting clockwise around the state diagram for the desired phase shifts we see that the contents of the dual LFSR for the channels are as follows: c_1 = 1000, c_2 = 0011, c_3 = 1100, c_4 = 1111, and c_5 = 1101. From the logic ones, the following exclusive-OR connections are needed to produce the five channels: $c_1 = x^1$, $c_2 = x^3 \oplus x^4$, $c_3 = x^1 \oplus x^2$, $c_4 = c_2 \oplus c_3$, and $c_5 = c_3 \oplus x^4$. The resultant exclusive-OR network is shown in Figure 10.10 along with the sequence of patterns produced by the LFSR and phase shifter to verify that the correct phase shift was achieved for each channel, where the gray box indicates the start of the sequence with respect to x^1.

Obviously, fewer logic 1s in the contents of the dual LFSR for a given phase delay will result in fewer exclusive-OR gates in the phase shifter implementation. This translates to lower area overhead and performance penalty. Therefore, if a range of phase shifts is acceptable for a given channel, the particular phase delay within that range that corresponds to the fewest logic 1s in the contents of the dual LFSR would be the most desirable choice for the phase shifter implementation. Since channels may share some of the exclusive-OR gates, the selection can be made considering all channels to reduce the overall exclusive-OR gate count and/or the maximum delay through the exclusive-OR network. Design of phase shifters that seek to minimize the number of exclusive-OR gates by allowing a range of phase shifts for each channel have resulted in an average exclusive-OR count of about one gate per channel with an average performance penalty of about two exclusive-OR gate delays per channel [217].

The area overhead of about one exclusive-OR gate per channel puts the LFSR with phase shifter in the same range as a CA register in terms of area overhead. A statistical comparison of LFSRs with phase shifters and CA registers in terms of the properties of their patterns that are favorable to high fault coverage in scan BIST applications show that they are comparable [216]. In fact the LFSRs with phase shifters perform slightly better than CA registers without phase shifters. Since only exclusive-OR gates need to be added to create new channels, the LFSR with phase shifter can be constructed from fewer flip-flops than a CA register without a phase shifter but designed for the same number of scan chains. In this case, the CA register must contain a stage for every scan chain; in fact the CA register should contain slightly more stages than the number of scan chains to avoid the structural dependencies observed near the ends of the CA register due to null boundary conditions. However, phase shifters can also be designed for CA registers to avoid the structural dependencies at the ends of the register and to facilitate implementation of CA registers with fewer stages than the number of scan chains [186]. The design procedure for implementing a phase shifter for a CA register is identical to that of the LFSR based algorithm given above with the exception that Step 1 is eliminated and the original CA design is used for the simulation in Step 2.

Phase shifters do a good job of removing structural dependencies but, by themselves, do not eliminate the problem of linear dependencies in test patterns produced by a single channel. Therefore, the size of the LFSR or CA should be chosen to help minimize the linear dependency problem. In addition, phase shifters can be combined with other techniques, such as test points [107] and bit manipulation to reduce the problems of linear dependencies. Techniques for implementing bit flipping [132] or bit fixing [119] have been used in conjunction with phase shifters. The procedures for implementing these techniques are the same as described in the previous subsections but the fault simulations are performed after the phase shifter has been designed. The resultant bit flipping or bit fixing logic is incorporated at the output of the phase shifter.

10.3.5 Multiple Capture Cycles and Partial Scan

An interesting observation has been made that fault coverage can sometimes be increased by incorporating multiple capture cycles for each scan cycle [152]. In this case, a scan cycle refers to shifting the test patterns from the LFSR (and phase shifter) into the scan chains while a *capture cycle* refers to the clock cycle where the scan chain is put into the system mode of operation to apply the test patterns and 'capture' the output response of the CUT. Normally a single capture cycle is associated with each scan cycle. However, it was observed that fault coverage could be improved and the total number of scan cycles could be reduced by applying multiple capture cycles after each scan cycle. As a result of the multiple capture cycles, the test patterns being applied to the combinational logic after the first capture cycle are a result of the output response of the combinational logic to the previous test patterns. This is similar to Circular BIST where the test pattern is the compacted output response. The test patterns are not coming directly from the TPG since the test vectors are being modified by the CUT such that structure and linear dependencies are modified as well. An additional 1% to 1.5% in total fault coverage was obtained with only 60% of the total number of scan cycles with this method [152]. The question of how many capture cycles should be used for each scan cycle can be determined through fault simulation. In the results reported, different numbers of multiple capture cycles were applied for different scan vectors during the same BIST sequence. This complicates the design of the test controller since the Scan Mode control signal is now not the simple single clock cycle, carry-out signal generated by the Shift Control counter. However, in the results reported, the BIST controller area overhead increased only by about 1% to account for varying multiple capture cycles [152].

Partial scan design-based BIST without multiple capture cycles can be used to reduce test time since there are fewer flip-flops in the scan chain [311]. However, in some of these implementations it has been observed that partial scan-based BIST gives better fault coverage than full scan-based BIST. Choosing the "right" flip-flops to leave out of the scan chain can provide improvements to fault coverage with fewer scan cycles. The results reported are similar to those of the multiple capture cycle; about 1% to 1.5% increase in fault coverage was observed with a total reduction in test time on the average of 65% [311]. The advantage of this technique is that it does not involve modifications to the BIST controller other than shortening the Shift Count to account for the shorter scan chains.

10.4 Benefits and Limitations

The main advantages of scan-based BIST lie in the advantages of scan design-based DFT. Those include the high level of automatic synthesis of scan chain along with the low area overhead and performance penalties. The area overhead is summarized for the various components of the basic scan-based BIST approach discussed in this chapter in terms of the approximate number of flip-flops, exclusive-ORs gates, 2-to-1 multiplexers (MUXs), and other elementary logic gates needed to implement each component. Note that the MISR does not include circuitry for holding the resultant signature at the end of the BIST sequence. The performance penalty is basically the same as that of scan design, a 2-to-1 multiplexer delay at the input to every flip-flop. Another important feature of scan-based BIST approaches is the fact that partial scan design can be used to lower area overhead and/or to avoid performance penalties in critical timing paths. The high degree of design automation for scan design has been carried over to scan BIST approaches. The simplicity of the basic scan design approach has facilitated development of VHDL synthesis for STUMPS that does not include phase shifters [78].

Scan BIST has been successfully applied to some very large chips achieving greater than 95% fault coverage [107]. The devices reported range in size from 200,000 to 750,000 gates with 9,000 to 41,000 flip-flops in the scan chains. The basic architecture for these devices is the STUMPS architecture with 80 to 128 scan chains. In all cases a 31-bit LFSR was used in conjunction with a phase shifter design using the procedure described in the previous section. The MISRs used in these devices varied between 31 and 80 bits and concentrators were used when the number of scan chains was larger than the number of MISR bits. In all devices, test point insertion was incorporated based on analysis of the controllability and observability of the circuits. Controllability test points included multiplexers and gates as in the case of ad-hoc DFT. Observability test points included the fan-out of internal nodes to parity tree-based concentrators. The number of test points inserted into these devices ranged from 50 to 600 controllability test points and 70 to 1200 observability test points. While the number of test points seems very high, it appears that many of these were used to bypass regular structures such as RAMs during the BIST mode of operation.

TABLE 10.2 Area overhead of scan-based BIST.

BIST Component	Area Overhead			
	Flip-Flops	**Exclusive-ORs**	**MUXs**	**Gates**
Scan Chain	0	0	N_{FF}	0
Input Isolation	0	0	N_{PI}	0
LFSR-TPG	$\lceil \log_2(FF_{max} \times N_{SC}) \rceil$	3	0	$\lceil \log_2(FF_{max} \times N_{SC}) \rceil$
Phase Shifter	0	$\lceil \log_2(FF_{max} \times N_{SC}) \rceil$	0	0
MISR	N_{SC}	$N_{SC} + 3$	0	N_{SC}
Shift Control	$\lceil \log_2(N_{SC}) \rceil$	$\lceil \log_2(N_{SC}) \rceil$	0	$\lceil \log_2(N_{SC}) \rceil$
Pattern Control	$\lceil \log_2(N_{TV}) \rceil$	$\lceil \log_2(N_{TV}) \rceil$	0	$\lceil \log_2(N_{TV}) \rceil$

where: N_{FF} = number of flip-flops in CUT replaced by scan flip-flops
N_{PI} = number of primary input to CUT requiring input isolation
FF_{max} = number of flip-flops in longest scan chain
N_{SC} = number of scan chains
N_{TV} = number of scan vectors to be applied

The main limitations of scan BIST can also be seen in the applications described above. Specifically, adding an LFSR to both ends of a scan chain for TPG and ORA functions is not typically sufficient to obtain high fault coverage. The problems of structural and linear dependencies must be addressed with additional logic and design time if these problems have a sufficient impact on the fault coverage. While phase shifters are easy to design, determining the proper phase shift to design for is not so easy. Furthermore, the problems of linear dependencies are not as easy to address as structural dependencies. Regardless of these limitations, scan BIST has become so popular in recent years that it is often referred to as Logic BIST, even though this term is inaccurate since there are many BIST approaches for general sequential logic.

CHAPTER 11

Non-Intrusive BIST

The BIST approaches presented for general sequential logic in the previous four chapters are all structured, embedded BIST architectures. They use existing flip-flops to create TPG and ORA functions or, in the case of scan BIST, points of controllability and observability for application of test vectors and capture of output responses. As a result, all of the BIST approaches discussed thus far incur performance penalties in the CUT because of the embedded nature of the BIST approach. The basic BIST architecture illustrated in Figure 1.4 is separate from the CUT and, therefore, does not impose performance penalties on the CUT. In this chapter, we explore non-intrusive BIST architectures and investigate the applications where these architectures are good candidates.

11.1 Non-Intrusive BIST Architectures

The basic BIST architecture discussed in Chapter 1 and illustrated in Figure 11.1a is a separate, test-per-clock, BIST architecture. Since the TPG and ORA functions are external from the CUT, they can be used to test multiple CUTs as shown in Figure 11.1b. This can significantly reduce the area overhead. Non-intrusive BIST architectures (also referred to as non-invasive BIST) do not impose performance penalties on the CUT like embedded BIST approaches. The only performance penalty is additional set-up time on the primary inputs due to the input isolation multiplexers.

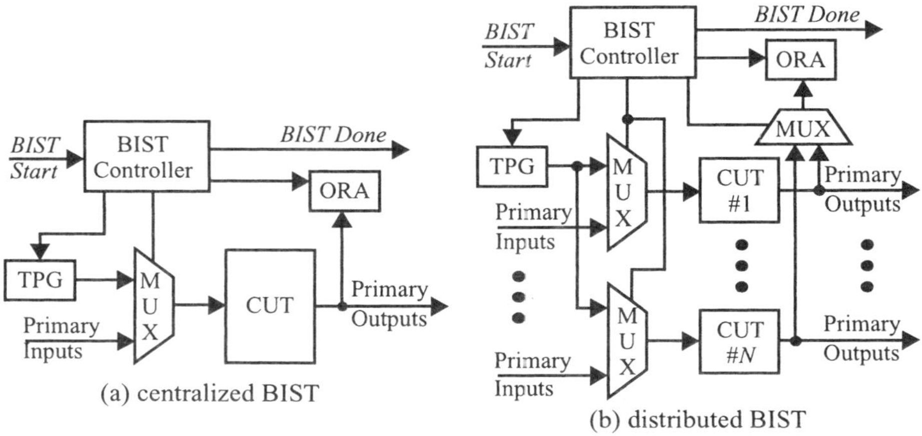

FIGURE 11.1 Non-intrusive BIST architectures.

There is additional clock-to-output delay due to the fan-out of the primary outputs to the ORA. The CUT in embedded BIST approaches incurs performance penalties not only from the gate delays of the embedded BIST circuitry, but also from the fact that in VLSI devices, the CUT will be spread out to make room for the embedded BIST circuitry. This means that wire lengths for routing will be longer with more resistance and capacitance to increase the routing delays. In VLSI applications of the non-intrusive BIST approaches, it is possible that the layout of the CUT will not be affected by the separate BIST circuitry and, as a result, will not incur additional performance penalty. Obviously, the chip area will be larger due to the BIST circuitry, which will in turn increase the routing lengths at the primary inputs and outputs, but the maximum operating frequency of the CUT may not be effected. As a result, non-intrusive BIST approaches are good candidates for high speed applications [266].

11.1.1 Board-Level Implementations

The first approach using this architecture was a board-level testing approach that used an LFSR for the TPG function and single input SAR for the ORA function [36]. As a result, a multiplexer was placed at the input to the SAR in order to select the primary output to be compacted for a given test session. The BIST sequence was executed once for each primary output. The approach was later implemented on-chip with a MISR compacting all primary outputs in parallel [229]. With the advent of Boundary Scan, an obvious evolution of this architecture was to incorporate the TPG and ORA functions in the Boundary Scan cells to further reduce the area overhead [93]. For

Boundary Scan cells capable of INTEST, the flip-flops and multiplexers are already in place for use by the TPG and input isolation function. As a result, a small amount of additional logic is needed to construct either an LFSR or CA register to supply test patterns at the primary inputs during the BIST mode. Similarly, the flip-flops of the Boundary Scan cells at the primary outputs can be used to construct a MISR for output response compaction with the Boundary Scan chain. This interface provides perfect system-level access to retrieve the final signature at the end of the BIST sequence. During the BIST mode, the Boundary Scan cells would need to be clocked by the same clock being supplied to the CUT.

Another variation of the non-intrusive BIST architecture is for PCBs that incorporate programmable logic devices, either CPLDs or FPGAs. In these applications, the programmable logic devices that are in-system re-programmable (ISR) can be reprogrammed during system-level testing to replace the normal system function with TPG and ORA functions to provide off-line BIST of VLSI devices on the PCB that do not incorporate BIST [234]. This is illustrated in Figure 11.2 for two possible cases. In the first case, the programmable logic device has full connectivity to the CUT, as illustrated in Figure 11.2a. As a result, the TPG, ORA and BIST controller can be programmed into the FPGA or CPLD to test the external CUT. Notice that there is no need for input isolation since the reprogramming of the FGPA or CPLD to replace the system function effectively isolates the CUT from unknown system data by deleting those normal system connections and replacing them with direct connections to the TPG. In the second case, multiple programmable logic devices exist on the board with

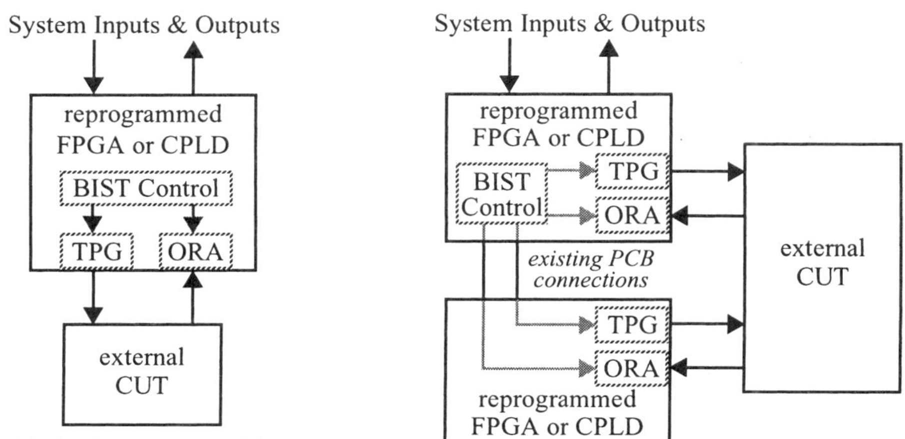

(a) single programmable logic device

(b) multiple programmable logic devices

FIGURE 11.2 Programmable logic-based board-level BIST.

connections to the circuitry to be tested, as illustrated in Figure 11.2b. In this case the TPG and ORA functions must reside in multiple devices and the BIST control must be coordinated across the multiple TPGs and ORAs. This requires PCB connections to exist between the multiple programmable logic devices in order to route the control signals during the BIST sequence. Alternatively, the BIST control functions can be replicated in the programmable logic devices. There will still need to be coordination between the multiple controllers to synchronize the BIST sequences to ensure reproducible results from one execution of the BIST sequence to the next.

Once the CUT has been tested, the programmable logic device(s) can be reprogrammed to perform the intended system function. As a result, there is no area overhead or performance penalties associated with this BIST approach other than storage of the configuration(s) to be downloaded into the programmable logic device(s) to create the TPG, ORA, and BIST controller functions. Most current FPGAs and CPLDs provide access to the configuration memory through the Boundary Scan interface, resulting in a common programming interface to all programmable logic devices in the system [196]. For those FPGAs and CPLDs that provide access to the core programmable logic and interconnect, the Boundary Scan interface can also be used for initiating the BIST sequence and retrieving the BIST results at the completion of the BIST sequence. For those devices that do not provide access to the core programmable logic and interconnect, the Boundary Scan cells themselves or the configuration memory can be used to initiate the BIST sequence and retrieve the results for pass/fail determination of the BIST [100]. BIST approaches for the FPGAs and CPLDs themselves will be discussed in Chapter 13.

Another application of reprogramming FPGAs/CPLDs during off-line testing is to program the device to test chips with scan chains. This could be performed in a manner similar to either the Random Test Socket approach illustrated in Figure 10.4 or the STUMPS approach illustrated in Figure 10.5. In either case, additional logic can be programmed into the FPGA(s) and/or CPLD(s) to perform a phase shifter function and/or bit manipulation function of the test patterns as they enter the scan chain(s). There is no area overhead associated with this technique in terms of logic since the programmable logic device will be reprogrammed with the intended system function after testing is complete. However, interconnections on the PCB must be incorporated in order to connect the FPGA(s) and/or CPLD(s) to the scan chains for the test mode of operation. This would require that I/O pins of the programmable logic device(s) be reserved for these connections to the scan chains. These connections include *Scan Mode* and *Scan In* inputs as well as *Scan Out* outputs to facilitate this off-line BIST capability. Regardless, programmable logic devices offer a number of unique opportunities to perform off-line testing at the PCB and system-level without area overhead or performance penalties. These opportunities should be investigated whenever programmable logic devices are incorporated on a PCB or in a system.

11.1.2 Device-Level Implementations

Later applications of the non-intrusive BIST architecture combined the TPG and ORA functions, as illustrated in Figure 11.3, by using the compacted output responses for the test patterns to be applied to the CUT. In one approach, a CA register was used for output response compaction and simultaneous test pattern generation [64]. In another approach, a MISR was used as the combined ORA/TPG function [266]. Since the TPG and ORA functions are separate from the CUT, all flip-flops used for the TPG and ORA directly contribute to the area overhead. Therefore, the combined TPG/ORA function reduces area overhead. For large CUTs the area overhead is much less than that of any of the BIST approaches that have been discussed in the previous chapters. When compared to a separate LFSR-based TPG and MISR-based ORA, the CA-based combined TPG/ORA achieved high fault coverage for many circuits, but not for all circuits [64]. The MISR-based combined TPG/ORA did not perform as well as the separate LFSR-based TPG and MISR-based ORA in terms of fault coverage for most circuits, although there are circuits for which the MISR-based combined TPG/ORA does provide higher fault coverage [234]. In fact, the fault coverage obtained by any of these non-intrusive BIST approaches is circuit dependent and can be low, in some cases well under 50%.

In order to use this device-level BIST approach during system-level testing, compaction of the output data must be avoided until the CUT is initialized. This can most easily be accomplished by incorporating blocking gates between the output of the CUT and the inputs to the MISR or CA register, as illustrated in Figure 11.3. The blocking gates are the same gates used for input isolation in Figure 6.2. They force the MISR to operate as an LFSR-based TPG until the BIST controller enables the valid output responses of the CUT to pass through the blocking gates to the MISR. The size of the MISR or CA register can be chosen to be either the maximum of the number of pri-

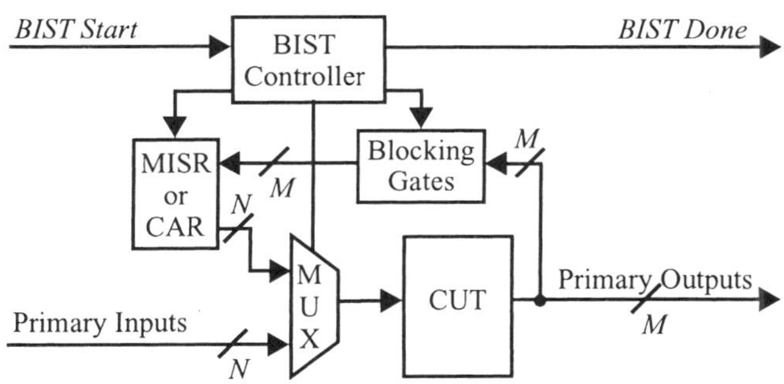

FIGURE 11.3 Modified non-intrusive BIST architectures.

mary inputs, N_{PI}, or the number of primary outputs, N_{PO}. However, if $N_{PO} > N_{PI}$ and N_{PO} is very large, the size of the MISR or CA register can be chosen to be smaller with concentrator circuits used to combine CUT outputs prior to entering the blocking gates.

Many VLSI designs can be partitioned into data path circuitry and control circuity that controls the processing of data through the data path. These non-intrusive BIST approaches tend to be most effective for testing the data path circuitry as opposed to the control circuitry. Basic architectures are illustrated in Figure 11.4 for simplex (one-way) and duplex (two-way) data path implementations. All of the BIST circuitry of Figure 11.3 is assumed to be contained within the block denoted 'BIST Circuitry'; this includes the blocking gates, the MISR or CA register, the BIST controller, and the input isolation multiplexers. The *BIST Start* and *BIST Done* signals can be connected to internal registers in the control circuitry for the device such that the BIST can be initiated by the system through the processor interface. For the simplex data path architecture, the outputs of the data path circuitry are looped back to the blocking gates of the BIST circuitry as shown in Figure 11.4a. In the duplex data path architecture, the outputs of both data paths can be looped back to the inputs of the other data path. A loopback multiplexer is used at the input to the data path that does not incorporate the BIST circuitry as shown in Figure 11.4b. This helps to minimize the area overhead for the overall BIST approach since only one 2-to-1 multiplexer is needed for each data input rather than duplicating the BIST circuitry. In both architectures, data path control signals coming from the control circuitry can be set via the proces-

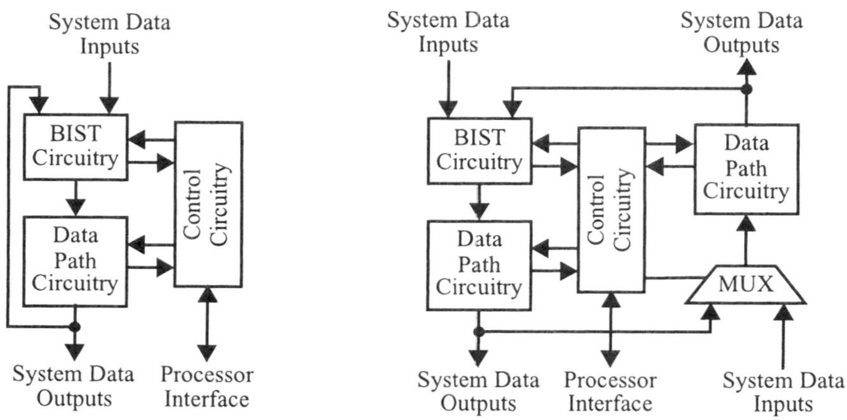

(a) simplex data path (b) duplex data path with loopback

FIGURE 11.4 Non-intrusive BIST architecture for data path structures.

sor interface prior to execution of a BIST sequence. The control signals are then set to other logic values prior to execution of subsequent BIST sequences. This not only reduces the area overhead of the BIST approach while testing the interface between the control circuitry and data path circuitry, but can also eliminate the need for BIST circuitry in the control portion and its associated performance penalties in some applications.

Control circuitry is often in the critical timing paths of the design since the control signals are generally set through a low-speed, asynchronous system software interface. As a result, the control circuitry is not always optimized for performance, with the performance optimization effort devoted to the design of the data path circuitry. The reduction in area overhead and avoidance of performance penalties by not putting BIST in the control circuitry is offset by longer testing time since multiple BIST sequences must be performed to test various logic values on the control signals. While the multiple executions of the non-intrusive BIST approach do test a considerable portion of the control circuitry, the fault coverage achieved for the control circuitry is generally less than that of the data path circuitry. During system-level testing, the control circuitry is typically easier to test than the data path circuitry via system diagnostic software since it is more controllable and observable. However, for manufacturing testing, other BIST approaches such as Circular BIST or scan-based BIST approaches can be applied to the control circuitry to provide higher fault coverage [266].

11.1.3 System-Level Implementations

The non-intrusive BIST architecture can be used in conjunction with loopback multiplexers in other devices or on other PCBs to provide system-level testing capabilities. This is illustrated in Figure 11.5 for a system-level implementation of such a BIST approach [289]. The unit to be tested by the BIST approach was a line unit for a switching system supporting Integrated Services Data Network (ISDN) capabilities. In this particular application, the intent of BIST was system-level testing and not device-level testing.[1] The unit consisted of 532 PCBs with only two VLSI devices residing on four PCBs containing the BIST circuitry, as shown in Figure 11.5b. The remaining VLSI devices and PCBs contained loopback multiplexers to accommodate

1. This is an unusual goal for most BIST implementations, but quite valid and quite effective in this particular application. This BIST implementation was initiated by system diagnosticians who were concerned about the difficulty of locating faults in the data path portion of the unit due to their lack of ability to insert/extract test patterns into/from the data paths during diagnosis. The control portions of the unit were easy to diagnose since there was good access through the system software interfaces.

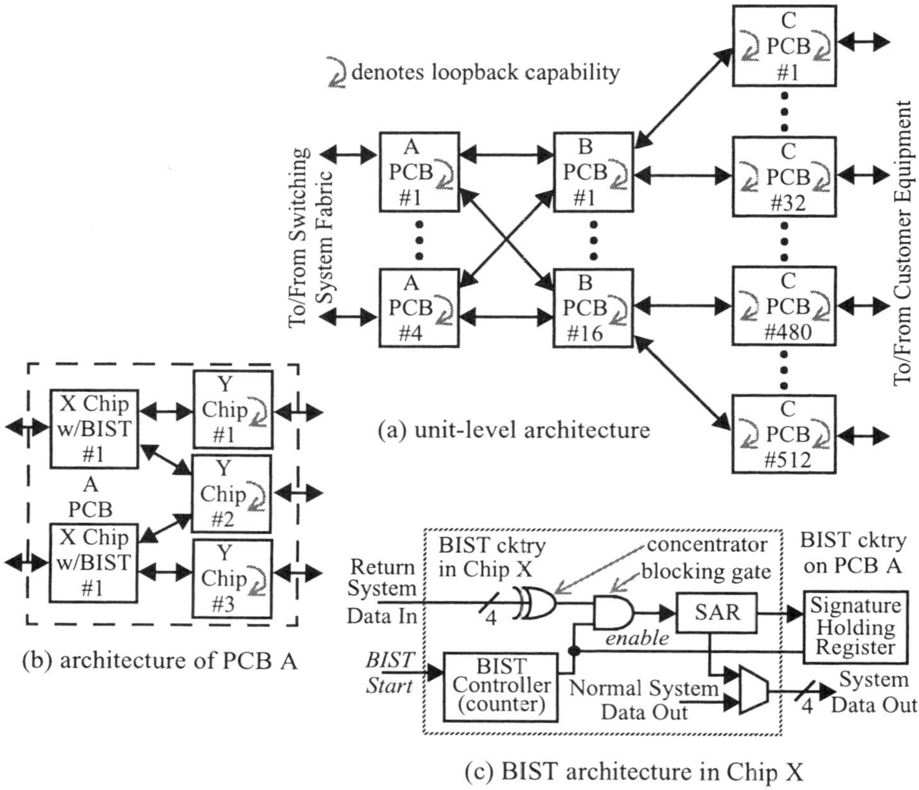

FIGURE 11.5 Non-intrusive BIST architecture for system-level testing.

the loopback of test data at various points in the unit. All PCBs and VLSI devices contained duplex (two-way) data path architectures, indicated by the double ended arrows in the figure. The loopback capability returns the outgoing test data on the incoming data paths. While the PCBs labeled 'A' and 'B' contained loopback capabilities only at the output of the PCB, the PCBs labeled 'C' contained loopback capabilities at both PCB inputs and outputs. This facilitated fault location to backplane wiring between any two PCBs in the unit [289].

The architecture of the BIST circuitry contained within each chip labeled 'X' is illustrated in Figure 11.5c. Since the VLSI device labeled 'X' did not contain a processor interface for retrieving the final BIST signature, an external shift register on the PCB served as the Signature Holding Register. This holding register was enabled for shifting in the signature whenever the SAR was compacting the output responses of the unit during the BIST sequence and disabled for holding the signature after the BIST sequence. The holding register could then be read by system software to retrieve the final signature after completion of the BIST sequence. By activating the loopback capabilities at the various points in the unit, faults could be isolated to a given PCB or set of interconnections between PCBs. The BIST controller used the blocking gate to prevent output responses from being compacted until the unit (including both outgoing and returning data paths) was completely initialized with test data from the BIST circuitry. Therefore, the length of time that the blocking gate prevents data from entering the SAR is determined by the maximum number of clock cycles required for test data from the SAR (which operates as an LFSR during the initialization sequence) to reach the most distant loopback multiplexer and return to the SAR.[1] At that point in time, the BIST controller activated the *enable* signal (a logic 1 in the design in Figure 11.5c) and the output responses entered the SAR for compaction, with the resultant signatures being used for the test patterns sent to the unit under test. At the same time the output of the SAR began to shift into the Signature Holding Register. At the end of the BIST sequence, the *enable* signal was deactivated by the BIST controller to freeze the resultant signature in the holding register.

Due to the large amount of circuitry under test and the many combinations of data paths, simulation of the complete unit was not possible at that time. Therefore, the 'good circuit' signatures for the various combinations of data paths were determined by running BIST sequences in a fault-free system. These were subsequently used by diagnostic software for pass/fail determinations. During system-level testing, various combinations of data paths were tested for a given BIST sequence with the loopbacks activated at the outputs of the PCBs labeled 'C'. In this way the complete data path was tested from chip 'X' to the outputs of the PCB labeled 'C'. If a failure was detected, different combinations of loopbacks were activated working back toward the output of the PCB labeled 'A' in order to determine the location of the fault(s). Once designed and implemented, the BIST then became part of the manufacturing unit-level test process. A conservative analysis determined that the cost of manufacturing unit-level testing was reduced by 10% [289].

11.1.4 Vertical Testability Implementations

Another modification to the basic non-intrusive BIST architecture provides vertical testability as well as a number of significant improvements to system-level testing

1. While the BIST controller design worked well for this application, we later realized that if the initialization sequence had been designed to last longer (or to be programmable) the BIST circuitry could have been used to test all the way out to the customer equipment as well.

that have not been supported by other BIST approaches. The modification requires only one additional set of multiplexers as shown in Figure 11.6. The additional set of 2-to-1 multiplexers facilitate selection of either the output of the CUT or the system input data as the data to be compacted in the MISR. This selection of system input data for compaction in the MISR facilitates additional features for system-level testing [266]. The system application in which this BIST approach was originally implemented was a high-speed, self-routing, broadband switching system [303]. The system consisted primarily of six full custom and standard cell VLSI devices that were used a total of 27 times per PCB with a total of 32 PCBs in the fabric of the switching system for a total of 864 VLSI devices in the system. Due to the self-routing capability of the switching system, most of the circuitry of these six devices comprised the data path as opposed to the control portion of the system. The data path architectures consisted of simplex data paths in four of the VLSI devices and duplex data paths in two of the VLSI devices with 8, 16, or 36 data inputs and outputs per chip, depending on the function of the particular chip. While the data paths were high-speed, full-custom implementations, all six of the VLSI devices had low speed control circuitry implemented using standard cells. The area overhead of the BIST approach ranged between 2% and 4% for each of the six devices. The *BIST Start* was supplied to all devices by a board-level control register so that the BIST sequence in all devices could be synchronized to run in parallel. The *BIST Done* for each device could be polled by system software during execution of BIST through the software interface to the control portion of each chip.

The gate-level design of the BIST circuitry is illustrated in Figure 11.7 for a single data path bit along with a table of the modes of operation. This bit-sliced design can be inserted at each input to the data path portion of the circuit. The only circuitry not shown is the exclusive-OR gates used to construct the characteristic polynomial of the

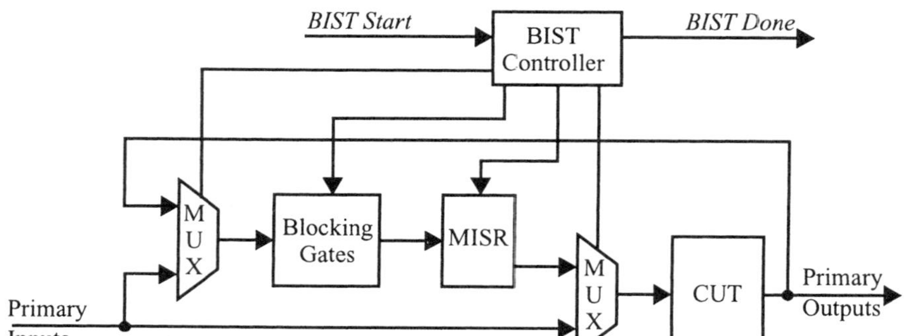

FIGURE 11.6 Modified non-intrusive BIST architecture for system-level testing.

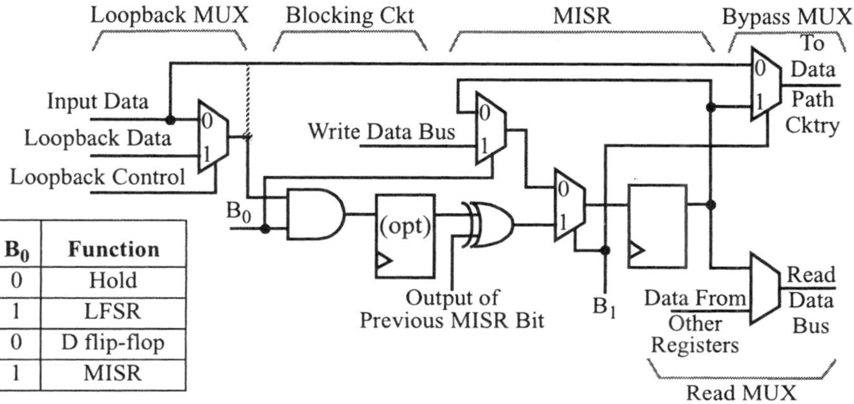

B_1	B_0	Function
0	0	Hold
0	1	LFSR
1	0	D flip-flop
1	1	MISR

FIGURE 11.7 Bit-sliced design of BIST circuitry.

MISR, loopback multiplexer control register bits, and the BIST controller. For high speed applications where the number of gate delays that can be allowed between flip-flops is limited, additional flip-flops can be added to this pipe-lined structure, such as the flip-flop shown at the output of the blocking circuit (denoted 'opt' for optional). The connection to the bypass multiplexer for normal operation can be taken directly from the system inputs or from the output of the loopback multiplexer, as shown by the 'dashed' connection in the figure. This latter connection can provide additional diagnostic or other system capabilities beyond that of the BIST feature, depending on the system needs.

The operation of the BIST circuitry is illustrated in the associated table as a function of the mode control signals B_0 and B_1. Note that there is no initialization mode for the MISR in this BIST design, instead the MISR/LFSR is initialized by writing the seed values in the register when B_0=1 and B_1=0. For large BIST circuits supporting many data path bits, this facilitates reseeding the LFSR during subsequent BIST sequences for improving fault coverage [240]. The ability to hold the final signature at the end of the BIST sequence is inherent in the design when B_0=B_1=0. The operation of the BIST circuitry as an LFSR (for test pattern generation only) or as a MISR (for output response compaction as well as simultaneous test pattern generation) is used in conjunction with the loopback multiplexer control signals to perform various manufacturing and system-level testing modes.

During manufacturing testing, the data path circuitry of the device is tested by setting all loopback multiplexers to select their loopback data, initializing the MISR with a seed value, and initiating the BIST sequence. At the beginning of the BIST sequence, the output response from the loopback data is disabled by the blocking gate to avoid unknown system data from entering the MISR. During this initialization sequence, the MISR performs as an LFSR generating pseudo-random test patterns based on the seed value that was written into the MISR prior to initiation of the BIST sequence. This gives the ability to choose seed values that provide the highest fault coverage. The initialization sequence proceeds for some predetermined number of clock cycles factored into the BIST controller design. Once the initialization sequence has completed, the blocking circuit is enabled to allow the output responses from the loopback data to enter the MISR for compaction. From that point in time, the compacted signatures are used as the test patterns applied to the data path. At the completion of the BIST sequence the BIST controller returns the BIST circuitry to the normal system mode of operation which holds the resultant signature until it can be read by system software for pass/fail determination of the BIST. Multiple BIST sequences can be executed with various seed values as well as with various settings of control values in the control circuitry in order to obtain high fault coverage during device-level testing.

This same test procedure can be used during manufacturing device-level testing without activating the loopback multiplexers such that all the system data inputs are selected to enter the MISR for data compaction. With this test mode in conjunction with a test fixture that connects the data path output pins of the device to the data path input pins of the device, high-speed testing of the I/O of the device can be performed on a low speed test machine using an external oscillator (built on the test fixture) for the high-speed clock.

During system-level testing of a PCB board containing VLSI devices that incorporate this BIST approach, a complete board-level test can be performed as illustrated in Figure 11.8a. The loopback multiplexers of the VLSI device(s) connected to input pins of the PCB are set to select loopback data while all other loopback multiplexers in all other VLSI devices are set to select normal system input data. The MISRs in the various VLSI devices are written with seed values and the BIST sequence is initiated. In this case where multiple VLSI devices will be executing their BIST sequences in parallel as well as in conjunction with each other, it is important that the *BIST Start* signal be synchronized so that all chips start their BIST at the same time to ensure that the board-level BIST sequence is reproducible from one BIST execution to the next.

The initialization sequence proceeds as in the case of device-level testing by blocking unknown system data from entering the MISR until initialization of all data paths on the PCB is complete. This means that the initialization sequence must be designed based on the worst case number of clock cycles for the longest data path on the PCB as opposed to the number of clock cycles for the data path through a particular VLSI device. Therefore, we once again encounter the case where system-level testing issues must be considered early in the design process in order to produce an effective BIST

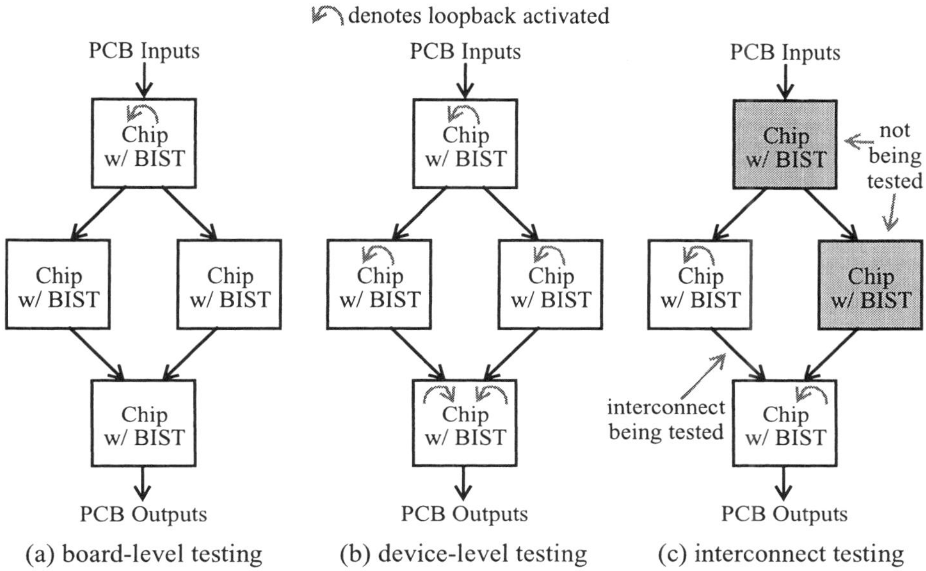

FIGURE 11.8 System-level testing features.

controller design that can be used for all levels of testing. At the completion of the initialization sequence, the data at the system data inputs is allowed to proceed through the blocking gate to be compacted by the MISR during the remainder of the BIST sequence. Once the BIST sequence is complete, the BIST controller in each device returns the BIST circuit to the system mode of operation and the resultant signatures are held within each VLSI device. All signatures are then read to determine the pass/fail status of the data path circuitry on the PCB. As in the case of device-level manufacturing testing, multiple BIST sequences can be executed with various seed values as well as with various settings of control values in the control circuitry.

In the event that one or more of the signatures are incorrect compared to the known fault-free board-level BIST signatures, additional BIST sequences can be executed to perform diagnosis for location of the fault(s). There may be some diagnostic information provided by the initial board-level faulty signature(s). However, since the MISR is also performing as the TPG, a faulty signature in one MISR will result in incorrect test patterns being produced and applied to circuity down stream in the data path. Therefore, it is best to use the device-level system test illustrated in Figure 11.8b. In this test, all loopback multiplexers are set to select loopback data such that only circuitry within each VLSI device is being tested by its own BIST circuitry. A failing signature at the completion of this test would identify the individual VLSI device as being faulty. If all device-level signatures are good for all BIST sequence executions, then the fault must reside in the interconnect or in the I/O buffer circuitry. Boundary Scan can then be used for diagnosing these faults.

If Boundary Scan is not implemented, the BIST circuitry can be used to complete the diagnosis as illustrated in Figure 11.8c. In this case, interconnect testing can be performed by executing BIST in two VLSI devices at a time that have common interconnections. The loopback multiplexers in the driving device are set to select loopback data while the loopback multiplexers in the receiving device are set to select normal system data. In the case where a given VLSI device receives data from multiple devices, the loopback multiplexers should be grouped with a separate loopback control bit for each driving VLSI device. In this way, only one group of loopback multiplexers is set to select system input data corresponding to the interconnect to be tested, while all other loopback multiplexers are set to select loopback data. The diagnostic resolution can be increased (even to the point of a single wire) by incorporating additional independent loopback control register bits, but this is at the expense of increased area overhead. This same interconnect testing strategy can be extended to testing backplane wiring or cables between PCBs [266].

11.2 Benefits and Limitations

The primary benefit of non-intrusive BIST approaches is low area overhead with very little performance penalty. The area overhead for the various approaches discussed in the previous section is summarized in Table 11.1 in terms of the number of flip-flops, exclusive-OR gates, multiplexers, and elementary logic gates (denoted 'Gates' in the table). Note that the area overhead includes logic to hold the final signature at the end of the BIST sequence (this was not included in the area overhead of BIST approaches in the previous chapters). The number of gates required to implement the BIST controller is not included since that design is more of a function of the application of the BIST at the system-level. Application of non-intrusive BIST approaches in production chips have provided greater than 90% fault coverage with 1.5% to 4% area overhead [266]. There is only a single multiplexer delay added to the normal system operation and this occurs only at the primary inputs. There is no area overhead or performance penalty associated with the BIST approaches that reprogram FPGAs and/or CPLDs during off-line testing to implement BIST to test other devices on the PCB or in the system. This is because the BIST circuitry does not exist during normal system operation since the FPGA/CPLD is reprogrammed after BIST to reinsert the system function logic.

The main problem with non-intrusive BIST compared to other BIST approaches is that fault coverage can be very low for some circuits. Without fault simulation, it is difficult to determine which circuits will achieve high fault coverage and which ones will not. There are two techniques that can be used to help improve fault coverage in non-intrusive BIST applications. Some global control signals such as global reset signals can inhibit fault detection when these signals are frequently activated by pseudo-random test patterns. Similarly, other control signals, such as a clock enable, can inhibit fault detection if not activated frequently enough. In such cases, these control signals should be driven from combinational logic to create weighted patterns for more or less frequent activation of the control signal as needed. Another technique is to partition the CUT into data path and control portions with non-intrusive BIST applied to the data path portion and a different BIST approach applied to the control portion.

TABLE 11.1 Area overhead of non-intrusive BIST approaches.

BIST Approach	Area Overhead			
	Flip-Flops	**Exclusive-ORs**	**Multiplexers**	**Gates**
Figure 11.1a	$N_{PI}+N_{PO}$	$N_{PO}+6$	$N_{PI}+N_{PO}$	$N_{PI}+N_{PO}$
Figure 11.1b	$N_{PImax}+N_{POmax}$	$N_{POmax}+6$	$N_{PItot}+2N_{POmax}$	$N_{PImax}+N_{POmax}$
Figure 11.3	$max(N_{PI},N_{PO})$	$N_{PO}+3$	$max(N_{PI},N_{PO})+N_{PI}$	$max(N_{PI},N_{PO})+N_{PO}$
Figure 11.5	$max(N_{PI},N_{PO})$	$N_{PO}+3$	N_{PI}	$max(N_{PI},N_{PO})+1$
Figure 11.6&7	$max(N_{PI},N_{PO})$	$max(N_{PI},N_{PO})+3$	$4max(N_{PI},N_{PO})$	$max(N_{PI},N_{PO})$

where: N_{PI} = number of primary inputs to CUT
N_{PO} = number of primary outputs of CUT
N_{PImax} = maximum number of primary inputs to a CUT
N_{POmax} = maximum number of primary outputs from a CUT
N_{PItot} = total number of primary inputs to all CUTs

Design automation is easy due to the simplicity of the implementation of the general non-intrusive BIST approach [328]. While partitioning the CUT into data path and control portions can be difficult from a general netlist description of the design, the determination for the use of weighted pseudo-random patterns for global control signals can be automated [234]. This is accomplished by performing an audit of the fan-out of the various primary input signals within the design; those primary inputs with the highest fan-out are the most likely candidates for global resets, presets, clock enables, etc. Weighted pseudo-random test patterns can then be used for those global primary input signals to maximize the fault coverage. Test point insertion can also be used to improve the fault coverage obtained with non-intrusive BIST approaches [77]. Of course this somewhat negates the advantages of non-intrusive BIST since the test points are intrusive in most cases.

Most BIST implementations focus on device-level applications for manufacturing testing, though sometimes they are used in system-level testing as well. The non-intrusive BIST techniques described in this chapter illustrate a key point: consideration of system-level testing issues early in the design process can provide significant benefits for system-level use of BIST. Yet, the additional area overhead and design effort to make the system-level use possible and profitable are minimal. Therefore, a key part of any BIST implementation should be to consider the system application(s) for which a given device is intended, the potential life cycle of that device, and what could be done to improve the value of the BIST for the system application(s) and life-cycle of that device.

11.3 The Clock Enable Problem Revisited

Now that we have seen the various BIST approaches for general sequential logic, we return to the clock enable problem discussed at the end of Chapter 6. Most of the BIST approaches in the previous chapters ignore this problem. If we replace flip-flops A through D in Figure 6.5 with BIST or scan flip-flops and let the *Enable* signal be driven by test data, the delay of the combinational logic between flip-flops C and D will lead to failing BIST results in fault-free circuits as the delay varies due to processing, temperature, and voltage variations. Therefore, the embedded BIST approaches (BILBO, pseudo-exhaustive BIST, Circular BIST, and scan BIST) treat the multiplexer used for the clock enable function as part of the combinational logic network being tested with no regard to the implications of the large combinational logic delays that may exist. The non-intrusive BIST approach is the one exception in that it facilitates maintaining clock enable signals in their normal system mode of operation during the BIST sequence such that they are active every N clock cycles.

This clock enable problem can be overcome in the embedded BIST approaches by partitioning the circuit such that the non-clock-enabled flip-flops and the clock-enabled flip-flops are in two different timing regions. As a result, they are treated as two different CUTs with input isolation between them. The clock-enabled flip-flops must then be kept in tact with the additional BIST logic added to the input of the clock-enable multiplexer. In addition, the *Enable* signal would be maintained at its normal rate of repetition and cannot be driven by pseudo-random patterns. In this manner, new test data is applied every N clock cycles allowing sufficient time for the combinational logic delays.

CHAPTER 12

BIST for Regular Structures

Regular structures typically consist of a 2-dimensional array of identical, densely packed cells surrounded by dedicated I/O blocks of circuitry that provide access to the array, as illustrated in Figure 12.1. Due to the regularity of the array, algorithmic approaches to testing the array are often the most suitable for efficient testing time and cost, as well as achieving maximum fault coverage. In many cases, the close packing of the cells in the array, intended to obtain a high density, leads to faults that are not typically of concern in general sequential logic. Therefore, fault modeling and test algorithms for regular structures tend to be different, and sometimes more complex, than for general sequential logic. As can be expected, BIST approaches for regular structures tend to be different than those for general sequential logic.

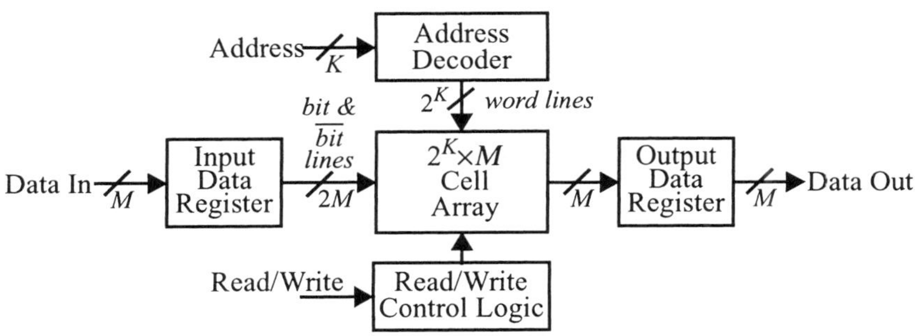

FIGURE 12.1 Basic block diagram for regular structures.

The structure depicted in Figure 12.1 has a write operation feature as indicated by the Read/Write Control Logic. As a result, this block diagram could represent a Random Access Memory (RAM) or Register File structure. Minor modifications would be needed to create a First-In First-Out (FIFO) or a First-In Last-Out (FILO, but more often referred to as a *stack*) memory structure by adding counters for the read/write pointer logic. Multiple port RAMs can be constructed from this basic block diagram by incorporating additional Address Decoders as well as additional Data Input and Output Registers. Similarly, the basic structure can be modified to represent a Read Only Memory (ROM) or Programmable Logic Array (PLA) by removing the Read/Write Control Logic and the Input Data Registers. Since the ROM and PLA represent a subset of the circuitry involved in the memory structures with write capabilities, we will consider the fault models typically associated with RAMs and memory structures of that type.

12.1 Regular Structure Fault Models

There are two basic types of fault models that are associated with regular structures: classical and non-classical faults. While the classical faults include the fault models discussed in Chapter 2, the non-classical fault models are more specific to regular structures and are a result of the dense packing of the cells in the array. Classical faults models include [50]:

- ***Cell array faults*** result when the cell contents are stuck-at-0 or stuck-at-1. To detect these faults, one must test each cell with both logic 0 and logic 1 values.
- ***Addressing faults*** result when a wrong address is selected due to a fault in the address decoder logic. To detect these faults, one must test all addresses with unique data.
- ***Data line faults*** result when input and output data registers (or bit and $\overline{\text{bit}}$ lines, where $\overline{\text{bit}}$ is the complement of the bit line) have faults which prevent correct data from being written into or read from the cells in the array. To detect these faults, one must pass both logic 0 and logic 1 values through every data input and output.
- ***Read/write faults*** are the result of stuck-at faults on the read/write control lines or faults in the read/write control logic that prevent read or write operations in the cell array. To detect these faults, one must write and read all cells.

Non-classical fault models for regular structures such as RAMs include [319]:

- ***Transition faults*** result when a cell cannot undergo a 0→1 transition or cannot undergo a 1→0 transition. To detect these faults, one must test both transitions.

- ***Data retention faults*** result in a cell losing its contents after a certain period of time. To detect these faults, one must read data after a period of no activity (no read or write operation).
- ***Destructive read faults*** result in a read operation changing the contents of a cell. These faults are sometimes referred to as *read disturb faults*. To detect these faults, one must perform multiple reads of the cells in the array after data has been written. Those reads should be performed for both logic 0 and logic 1 values.
- ***Pattern sensitivity faults*** result in the contents of a given cell being affected by contents of other cells. A bridging fault between cells would be one example of a fault causing pattern sensitivity. To detect these fault, one must surround the cell under test with specific logic values in adjacent cells and verify that the contents of the cell under test are not changed.
- ***Coupling faults*** result in the contents of a given cell being affected by operations on other cells. Coupling faults and pattern sensitivity faults seem similar; the basic difference is that the pattern sensitivity fault is a function of the static contents of neighboring cells while coupling faults are the result of transitions in other cells due to write or read operations. There are several types of coupling faults and their detection is a function on the type of coupling [319].

The logic values required to detect pattern sensitivity in a given cell depend on the physical layout of the cell array. For example, a checker board pattern in the cell array will put the opposite logic values in directly adjacent cells as illustrated in Figure 12.2a. However, diagonally adjacent cells will have the same logic values. Therefore, one must consider the nature of the possible pattern sensitivity and apply patterns with respect to the suspected adjacency pattern sensitivity. This is complicated by the fact that in large RAMs the bits of multiple words are often interleaved; for example, the ith bit from multiple words may be grouped together as opposed to all the bits associated with a single word being grouped together. Furthermore, a static RAM cell contains a bit (*B*) and a $\overline{\text{bit}}$ ($\overline{B}$) subcircuit. The pattern sensitivity may be with respect to the bit or the $\overline{\text{bit}}$ circuitry. In some cases the physical layout of the cell array may be such that the orientation of a cell is 'mirror-imaged' with respect to adjacent cells in the array; a simple example of this is illustrated in Figure 12.2b. In

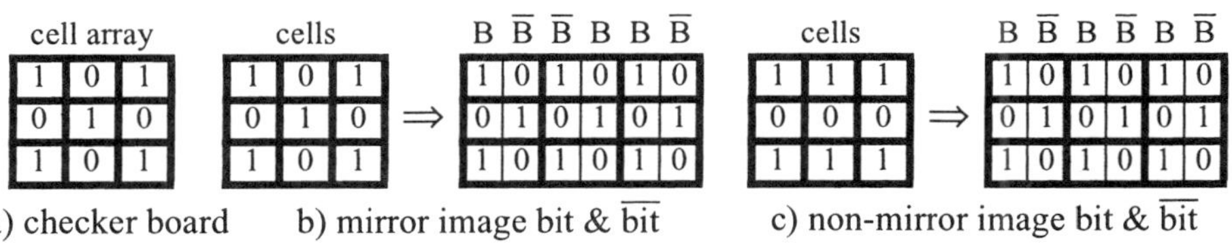

FIGURE 12.2 Checker board test patterns for pattern sensitive faults.

these cases the checker board applied to the complete cell may hold for the bit (B) and $\overline{\text{bit}}$ ($\overline{\text{B}}$) portions of the cell as well [118]. However, in the case of a non-mirror imaged layout, a different pattern may be required in the cell array to obtain the checker board at the bit and $\overline{\text{bit}}$ level, as illustrated in Figure 12.2c.

As a result of the specific fault models associated with regular structures, particularly the non-classical fault models, it is preferable to develop BIST implementations based on test algorithms that are known to provide high fault coverage for these various fault models. While fault simulation of the non-classical faults may be impractical during an ASIC development, most of the test algorithms that have been developed are shown to detect all of the classical fault models and varying degrees of the non-classical fault models [50]. As a result, the obvious choice for the TPG design would be a counter and/or FSM-based TPG in order to produce the algorithmic test patterns that will detect both classical and non-classical faults, as was illustrated for the March Y algorithm example in Figure 4.1. When a test algorithm produces a known expected output response for a regular structure, a comparison-based ORA can be used, as discussed and illustrated for the March Y algorithm example in Figure 5.2. However, in read-only regular structures, such as ROMs and PLAs, a MISR or an accumulator is the better choice.

12.2 RAM BIST Architectures

Some example RAM test algorithms that are reasonably straightforward for BIST implementations are given in Table 12.1 [319]. These algorithms include variations of the Modified Algorithmic Test Sequence (MATS) and March tests including the March Y algorithm used for the FSM-based TPG design in Chapter 4. Of the test algorithms given in the table, the March C- algorithm has been observed to provide the highest fault coverage [327]. A more detailed breakdown of the fault detection capabilities of these test algorithms is summarized in Table 12.2 in terms of the type of faults detected by the tests. Data retention faults can be detected by inserting a delay between each March sequence (where a March sequence is denoted by the up or down arrow in Table 12.1) [319]. There are additional fault models and other test algorithms for multi-port memories [321].

The test algorithms as shown are intended for RAMs with one data bit per word. However, they can also be used to test RAMs with multiple bits per word. In some cases, this requires possible modification to the algorithm to increase fault detection capability for coupling and pattern sensitivity faults within the bits of the words of the RAM [323]. In the case where the bits of different words are interleaved, the March algorithms can be used as shown, with all 0s and all 1s for the data patterns written into and read from the RAM to detect coupling faults (for those March test algorithms that detect coupling faults as indicated in Table 12.2) [323]. In the case where the bits of a given word are physically adjacent in the RAM, modifications can be made to the algorithms to ensure coupling and pattern sensitivity fault detection by the March test.

TABLE 12.1 Example RAM test algorithms.

Algorithm	Algorithm Sequence	Test Length
MATS	⇑⇓(w0) ⇑⇓(r0,w1) ⇑⇓(r1)	4N
MATS+	⇑⇓(w0) ⇑(r0,w1) ⇓(r1,w0)	5N
MATS++	⇑⇓(w0) ⇑(r0,w1) ⇓(r1,w0,r0)	6N
March X	⇑⇓(w0) ⇑(r0,w1) ⇓(r1,w0) ⇑(r0)	6N
March Y	⇑⇓(w0) ⇑(r0,w1,r1) ⇓(r1,w0,r0) ⇑⇓(r0)	8N
March C-	⇑⇓(w0) ⇑(r0,w1) ⇑(r1,w0) ⇓(r0,w1) ⇓(r1,w0) ⇑⇓(r0)	10N

where: w0 or w1 - indicates writing all 0s or all 1s, respectively
r0 or r1 - indicates reading and expecting all 0s or all 1s, respectively
⇑ - indicates addressing entire RAM in ascending order
⇓ - indicates addressing entire RAM in descending order
⇑⇓ - indicates addressing entire RAM in ascending or descending order
N = the number of words in the RAM

TABLE 12.2 Fault detection for RAM test algorithms.

	Faults Detected			
Algorithm	**Stuck-at**	**Address**	**Transition**	**Coupling**
MATS	yes	some	no	no
MATS+	yes	yes	no	no
MATS++	yes	yes	yes	no
March X	yes	yes	yes	some
March Y	yes	yes	yes	some
March C-	yes	yes	yes	yes

The modifications consist of running the algorithm multiple times with a different *data background sequence* during each execution. A data background sequence is a pattern that can be used to detect coupling and pattern sensitivity faults depending on the physical layout of the RAM. Therefore, the collection of all data background sequences should then provide testing independent of the physical layout of the RAM at the expense of multiple executions of the BIST sequence with a different data background sequence used during each BIST execution. The data background sequences for various size RAM words are given in Table 12.3. Data background sequences for larger words should be apparent from the examples that are given in the table. Each data background sequence is used to replace the 'all 0s' in the **w0** and **r0** of the algorithms from Table 12.2 with the inverse pattern replacing the 'all 1s' in the **w1** and **r1** of the algorithm.

TABLE 12.3 Example data background sequences for March RAM tests.

Bits/Word	Data Background Sequences			
2	00	01		
4	0000	0101	0011	
8	00000000	01010101	00110011	00001111

One of the early BIST architectures for embedded RAMs in VLSI devices is illustrated in Figure 12.3 [118]. The approach uses a counter for the TPG and test controller, and a MISR for the ORA. The test algorithm (also given in the figure) produced by the BIST circuitry is an $8N$ algorithm where N is the number of data words in the RAM; $N=2^K$ where K is the number of address bits. By using the address as the data, a unique data word is written into each address location. For cases where $M>K$, the K address bits were repeated to obtain M data bits. By inverting the address when used as data during the second write cycle, all RAM cell locations are tested for all stuck-at

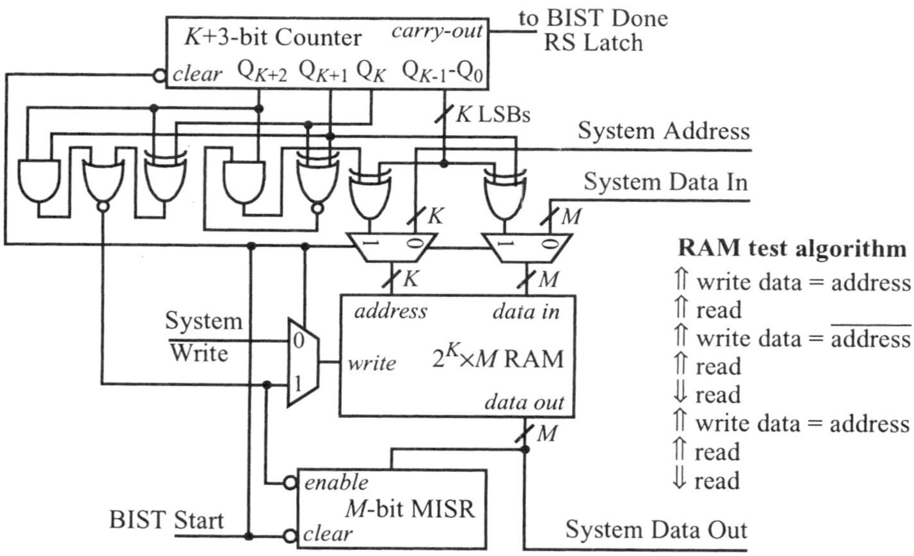

FIGURE 12.3 Early BIST architecture for embedded RAMs.

faults. There are three write cycles in the algorithm where the third write cycle returns the cells to their original data values such that transition faults are detected. There are two back-to-back read cycles of all address locations to test for destructive read faults. The address locations are read in reverse order during the second read cycle to overcome error cancellation in the MISR. Data retention faults were detected during manufacturing tests by stopping the clock for a specified period of time and then continuing with the BIST sequence. For example, RAMs that were tested at 8MHz, with a complete cycle through all addresses in 125 microseconds, incorporated a 2.5 millisecond pause in the clock for the data retention test.

This BIST approach was implemented in the standard cell logic area of six different production ASICs to test various RAM sizes ranging from 4 Kbits to 16 Kbits with a total area overhead ranging from 3% to 5% for the six devices [118]. It should be noted that all BIST logic was constructed from existing counters and registers in the ASICs with no additional flip-flops needed to implement the BIST circuitry, as opposed to the dedicated logic implied by Figure 12.3. The active high *BIST Start* signal goes to the active low *clear* inputs of the counter and MISR for initialization at the beginning of the BIST sequence. In each of the six devices, the RAMs contained 16-bits per word and, as a result, 16-bit registers were used for the MISR. It is important to disable the MISR during write operations if the RAM output data is unknown to prevent non-reproducible BIST results; this is illustrated by the active low *enable* to the MISR in Figure 12.3 where the write enable signal is assumed to be active high.

The main problem with this test algorithm was caused by the use of the RAM address as the test data where adjacent bits in the RAM may have the same logic values. As a result, this BIST approach failed to detect some pattern sensitivity faults encountered during production wafer and device-level testing. Because pattern sensitivity faults were actually observed in the manufactured chips that were not detected by the BIST approach, a new test algorithm was investigated. Based on the physical layout of the RAMs produced by the RAM layout generator, which included interleaving of bits for multiple words, a modified March test algorithm was developed [118].[1] The test algorithm is given in Table 12.4 to illustrate modifications that can be made to make the March test algorithms efficient for detecting pattern sensitivity faults for a specific physical layout.

A new BIST design was developed based on the modified March test algorithm and this new BIST approach used a comparator-based ORA [118]. As it turned out, the

1. This RAM layout generator was used to construct embedded RAMs for ASICs at Bell Labs in the mid-1980s. This was the same RAM layout generator into which the BIST with the improved test algorithm was later incorporated.

new BIST implementation was slightly more area efficient than that of the counter/MISR-based BIST approach in Figure 12.3. In fact, the area overheads with this approach for the same six devices were in the range of 2.5% to 4.5% when the BIST circuitry was implemented in the standard cell logic area of the devices. The only problem with this new approach was that the comparator was not completely tested by the BIST sequence and additional test vectors had to be applied after the BIST sequence to completely test the ORA.

TABLE 12.4 Modified March test for a specific interleaved RAM.

Address	RAM Operation
⇑	if (address mod 8) < 4, write 0s, else write 1s
⇑	read
⇑	read
⇑	if (address mod 8) < 4, write 1s, else write 0s
⇑	read
⇓	read
⇓	if (address mod 8) < 4, write 0s, else write 1s
⇓	read

Eventually, this new BIST approach was incorporated in the RAM layout generator with the TPG and test controller combined with the Address Decoder, Input Data Registers, and Read/Write Control Logic while the comparator was combined with the Output Data Registers as illustrated in Figure 12.4. The area overhead for the RAMs themselves was approximately 10% for 4 Kbit RAMs, 5% for 8Kbit RAMs,

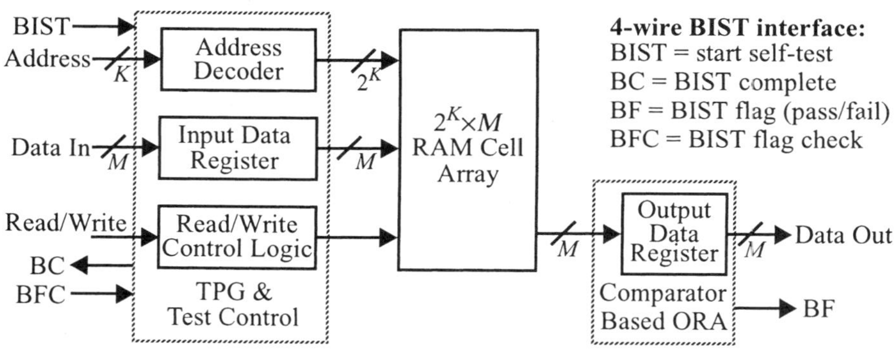

FIGURE 12.4 Basic block diagram for regular structures.

and 2.5% for 16 Kbit RAMs. The area overhead for the six devices discussed above, however, ranged from 1% to 1.5% when implementing the RAM generator-based BIST approach. This represented a significant reduction in the area overhead for RAM BIST because the generator provided greater area optimization over the standard cell layout. More importantly, designers didn't need to have a complete understanding of the classical and non-classical fault models, RAM test algorithms, or the architecture of the BIST approach; designers only needed to understand the interface to the BIST circuitry for access and execution during wafer, device, board, and system-level testing. Since the BIST approach used a comparator-based ORA, there was only a single *Pass/Fail* indication at the end of the BIST sequence which was given by the *BIST Flag* (BF) signal. For timing the BIST sequence and to indicate when it had completed, a *BIST Complete* (BC) signal was incorporated in the design. The *BIST Start* input was simply denoted 'BIST' as illustrated in Figure 12.4. Since the comparator-based ORA was not completely tested as part of the normal BIST sequence, an additional BIST sequence was incorporated to test the comparator; this additional comparator testing BIST sequence was initiated using the *BIST Flag Check* (BFC) input.[1]

12.3 ROM and PLA BIST Architectures

Since ROMs and PLAs are read-only structures, a simple but effective BIST technique is to apply exhaustive test patterns to the inputs while compacting the output responses. The TPG can be a counter or an LFSR with the additional logic to generate the all 0s state. An LFSR or CA-based MISR is a good choice for the ORA as long as steps are taken to minimize signature aliasing. The basic architecture for this approach is illustrated in Figure 12.5. In this implementation, the TPG produces 2^{N+1} test patterns where the toggle flip-flop inverts the inputs to the ROM or PLA during the second set of 2^N test patterns. As a result, the ROM or PLA is read in the opposite direction to assist in reducing signature aliasing and error cancellation in the MISR [271]. If an LFSR is used for the TPG, the *carry-out* signal can be produced by the NOR logic used to facilitate generation of the all 0s test pattern [103]. If a counter is

1. During the development of this BIST approach, we debated the use of a comparator versus a MISR. The choices were additional test vectors for the comparator test versus a low probability of signature aliasing using the MISR. Those who favored the comparator assured the MISR fans that designers would be happy to write and apply test vectors to test the comparator. When we surveyed Bell Labs designers, they replied just as MISR fans had predicted; the designers unanimously said "We are not writing test vectors for your comparator. If you put it in, then you test it!" As a result, the BIST Flag Check sequence was added to the BIST circuitry to test the comparator.

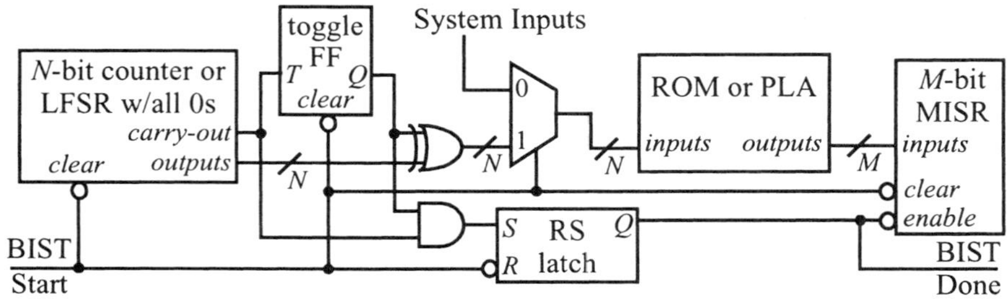

FIGURE 12.5 BIST architecture for ROMs and PLAs.

used for the TPG, on the other hand, an N+1-bit counter could be used instead of the toggle flip-flop with the Nth bit used to control inversion of the test patterns being applied to the ROM or PLA. The *carry-out* from the N+1 bit counter would then be used to set the *BIST Done* RS latch. A final consideration for signature aliasing would be the number of outputs, M, from the ROM or PLA. If M is small, say M<16, a larger MISR can be used to reduce the probability of signature aliasing; in this case, only those MISR bits that are compacting an output from the ROM or PLA would have an exclusive-OR gate incorporated (aside from the exclusive-OR gates used to implement the characteristic polynomial).

A variation on this architecture which further reduces the probability of aliasing when testing ROMs is to add an extra output bit to the ROM and program that bit to correspond to the quotient that is being shifted out of the MISR as a result of the polynomial division [348]. During the BIST sequence, the last bit of the MISR is compared to the additional ROM output and any mismatches are latched in a *Pass/Fail* indicator flip-flop. By checking the *Pass/Fail* indication as well as the signature contained in the MISR at the end of a BIST sequence, the probability of signature aliasing is greatly reduced. However, since the BIST sequence consists of a single pass through the ROM address space, there still exists the possibility of error cancellation in the MISR. The additional output of the ROM is programmed to be the fault-free circuit quotient exiting in the MISR for a specific address ordering when accessing the ROM. Therefore, this bit cannot be used during a second pass through the ROM in the opposite direction if a second pass is desired to overcome error cancellation in the MISR [348].

Accumulators are often used for the ORA when implementing BIST for ROMs (this is typically referred to as *checksum* for ROMs). In this case, there is probably no need to address the ROM in the reverse direction since this will not reduce error cancellation in the accumulator. While this reduces the area overhead of the TPG by 1-bit in the counter implementation or by the toggle flip-flop in the LFSR implementation, the area overhead of the accumulator compared to the MISR is greater than the savings of the extra flip-flop in the TPG. In addition, the aliasing probability in accumulators is reduced by increasing the precision (for example, going from a single precision to a double precision accumulator) which can significantly increase the area overhead [242].

The BIST architecture of Figure 12.5 works well for ROMs and many PLAs but there are cases where the number of PLA inputs is too large for a counter or LFSR-based TPG since the 2^{N+1} clock cycle test time would be impractical. ROMs are limited in size in terms of the number of address bits, N, as all 2^N words are implemented in the ROM. The number of product terms in a PLA, on the other hand, is always less than 2^N and in most cases much less than 2^N. As a result, N can be large (greater than 35 or 40) without making the size of the PLA impractical. Therefore, one can either design an FSM-based on deterministic test patterns that will detect faults in PLAs [41] or one can design the PLA for embedded BIST techniques that have been proposed [309]. Either approach should include tests for missing and extra cross-point faults that are specific to PLAs in addition to the normal stuck-at and bridging faults [50].

Most embedded BIST approaches for PLAs incorporate a shift register at the inputs to the PLAs where the bit and the complemented $\overline{\text{bit}}$ lines are decoded from the input line [99]. Another shift register is incorporated along the product term lines to facilitate activation of each product term independently and one at a time. During the BIST sequence, the normal inputs to the PLA are disabled such that the shift registers can control the lines. The basic idea behind the shift registers is that while a given product term line is activated by a bit in the product term shift register, the bit and $\overline{\text{bit}}$ lines are activated via the input shift register to test for missing and extra cross-points. This sequence of tests also results in a walking pattern across the bit and $\overline{\text{bit}}$ lines as well as across the product term lines which will detect bridging faults between adjacent lines. The shift registers incorporate multiplexers and decoders for activating product term lines and bit/$\overline{\text{bit}}$ lines. In addition to the shift registers, additional inputs and outputs are added to the PLA to facilitate something similar to parity bits on the product term lines [309]. These additional inputs and outputs are used only during testing. Finally, parity checking circuitry is incorporated at the outputs of the PLA. Therefore, the shift registers act as the TPG while the parity checking circuitry performs the ORA function in conjunction with the additional input and output lines that control the parity sense. As a result, the area overhead and performance penalties are incorporated within the PLA, as opposed to the non-intrusive BIST approaches to PLA testing such as the one illustrated in Figure 12.5. This makes embedded PLA BIST approaches good candidates for incorporation in automatic PLA layout generators as opposed to designers attempting to manually implement the BIST approaches in the PLA.

12.4 Bypassing Regular Structures During BIST

Some regular structures (RAMs and other memories with write capabilities) often pose as a blocking function to the test patterns and CUT output responses during BIST sequences for general sequential logic. This is due to the controllability and/or observability of the CUT being deteriorated when the logic drives the inputs to the regular structure and/or the regular structure outputs drive the logic. As a result, fault coverage is lower, unless the regular structures can be bypassed. In some RAMs, the data being written to a given address will pass through to the output data port such that holding the RAM in the write mode of operation allows the test patterns and/or CUT output responses to pass through the RAM as if it were simply wires connecting the input data port to the output data port. If this option is available, it facilities a low area overhead solution to the problem of bypassing the RAM data inputs (of course the address inputs pose another problem). Alternatively, test points can be inserted, as illustrated in Figure 12.6, to bypass regular structures during BIST of the general sequential logic [107]. Some regular structures implement scan design-based DFT in I/O blocks of the regular structure. The object is to allow independent test of the embedded regular structure and the general sequential logic of the ASIC. These scan chains can sometimes be used to extract sequential logic output responses seen at the inputs to the regular structure as well as to insert test patterns at the outputs from the regular structure using the scan chain. When regular structures are bypassed, it is important to remember to test the interface between the regular structure and the general sequential logic.

Note that read-only structures like ROMs and PLAs do not pose a problem since the data they contain is known and since they operate like combinational logic rather than memory elements. Therefore, they need not be bypassed during BIST of the general sequential logic. If the address and input data lines to the RAM are driven directly by BIST flip-flops, bypassing the RAM may not be necessary. This is particularly true in the case of Circular BIST and scan-based BIST approaches since the shift register nature of these two approaches will naturally carry CUT responses around the RAM.

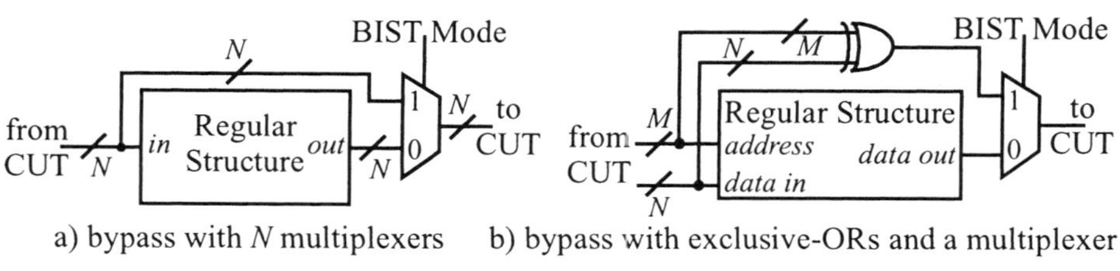

FIGURE 12.6 Test points for bypassing regular structures.

Similarly, the outputs of the RAM may not have to be isolated from the sequential logic if the RAM is initialized to a known state prior to execution of the logic BIST. One convenient way of initializing the RAM is to run the RAM BIST first, with the data stored in the RAM as a result of its BIST sequence used as the initialized values for the logic BIST. In some cases, the RAM and logic BIST can be executed in parallel with the RAM output data feeding directly into the general sequential logic under test. In this approach, the RAM inputs will be isolated as a result of the RAM BIST TPG. However, if the logic BIST sequence is longer than the RAM BIST sequence, the RAM can be put back into the system mode of operation at the completion of its BIST sequence. This allows logic BIST data to pass through the RAM in order to test the interface between the general sequential logic and the RAM during the remainder of the logic BIST sequence.

12.5 Benefits and Limitations

Regular structures pose a different testing problem than that of general sequential logic in many respects. This is primarily due to the densely packed cells of the array. The addition of DFT logic within the cells is usually out of the question since the area overhead and performance penalties are too great as a result of the large multiplicative effect of the many cells. The close proximity of the cells leads to the occurrence of defects that result in non-classical fault models, such as pattern sensitivity and coupling faults. The different physical layout of a regular structure requires different test sequences and, as a result, different BIST approaches when compared to general sequential logic. As a result, specific test algorithms that are developed to detect both classical and non-classical faults should be used to test regular structures as well as to develop BIST approaches.

Regular structures offer unique BIST opportunities as well. For example, due to their regularity where cells are connected by abutment, parameterized generators have been developed for many types of regular structures in order for an array of almost any size to be generated quickly and automatically. Once the physical layout generators have been developed, these regular structures are used repeatedly in many chip designs, easily making the development of the generator a cost effective investment that benefits many products and projects. Similarly, the development of test algorithms for those structures that detect classical as well as non-classical faults represents another good investment. The result is reusable solutions where the test development costs are recovered many times over by various projects.

The most effective development of BIST for regular structures is to implement BIST in the dedicated I/O blocks of the regular structure and have the BIST automatically produced as part of the regular structure by the physical layout generator. When the BIST approach was implemented in the RAM layout generator at Bell Labs, we not only saw significant reductions in area overhead associated with BIST, but more importantly, there was a significant increase in the acceptance and number of implementations of BIST by designers. As a result, all embedded RAMs were tested with a known good test algorithm and a standardized interface to the BIST circuitry. Instead of needing to understand the test algorithms, the non-classical fault models, and evaluate various test algorithms for their fault detection capability, designers only needed to know the operation of the BIST interface. When new faults were observed in production devices, usually as a result of design rule shrinks, the BIST approach programmed in the physical layout generator was modified to test for those new faults. Therefore, all BIST implementations for the regular structures in the new design rules were updated and accurate. For example, the RAM BIST approach went from an 8*N* algorithm (given in Table 12.4) to a 10*N* algorithm and eventually to a 13*N* test algorithm as a result of various defects and fault behaviors encountered and observed in production devices. Two versions of the RAM generator (one with and the other without BIST) were eventually replaced by a single version with BIST as a result of most designers opting to use BIST. Therefore, incorporating BIST into a physical layout generator for regular structures is one of the most effective ways to design such structures. High quality tests are obtained for all regular structures with very little design and test development effort once the BIST approach has been incorporated into the generator. A simple interface to the regular structure BIST, such as the 4-wire interface of Figure 12.4, facilitates easy access to the BIST during all levels of testing, including system-level testing.

An alternative to the incorporation of BIST circuitry into a physical layout generator is to develop HDL models (VHDL or Verilog) for the BIST circuitry based on test algorithms specific to a given regular structure. These HDL models can be compiled into a library for easy incorporation with the designer's HDL models for the system being developed. The BIST circuitry will then be synthesized along with the system logic for layout in the standard cell logic portion of the chip to provide BIST for the regular structures in the device. While this approach is not as efficient in terms of area overhead when compared to BIST incorporated in a physical layout generator, it is efficient from the standpoint of design time, effort, and quality (fewer design errors with high fault coverage for both classical and non-classical faults). This HDL-based approach to BIST for regular structures does offer advantages over a generator-based approach in some applications. For example, when there are multiple regular structures of the same size, the BIST circuitry can be shared by those regular structures to reduce the area overhead that would be incurred with a separate BIST circuit for each regular structure.

CHAPTER 13

BIST for FPGAs and CPLDs

Field Programmable Gate Arrays (FPGAs) and Complex Programmable Logic Devices (CPLDs) might be considered to be regular structures similar to RAMs, ROMs, and PLAs. An FPGA or CPLD consists of an array of programmable logic blocks (PLBs) and programmable I/O cells interconnected by a programmable routing network. However, the re-programmability of these devices makes them more complex to test than the regular structures discussed in the previous chapter. The BIST approaches for general sequential logic could be incorporated in the system function to be programmed in the device during system operation. The area overhead and performance penalties associated with the traditional BIST approaches can be avoided by taking advantage of the in-system re-programmable (ISR) nature of many current FPGAs and CPLDs. ISR devices include flash memory and RAM-based devices, which are programmed by writing an on-chip configuration memory [196]. During off-line testing, the programmable logic device can be reconfigured to test itself [270]. Once testing is complete, the device can be reprogrammed to reinsert the desired system function. In this way, BIST is obtained with no area overhead or performance penalty to the system function. In this chapter, we explore BIST approaches that facilitate complete testing of programmable logic and interconnect resources in FPGAs and CPLDs. The result is a generic BIST for a generic device.

Because of the many different modes of operation and combinations of interconnection provided by FPGAs and CPLDs, complete testing will involve multiple configurations of the device. For use during system-level testing, these multiple configurations must be stored in the system memory along with the configuration for the intended system function of the programmable logic device. To minimize the system memory requirements as well as the test time (which is dominated by the device programming time), an important goal is to keep the number of BIST configurations to a minimum. All FPGAs or CPLDs in the system, that are of the same type and size, can use the same BIST configurations. All programmable logic devices can be tested in parallel. An advantage of ISR FPGAs and CPLDs is that if the locations of faults detected by the BIST can be determined, then the system function can be reconfigured to avoid the faults when the device is reprogrammed for system operation to provide fault tolerant operation [5].

13.1 Overview of FPGAs and CPLDs

FPGA and CPLD architectures have historically been quite different until recently. In their similarities, both contain an array of PLBs interconnected by a programmable routing network. FPGAs have traditionally consisted of an $N \times N$ array of small PLBs, as illustrated in Figure 13.1a, where N is in the range of 10 to 30 PLBs [196]. Traditional CPLD architectures have consisted of a $2 \times N$ array of large PLBs, as illustrated in Figure 13.1b, where N is in the range of 2 to 8 PLBs.

The architectures of the PLBs themselves were a major historical difference between FPGAs and CPLDs. The traditional PLB architecture for FPGAs is illustrated in Figure 13.2a. The typical structure of an FPGA PLB consists of 2 to 8 memory blocks that can function as look-up tables (LUTs) or RAMs, a flip-flop associated with each

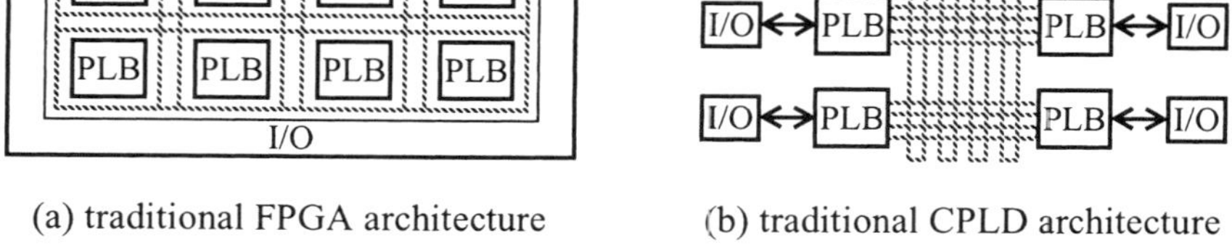

FIGURE 13.1 Traditional architectures for FPGAs and CPLDs.

LUT/RAM block, and multiplexing output logic. In the LUT mode of operation, each LUT could be used to define any combinational logic function of 3 to 4 inputs (denoted K in the figure) with the ability to combine LUTs to provide combinational logic functions of 5 or 6 inputs. The RAM mode of operation may be configured as various types of RAMs: synchronous, asynchronous, single-port, dual-port, etc. The typical LUT/RAM block also contains special-purpose logic for arithmetic functions such as counters, adders, multipliers, etc. The flip-flops can also be configured as latches and usually have programmable clock-enable, preset/clear, and data selector functions. The multiplexers in the FPGA PLBs allow signals to bypass the LUT and enter the flip-flop directly, allow the output of the LUT to bypass the flip-flop, or allow the output of the LUT to enter the flip-flop. As a result, an FPGA PLB can implement combinational and/or sequential logic functions. The PLB architecture for CPLDs, on the other hand, typically consists of a re-programmable PLA with 30 to 40 inputs and 8 to 16 outputs as illustrated in Figure 13.2b. Like FPGAs, CPLDs also have flip-flops that can be bypassed via the output multiplexers to facilitate implementation of combinational and/or sequential logic functions. While the PLB inputs and outputs connect directly to the programmable interconnect network in FPGAs, traditional CPLD PLB outputs are usually connected to dedicated programmable I/O cells at the periphery of the CPLD as well as the programmable interconnect network.

The programmable interconnect network for both FPGAs and CPLDs consists of wire segments that can be connected via *configurable interconnect points* (CIPs), sometimes referred to as programmable interconnect points (PIPs). The basic CIP structure consists of a transmission gate controlled by a configuration memory bit (Figure 13.3a). The logic value written into the configuration memory bit determines whether the two wire segments are connected or not. There are three basic types of CIPs which we will refer to as the *cross-point CIP* (Figure 13.3b), the *break-point CIP* (Figure 13.3c), and the *multiplexer CIP* (Figure 13.3d) [291]. Wire segments in

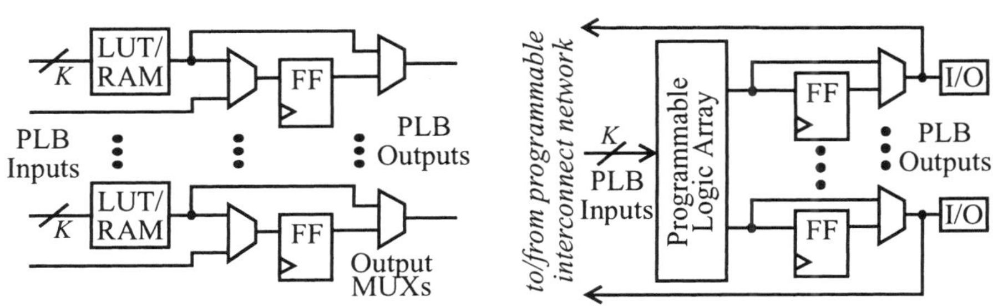

(a) traditional FPGA PLB architecture (b) traditional CPLD PLB architecture

FIGURE 13.2 PLB architectures for FPGAs and CPLDs.

the programmable interconnect network are bounded by these CIPs. There is also a *compound cross-point CIP* (Figure 13.3e) which is a combination of four cross-point and two break-point CIPs. While a cross-point CIP connects wire segments in disjoint planes (a horizontal segment with a vertical segment), a break-point CIP connects two wire segments in the same plane. Multiplexer CIPs have multiple input wire segments and a single output wire segment, where a given input wire segment selected by the configuration bits is connected to the output wire segment. Multiplexer CIPs come in two varieties, decoded and non-decoded multiplexer CIPs. In the decoded multiplexer CIP, a group of 2^k cross-point CIPs sharing a common output wire are controlled by k configuration bits; the decoding logic is incorporated between the configuration bits and the transmission gates. There is a configuration bit for each transmission gate in non-decoded multiplexer CIPs such that k wire segments are controlled by k configuration bits; usually only one of the configuration bits is active for any given configuration. Traditionally, the programmable interconnect network in FPGAs used either compound cross-point CIPs or a combination of cross-point CIPs and break-point CIPs. FPGAs incorporated multiplexer CIPs but these were usually located at the inputs to the PLBs. CPLDs on the other hand, used multiplexer CIPs almost exclusively. This was another major difference between traditional FPGA and CPLD architectures.

In recent FPGA and CPLD architectures, it is much more difficult to discern any differences in the architectures. The number of PLBs has grown in both FPGAs and CPLDs and the array size in both is $N \times M$ where N is not necessarily equal to M. In addition, PLBs are often clustered and grouped with the large embedded memories including RAMs, FIFOs, etc. The programmable interconnect network in recent FPGAs and CPLDs is primarily constructed from non-decoded multiplexer CIPs that

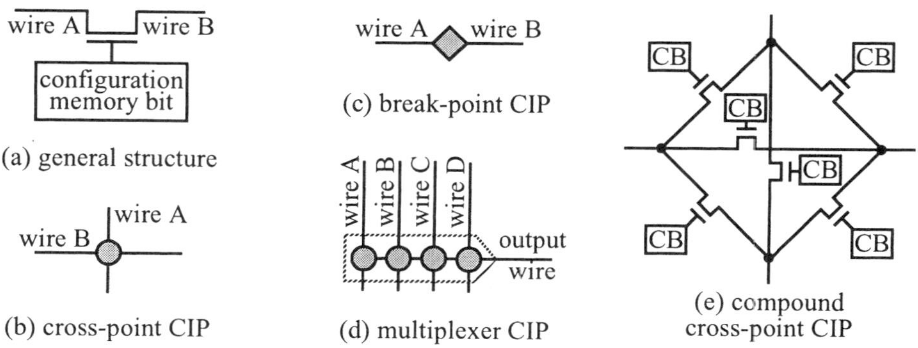

FIGURE 13.3 Configurable interconnect points.

are buffered to prevent signal degradation due to the series resistance (approximately 1KΩ) of each transmission gate the signal passes through. Most recent FPGAs and CPLDs contain large embedded RAMs (much larger than the RAMs used within FPGA PLBs). The principle difference that continues to exist between FPGAs and CPLDs is the architecture of the PLBs in which the combinational logic continues to be performed by smaller LUT/RAMs for FPGAs versus larger re-programmable PLAs for CPLDs.

13.2 Logic BIST Architectures

The basic idea underlying BIST approaches for the programmable logic resources in FPGAs and CPLDs is to configure groups of PLBs as TPGs and ORAs, and another group of PLBs as *blocks under test* (BUTs), as illustrated in Figure 13.4a. The BUTs are then repeatedly reconfigured so that they are tested in all of their modes of operation [280]. Each reconfiguration of the FPGA or CPLD to test a different PLB mode of operation is referred to as a *test phase*. A *test session* is a collection of test phases that completely test the BUTs in all of their modes of operation. Once the BUTs have been tested, the roles of the PLBs are reversed so that in the next test session the previous BUTs become TPGs or ORAs, and vice versa. If at least half of the PLBs are BUTs during each test session, only two test sessions are needed to test all PLBs in the FPGA or CPLD. This is illustrated in Figure 13.4b for an 8×*M* array of PLBs

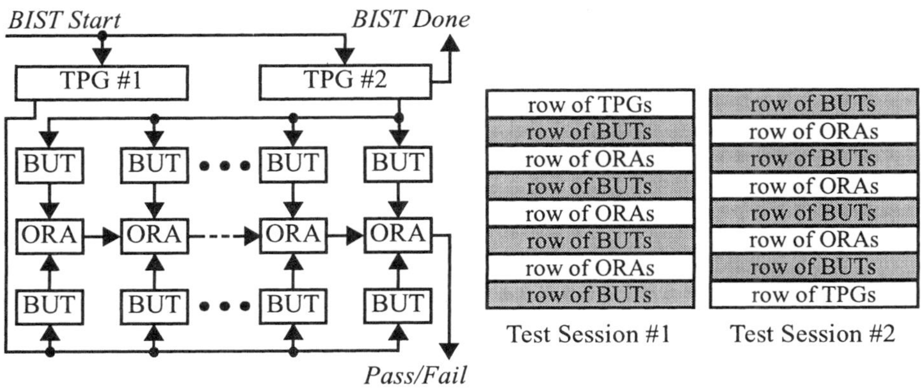

(a) detailed design (b) test sessions

FIGURE 13.4 Basic architecture for logic BIST for FPGAs.

[282]. Figure 13.4a corresponds to the first four rows of PLBs in test session #1 in Figure 13.4b.

Each test phase consists of the following steps: 1) reconfigure the FPGA or CPLD with a BIST configuration, 2) initiate the BIST sequence which includes initialization, test pattern generation, and output response compaction, and 3) read the BIST results from the ORAs. In step 1, the test controller (ATE for wafer/package-level testing; system or maintenance processor for board/system-level testing) interacts with the FPGA/CPLD under test to reconfigure the device by retrieving a BIST configuration from a storage medium and loading it into the configuration memory of the FPGA/CPLD. The test controller also initializes the TPGs, BUTs, and ORAs, and initiates the BIST sequence (via the *BIST Start* input in step 2), and reads the subsequent *Pass/Fail* results (step 3). After the test phase is complete, the test controller must reconfigure the FPGA/CPLD for the next test phase, or for its normal system function once the complete BIST session has been executed. Therefore, the normal system function configuration must be stored along with the BIST configurations. The application time for any given test phase is dominated by the FPGA/CPLD reconfiguration time. When diagnosis is incorporated for fault tolerant applications, the number of diagnostic configurations is important, since the total test and diagnosis time is a major factor in the system down time (system availability) and cost. As a result, an important goal of any BIST approach is to minimize the total number of configurations used for testing and diagnosis. This means minimizing not only the number of test phases but also the number of test sessions.

All BUTs are configured to have the same logical function and receive the same input test patterns from the two or more identical TPGs. Since all fault-free BUTs must produce the same output patterns, comparator-based ORAs can simply compare corresponding outputs from different BUTs. While comparator-based ORAs require additional testing, we must completely test the PLBs that are used for TPGs and ORAs in order to completely test all PLBs in the device. Therefore, this is one of the cases where the comparator-based ORA is the best choice. Figure 13.5 shows the structure of an ORA comparing four pairs of BUT outputs. The signal *BuOi* (*BdOi*) is the ith output from the BUT *up* above (*down* below) the ORA. The flip-flop stores the result of the comparison, and the logically ORed feedback loop latches any mismatch in the flip-flop. The flip-flops of all ORAs can be connected to form a scan chain (indicated by the connections between ORAs in Figure 13.4a) so that the *Pass/Fail* indications from each ORA can be shifted out at the end of the BIST sequence [100]. An earlier BIST approach recorded only one *Pass/Fail* result for every row of ORAs [282]. However, storing the test results of each ORA PLB significantly improves the diagnostic resolution achievable by this architecture. For an $N \times M$ array of PLBs, the number of ORA PLBs is $N_{ORA}=(NM/2)-N$.

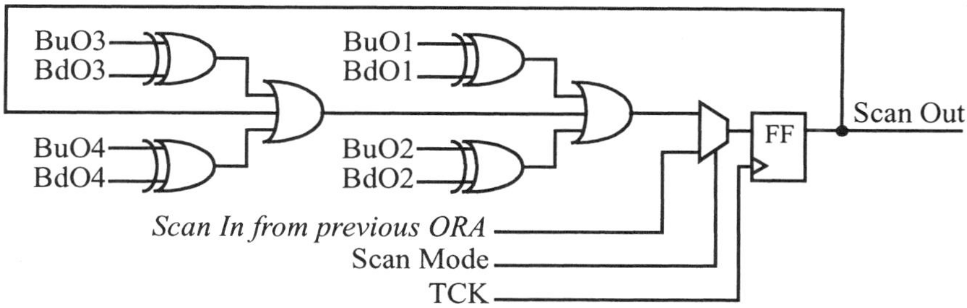

FIGURE 13.5 Comparison-based ORA for FPGA/CPLD BIST.

Two important features of this architecture help in testing the FPGA or CPLD. First, every BUT (except those in the first two and last two rows) is simultaneously compared with two other BUTs by two different ORAs. Second, the pair of BUTs being compared by each ORA is fed by two different TPGs. As a result, any single faulty PLB is guaranteed to be detected and located [8]. Although it cannot be guaranteed that the two test sessions illustrated in Figure 13.4b will detect all possible combinations of multiple faulty PLBs, it appears that the conditions that allow a group of faulty PLBs to escape detection are so restrictive, they are very unlikely to occur in practice [9]. But if needed, the BIST sequence can be enhanced to guarantee the detection of any combination of multiple faulty PLBs by adding the third and fourth test sessions shown in Figure 13.6. These are obtained by rotating the rows of TPGs, BUTs, and ORAs in test sessions 1 and 2 by 90° so the TPGs, BUTs, and ORAs now form columns. Hence, a group that escapes detection in the first two sessions is guaranteed to be detected in at least one of the additional two sessions [9].

This BIST architecture has a very regular, easily-scalable structure that can be automatically generated by algorithmic placement of the PLBs and routing of their inter-

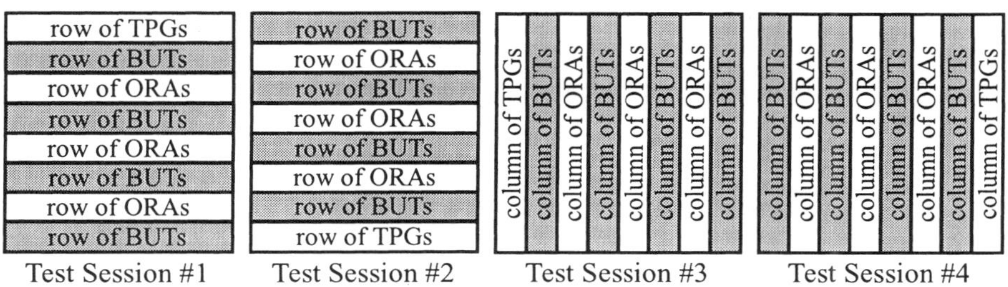

FIGURE 13.6 FPGA/CPLD logic BIST configurations for detection.

connections, based on the dimensions of the FPGA or CPLD array in terms of the number of PLBs in each direction, $N \times M$. Since an ORA compares the outputs of its two neighboring BUTs, all signals from BUTs to ORAs use only local routing resources. Global routing is used to distribute the patterns generated by the TPGs to the BUTs. Ignoring any fan-out limitations, adding rows and columns to an array of PLBs will just extend the length of the vertical and horizontal global lines fed by TPG outputs. Therefore, the usage of the global routing resources required for distributing the TPG patterns does not change with the FPGA/CPLD array size. When loading on the TPG outputs exceed fan-out limits in large arrays, the array can be divided into smaller arrays with the BIST architecture implemented separately in each array. For example, an 16×16 array can be divided into four 8×8 arrays with the four quadrants tested concurrently. The scan chains of each quadrant will be connected into a single scan chain, so that the BIST results retrieval time grows linearly with the size of the FPGA. The number of test sessions is independent of the size of the FPGA. Since all BUTs are tested in parallel, the BIST execution time is also independent of the size of the FPGA. In addition to the time for scanning out the BIST results, the only other testing time dependence on the size of the array is the reconfiguration time for each test phase.

Many recent FPGAs and CPLDs contain large embedded RAM arrays (much larger than the RAMs used within PLBs). To test such RAMs, an additional test session can be used in which the PLBs surrounding a RAM implement a large FSM to generate a March test sequence appropriate to test the embedded RAMs. All embedded RAMs may be tested in parallel, possibly sharing the same BIST controller. If the FPGA has several identical RAM modules, the TPG(s) can supply the RAMs with the same test patterns and a comparison-based ORA can be used to detect mismatches between corresponding outputs. From this point, we are ready to test the PLBs. Since this is where the major difference lies in recent FPGA and CPLD architectures, the BIST configurations must be customized to the particular PLB architecture, as we shall see in the following subsections.

13.2.1 LUT/RAM-based PLBs

There are many modes of operation to test in the PLBs. As a result, the TPGs may have different structures depending on the test patterns needed in different phases. Due to the limited number of inputs to a LUT and flip-flop, pseudo-exhaustive testing can be applied such that every subcircuit of a PLB is tested with exhaustive patterns in each one of its modes of operation. For the LUT/RAM block of an FPGA PLB, the first test should be the RAM mode of operation. This configuration is put as the first PLB test phase because it detects the most faults in the LUT/RAM blocks of the PLB [283]. If faults are detected during this first test phase, manufacturing testing can move on to the next chip to help minimize testing time and cost. Similarly, system-level testing will more quickly identify a faulty device for PCB replacement and repair. In this case, the TPGs are configured to apply a March test, which detects, among other faults, all the stuck faults in memory cells, as well as all faults in the address and read/write circuity of the RAM mode of operation. When the LUT mode of operation is tested, the LUTs in a PLB can be programmed with alternating exclusive-OR/exclusive-NOR functions for the LUT outputs during one test phase, and alternating exclusive-NOR/exclusive-OR functions during a second test phase. For any combinational logic function of n inputs, the TPG will apply all 2^n vectors. Since a counter mode of operation is a common feature of PLBs, a counter-based TPG is the most economical choice in terms of the number of PLBs required to construct the TPG, as opposed to an LFSR with additional logic necessary to generate the all 0s state.

The flip-flops (and other special logic for adders, multipliers, etc.) can also be tested with pseudo-exhaustive test patterns using a counter-based TPG. The total number of test phases required to completely test the PLB is a function of the PLB architecture. Typical numbers of test phases range from 9 to 16 per test session for a total of 18 to 32 test phases for complete BIST of the FPGA logic.

13.2.2 Fully Programmable OR-Plane PLA-based PLBs

There has been a great deal of research and development in mask programmable PLA testing over the past 20 years [12]. However, there has been surprisingly little work reported in the area of testing re-programmable PLAs [188]. Re-programmable PLAs are the major combinational logic component of PLBs in CPLDs, which can contain hundreds of large PLAs. Typical sizes of these PLAs include 35 to 40 inputs, as many as 80 product terms, and 8 to 16 outputs with product term sharing. The number of inputs prevents exhaustive testing. Therefore, effective testing techniques for re-programmable PLAs are necessary in order to ensure efficient testing time for a BIST implementation. All re-programmable PLAs require multiple test phases for complete testing such that the goal is to minimize the number of test phases as well as the number of test vectors that must be generated by the TPG for each test phase. In order to establish the TPG design, we must consider the test phases and their associated test vectors required to test the PLA.

A DFT technique was proposed for Electrically Erasable PLAs (EEPLAs) that adds extra circuitry to the PLA in order to detect single and multiple stuck-at faults (including line and transistor faults in the PLA) as well as certain bridging faults on input lines, product term lines, and output lines [220]. This extra hardware consists of one transmission gate for each input bit line, each input $\overline{\text{bit}}$ line, as well as two test control inputs to the PLA. These transmission gates affect the performance and increase the area of the PLA and are not included in any commercially available CPLDs. In an EEPLA with n inputs, m outputs, and p product terms, a complete test requires $max\{2np,mp\}$ test phases and one test vector associated with each test phase using this DFT technique. The testing time for this DFT technique was later improved by reducing the number of test phases while increasing the number of test vectors per test phase [188]. However, this improved approach still requires a large number of test phases which means hundreds of reconfigurations of the CPLD.

A more recent approach facilitates the systematic development of a minimum set of either two or four test phases (depending on the size of the PLA) and the associated set of test patterns for each phase without any modification to the re-programmable PLA [267]. These two or four test phases and their associated test vectors detect all single and multiple stuck-at faults among the cross-points, input lines, product term lines, and output lines, as well as wired-AND, wired-OR, and dominant bridging faults among the input lines, product term lines, and output lines. In addition, faults in the configuration memory and programming circuitry are also detected.

To illustrate this test procedure, we will consider an 8-input (N_0-N_7, with both bit and $\overline{\text{bit}}$ lines for each input), 8-product term (P_0-P_7), 4-output (M_0-M_3) PLA. This PLA can be completely tested with only two test phases. The first test phase is illustrated in Figure 13.7a where each product term line is programmed with a unique input bit ANDed together with the other 7 input $\overline{\text{bit}}$ lines. Note that in the test phases illustrated in the figure a '1' in the AND-plane implies the bit line cross-point is active and the $\overline{\text{bit}}$ line cross-point is inactive while a '0' in the AND-plane implies the bit line cross-point is inactive and the $\overline{\text{bit}}$ line cross-point is active. Similarly, a '1' in the OR-plane implies the cross-point is active while a '0' implies the cross-point is inactive.

The configuration pattern in the AND-plane indicates that one input bit and all other input $\overline{\text{bit}}$ lines are active in a given product term line during the first test phase. If the number of product terms is greater than the number of inputs ($p>n$), we would begin to repeat the cross-point pattern such that product term line $n+1$ would have the same pattern as product term line 1. The OR-plane is configured by assigning the first product term line to the first output line, the second product term line to the second output line, and so on. Note that a '1' ('0') in the OR-plane indicates that a given product term is connected (not connected) to the output line. If the number of product term lines is greater than the number of output lines ($p>m$), we begin to repeat the cross-point pattern in the OR-plane such that product term line $m+1$ is assigned the same pattern as product term line 1 as illustrated in Figure 13.7a. In the figure, only two product term lines are assigned to each output line and spaced such that each product term line is active for only one output line.

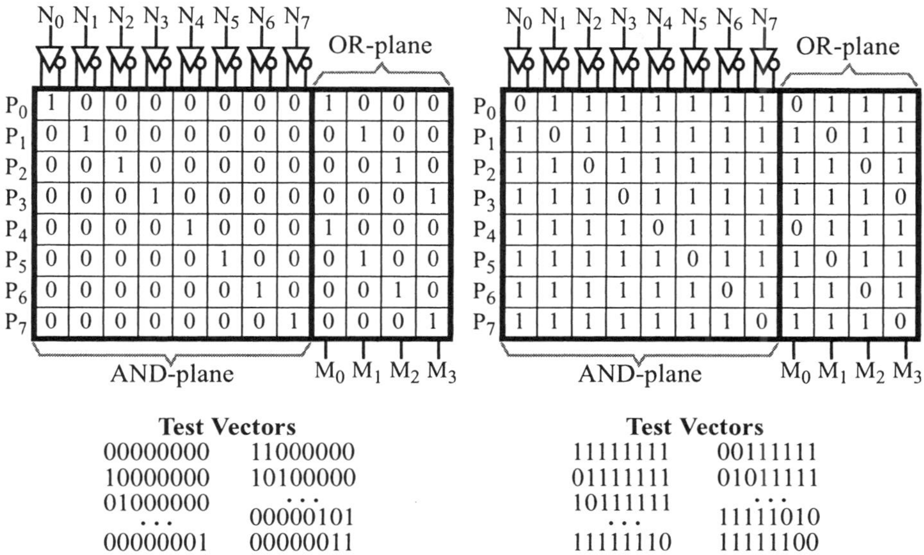

(a) Test Phase #1 (b) Test Phase #2

FIGURE 13.7 Re-programmable PLA BIST configurations.

The complete set of test vectors for the first test phase are also illustrated in Figure 13.7a. This set of test vectors includes the all 0s pattern, walking a 1 through a field of 0s, and all combinations of two 1s in a field of 0s. There is a total of $(n^2+n+2)/2$ test vectors in the set where n equals the number of inputs to the PLA. For our example PLA this would be a total of 37 test vectors as opposed to 256 test vectors for exhaustive testing. This reduction in the number of test patterns is even greater for larger PLAs. For example, a 39-input PLA would require only 781 test patterns for each test phase as opposed to 2^{39} test patterns for exhaustive testing.

The second test phase configuration, illustrated in Figure 13.7b, is simply the inverse of the first test phase configuration. Every activated cross-point from the first test phase is deactivated during the second test phase and vice versa. Similarly, the set of test vectors for the second test phase is the inverse of the test vectors for the first test phase. There are a total of $(n^2+n+2)/2$ test vectors in the set consisting of the all 1s pattern, walking a 0 through a field of 1s, and all combinations of two 0s in a field of 1s. These two test phases and their associated sets of test patterns can completely test any fully programmable OR-plane-based re-programmable PLA where $n \geq p$ [276].

For PLAs where $p>n$, two additional test phases are needed for a complete test of the OR-plane. The first two phases (shown in Figure 13.7a and b) and the associated test vectors completely test the programmable AND-plane for all stuck-at faults and bridging faults, regardless of the number of product terms, p. In addition, these two test phases detect most of the faults in the OR-plane. However, the repeated product terms prevent detection of some faults including some product term lines stuck-at-0 at the output line cross-points as well as some output line cross-points stuck-on. The third test phase detects the remaining output line cross-points stuck-on and is obtained by simply activating all cross-points in the AND-plane (both bit and $\overline{\text{bit}}$ in order to turn on all product term lines regardless of the input logic values) while deactivating all cross-points in the OR-plane. There is only one test vector for this test phase consisting of all don't cares where an output response of all 0s can be expected for the fault-free case. The fourth test phase detects the product term lines stuck-at-0 at the output line cross-points. This test phase configuration is obtained by assigning a unique non-zero binary value to each product term line, as illustrated in Figure 13.8a for a PLA where n=4, p=8, and m=4. Since there are 2^n-1 unique non-zero binary values available for product term assignment, this covers all possible sizes for a PLA (a structure with 2^n product terms is a ROM and not a PLA). For this test configuration, all cross-points are activated in the OR-plane. The set of test vectors consists of the binary count values from 1 to p.

A special case exists when $p > n$ and p is an integral multiple of n and m. In this case, the first test phase configuration must be modified so that when the OR-plane pattern is repeated, the activated cross-points are shifted by one (each time they are repeated)

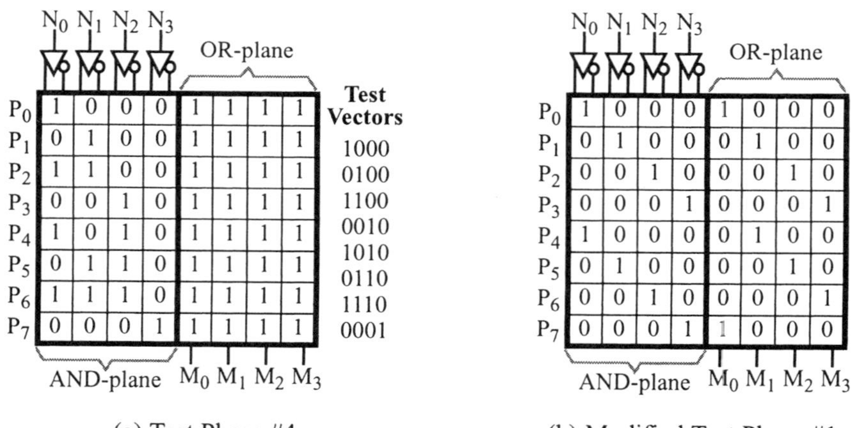

FIGURE 13.8 Special re-programmable PLA BIST configurations for case $p>n$.

as illustrated in Figure 13.8b. Here the objective is to avoid identical product terms being assigned to the same output line. The second test phase configuration for this special case is again the inverse of this modified first test phase configuration. The sets of test vectors for these two test phases remain unchanged along with the third and fourth test phase configurations and their sets of test vectors described above.

The TPG for the first two test phases could be designed as an n-bit FSM to generate the all 0s pattern, walking a 1 through a field of 0s, and all combinations of two 1s in a field of 0s for the first phase. A simple, but inefficient, TPG design is illustrated in Figure 13.9. It uses $2N$ flip-flops and generates N^2+N test vectors including the desired test vectors. In this TPG design the *BIST Start* signal (not shown in the figure) is used to reset the shift registers when it is inactive. The same FSM can be used for the second test phase with the outputs inverted (the OR gates become NOR gates) such that we apply the all 1s pattern, walking a 0 through a field of 1s, and all combinations of two 0s in a field of 1s. For PLAs where $p>n$, the TPG for the fourth test phase is a binary counter to count from 1 to p such that the counter is of size $\lceil \log_2(p) \rceil$ with logic 0s supplied to all other PLA inputs. Since the test vector for the third test phase is all don't cares, no TPG is needed for that test phase.

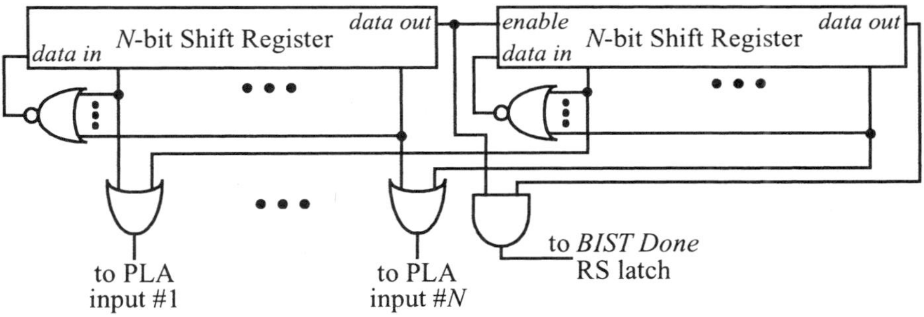

FIGURE 13.9 TPG for re-programmable PLA test phases #1 and #2.

13.2.3 Partially Programmable OR-Plane PLA-based PLBs

Many current CPLDs use partially programmable OR-planes for the PLA in the PLB. This facilitates limited product term sharing with fewer configuration bits required to implement the combinational logic functions. In some cases, certain product term lines may connect to a single output line. These product term lines will be referred to as unique product term lines. Partially programmable OR-plane-based re-programmable PLAs require a total of four test phases to completely test the PLA for all single and multiple stuck-at faults among the cross-points, input lines, product term lines, and output lines as well as wired-AND, wired-OR, and dominant bridging faults among the input lines, product term lines, and output lines [27]. These test phases and their associated test vectors are somewhat different from those of the fully programmable OR-plane-based PLA discussed above. The four test phases and their associated test vectors are illustrated in Figure 13.10 for a partially programmable OR-plane PLA with n=8, p=8, and m=4.

In the first test phase, the AND-plane is configured in the same manner as in the case of the AND-plane for the fully programmable OR-plane PLA discussed in the previous subsection. The OR-plane is configured such that every other cross-point is activated along each row and each column with the exception of unique product term lines. Every unique product term line cross-point is activated for this configuration as illustrated in Figure 13.10a. As a result, no two horizontally or vertically adjacent cross-points are activated with the exception of the OR-plane cross-points associated with unique product term lines. In order to completely test the AND-plane for all stuck-at faults and bridging faults, all cross-points associated with unique product term lines must remain active during the first two test phase configurations. The test vectors for the first test phase are the same as those for the first test phase of the fully programmable OR-plane PLA and include the all 0s patterns, walking a 1 through a field of 0s, and all combinations of two 1s in a field of 0s.

The second test phase configuration, illustrated in Figure 13.10b, is simply the inverse of the first test phase configuration with the exception that every cross-point associated with a unique product term line remains activated. Otherwise, every activated cross-point from the first test phase configuration is deactivated during the second test phase configuration and vice versa. Similarly, the set of test vectors for the second test phase is the inverse of the test patterns for the first test phase to obtain the set consisting of the all 1s pattern, walking a 0 through a field of 1s, and all combinations of two 0s in a field of 1s.

The third test phase is illustrated in Figure 13.10c. This test phase configuration is basically the same as the first test phase, except that half of the cross-points associated with unique product term lines are deactivated for testing. Every other cross-point associated with a unique product term line is deactivated in order to detect bridging faults for those unique product term lines as well as the cross-point stuck-on faults. The only test vectors needed for this test configuration are those that turn on the unique product term lines whose cross-points are deactivated in the OR-plane. These test vectors will be all 0s with a logic 1 on the input associated with a given unique product term line.

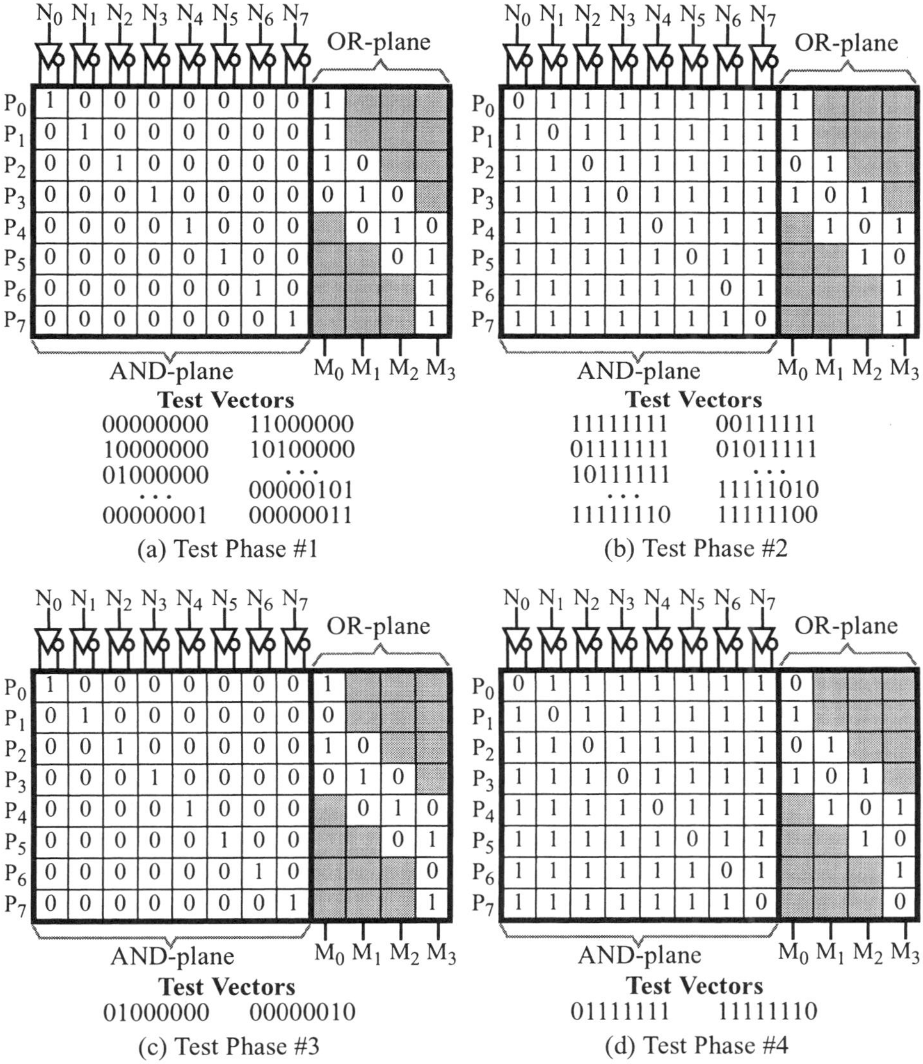

FIGURE 13.10 Partially re-programmable PLA BIST configurations.

The fourth test phase is illustrated in Figure 13.10d and is basically the same as the second test phase except that the other half of the cross-points associated with unique product term lines are deactivated. In other words, the fourth test phase is the inverse of the third test phase. Hence, those cross-points associated with unique product term lines that were activated in the third test phase are deactivated in the fourth test phase and vice versa. Again, the only test vectors needed for this configuration are those that turn on the unique product term lines whose cross-points are deactivated in the OR-plane. These test vectors will be all 1s with a 0 on the input associated with a given unique product term line.

A special case occurs when $p>n$ and one or more output lines connect to more product term lines than there are inputs to the PLA. In this case, those product term lines must be treated as though they were a part of a fully programmable OR-plane where $p>n$ and tested using the technique described in the previous subsection. However, the PLA can still be completely tested in four test phases.

The TPG design for the first two test phases for a partially programmable OR-plane PLA-based PLB would be the same as that for the first two test phases for a fully programmable OR-plane. While the TPG for the first and second test phases could be used for the third and fourth test phases, respectively, a much simpler FSM design could be used since the number of test patterns is much smaller than that required for the first two test phases. Assuming c unique product term lines, the TPG for test phases 3 and 4 will need to generate only $c/2$ test vectors per test phase.

Testing the flip-flop and other logic, such as output multiplexers and carry logic for adders, can be performed with pseudo-exhaustive test patterns once the PLAs have been tested since the number of inputs to the circuitry at the outputs of the PLA is small. In this case, the PLA can be programmed to act like wires such that inputs are routed through the PLA with no logical operation to achieve direct access to the flip-flops and output multiplexer circuitry. This is similar to the sensitized partitioning technique described in Chapter 8. This applies to both fully programmable and partially programmable OR-plane PLA-based PLBs.

13.3 Interconnect BIST Architectures

Wire segments and their associated CIPs within an FPGA or CPLD can be classified as either global or local routing resources. Global routing resources are used to interconnect non-adjacent PLBs and programmable I/O cells. Local routing resources are specific to a given PLB and are used to connect that PLB either to global routing resources, or to directly adjacent PLBs in some architectures. The fault models of interest for FPGA and CPLD interconnect networks include CIPs stuck-on and stuck-off, wire segments stuck-at-0 and stuck-at-1, open wire segments, and shorted wire segments (bridging faults). Detecting the CIP stuck-on and stuck-off faults also detects stuck-at-1 and stuck-at-0 faults in the configuration memory bits that control the CIPs.

Bridging faults are the more difficult faults to test in FPGAs and CPLDs. Without detailed physical layout information regarding adjacency between wire segments, one can only assume the "rough" physical relationship of wire segments and busses available in data books and graphical editors included in the FPGA/CPLD CAD tool suite. From this information, adjacent groups of wires can be determined that *may* have pair-wise shorts even though not every wire in one group is necessarily adjacent with every other wire in the other group [291]. This treatment allows layout-independent development of interconnect BIST configurations. This independence includes bus rotations which change the adjacency relationship among the wires of the same group. Obviously, the optimal approach to developing interconnect BIST configurations would be to begin with the physical design database and use capacitance extraction techniques discussed in Chapter 2 to establish which bridging faults are possible and also which have a higher probability of occurring [277].

To detect the various interconnect faults, the BIST must check that every wire segment and CIP is able to transmit both a logic 0 and a logic 1, and that every pair of wire segments that may be shorted can transmit both (0,1) and (1,0) values. This latter test case includes two wire segments that share a common CIP, since a CIP stuck-on fault creates a short between its two wire segments in the same manner as a bridging fault between those two wire segments. Applying logic 0 and logic 1 values at one end of a wire and observing the values at the other end, detects any stuck-open fault in any type of closed CIP (cross-point, break-point, or multiplexer) along the wire as well as any open or stuck-at fault affecting any wire segment composing the wire. Detecting a stuck-on CIP fault requires that opposite logic values be applied to the two wire segments associated with that CIP, so that the fault will yield an incorrect logic value on one of the two segments. A compound cross-point CIP can be completely tested in three test phases as illustrated in Figure 13.11 [298]. This requires that the opposite logic values are applied to the two wires used for signals as shown in the figure; note that the configuration bits with a logic 1 to turn on the transmission gate are denoted by a gray box while the configuration bits with logic 0 are denoted by white boxes.

A multiplexer CIP requires one test phase for each of its inputs. As a result, large multiplexer CIPs used for programmable routing resources in the more recent FPGA and CPLD architectures can easily dominate the number of test phases required for complete BIST of the programmable logic device. Both logic 0 and 1 values must be applied to the selected input to ensure that the cross-point CIP for that input is not stuck-off. The opposite logic values must be applied to the non-selected inputs to detect stuck-on CIP faults in the unselected inputs. The type of the multiplexer CIP,

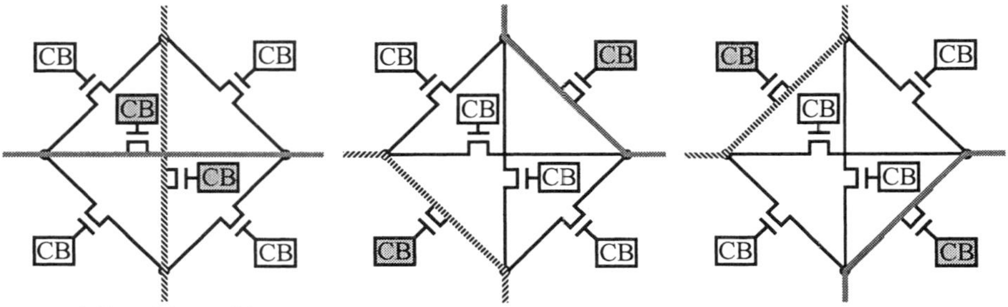

FIGURE 13.11 Compound cross-point CIP test phases.

decoded or non-decoded, also impacts the test patterns that must be applied in order to completely test all of the logic and circuitry of the multiplexer CIP [158]. For decoded multiplexer CIPs, the opposite logic values should be applied to all unselected inputs while applying logic values to the selected input, as shown in Table 13.1 for a 4-input decoded multiplexer CIP. This can lead to additional routing complexity in order to completely test the multiplexer CIP. In the case of non-decoded multiplexer CIPs, each input need only be tested during two test phases, once when that input is selected and again when that input is not selected. This reduces the routing complexity considerably if we test only one unselected input each test phase as illustrated in Table 13.1 for a 4-input non-decoded multiplexer CIP. Note that the choice of the unselected input to be tested during a given test phase is arbitrary, but all unselected inputs should be tested at least once during the set of test phases.

TABLE 13.1 BIST configurations and test vectors for 4-input multiplexer CIPs.

Test Phase	Decoded MUX CIP						Non-decoded MUX CIP							
	C_0	C_1	I_0	I_1	I_2	I_3	C_0	C_1	C_2	C_3	I0	I1	I2	I3
1	0	0	0	1	1	1	1	0	0	0	0	1	X	X
			1	0	0	0					1	0	X	X
2	1	0	1	0	1	1	0	1	0	0	X	0	1	X
			0	1	0	0					X	1	0	X
3	0	1	1	1	0	1	0	0	1	0	X	X	0	1
			0	0	1	0					X	X	1	0
4	1	1	1	1	1	0	0	0	0	1	1	X	X	0
			0	0	0	1					0	X	X	1

The strategy of one interconnect BIST approach is to configure a subset of routing resources (wire segments and CIPs) to form two groups of *wires under test* (WUTs) that receive identical test patterns from a TPG, while the values at the other end of the WUTs are compared by one or more comparison-based ORAs as illustrated in Figure 13.12 [291]. A WUT may be composed of several wire segments connected by closed CIPs. To check local routing resources, WUTs may also go through PLBs configured to pass the inputs of the PLB directly to the outputs. Note that without involving PLBs in configuring WUTs, one cannot test local routing and achieve a complete interconnect test. In Figure 13.12, the WUTs are shown by bold lines, and the activated (closed) CIPs are shown in gray, while the open CIPs are white. The first group of WUTs connects the wire segments S_1, S_2, S_3, S_4, and S_5, while the second group connects wire segments S_6, S_7, S_8, and S_9. To test the CIPs that are configured to be open for stuck-on faults, the opposite logic values must be applied to the wire segments on the opposite side of the open CIPs; these wire segments are S_{10}, S_{11}, S_{12}, S_{13}, and S_{14} in Figure 13.12.

The TPG applies exhaustive patterns to the WUTs, all possible 2^n test vectors to every group of n WUTs. Although this test is longer than walking patterns (which also provide a complete set of tests), the exhaustive patterns can be generated using a binary counter for the TPG which requires fewer PLBs than for an FSM designed to generate walking patterns. For reasonably small n, the application time for 2^n test vectors is still negligible compared to the reconfiguration time of the FPGA or CPLD. The TPG can be extended to a larger counter in order to provide test patterns to the opposite sides of open CIPs that isolate the WUTs from the rest of the interconnect in order to test these CIPs for stuck-on faults. As a result, the TPG must control any wire that may be shorted to some WUT (via a stuck-on CIP or a bridging fault) such that when the TPG drives a 0(1) on the WUTs, it also sets wire segments S_{10}, S_{11}, S_{12}, S_{13}, and S_{14} to 1(0) at least once during the BIST phase [291]. Similarly, the larger TPG is used to apply test patterns to groups of wires adjacent to WUTs A and B to detect bridging faults between these groups of wires.

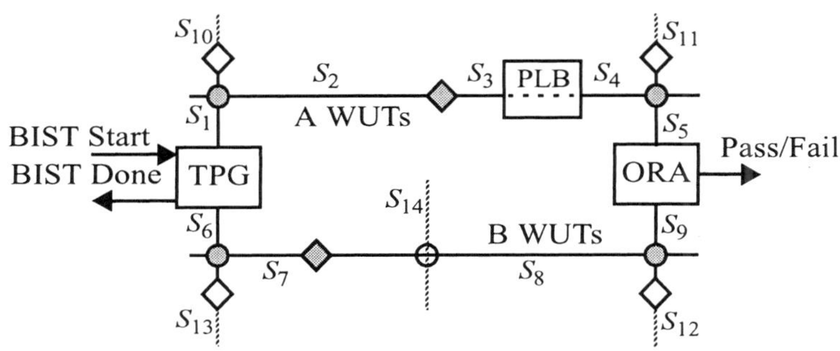

FIGURE 13.12 BIST structure for FPGA/CPLD interconnect resources.

When comparison-based ORAs are used, the only condition that will allow a fault to escape detection at the ORA would be identical (or equivalent) faults in the two compared sets of WUTs [102]. For example, each set of WUTs could have an open in its ith wire which could go undetected. This problem is overcome by testing every set of WUTs multiple times, to be compared each time with a different set of WUTs [281]. Alternatively, two sets of WUTs can be retested comparing different pairs of wires within the two sets of WUTs.

An alternative approach to comparison-based ORA and two sets of WUTs is to calculate parity over the set of WUTs and to transmit the parity over a separate wire to the ORA where it is compared to the parity that is regenerated for the set of WUTs at the ORA, as illustrated in Figure 13.13 [298]. The TPG consists of an N-bit binary counter and an N-bit parity generate circuit. The ORA consists of an N-bit parity regenerate and check circuit (discussed in more detail in Chapter 15) and an RS latch. This approach is capable of detecting any single fault in the interconnect network as well as any multiple faults that produce an odd number of bit errors in the WUTs as they enter the ORA. This approach helps to alleviate some of the routing congestion problems encountered when developing interconnect BIST configurations. Regardless of whether the BIST approach is comparison-based or parity-based, multiple groupings of TPG, WUTs, and ORA must be executed in parallel within the FPGA or CPLD in order to minimize the total number of BIST configurations and, as a result, the total testing time. The number of test phases required for complete interconnect testing typically ranges from 25 to 40 [281].

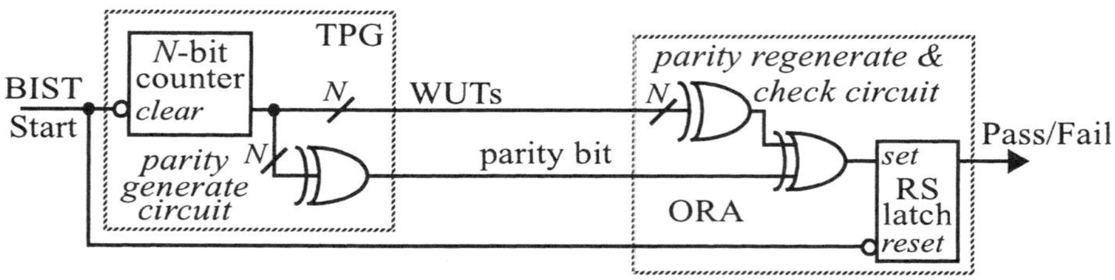

FIGURE 13.13 Alternative BIST structure for FPGA/CPLD interconnect resources.

13.4 Boundary Scan Access to BIST

Practically all ISR FPGAs and CPLDs feature a Boundary Scan interface that can also be used to access the configuration memory for programming the device. Some architectures provide user-defined access to the FPGA or CPLD core programmable logic and interconnect resources, which can be used to control the BIST sequence and the ORA scan chain. This allows FPGAs and CPLDs to be reconfigured for and to execute BIST during system-level testing without requiring additional I/O pins. For those FPGAs and CPLDs that do not feature user-defined access to the core logic and interconnect resources from the Test Access Port (TAP), the equivalent functionality could be implemented in the core logic for the BIST mode operation at the cost of four BIST-dedicated I/O pins and associated connections on the PCB for board and system-level testing. However, other techniques discussed in this section can provide the required Boundary Scan access without the need for additional I/O pins.

Reconfiguring the FPGA or CPLD for the BIST test phases, initiating the BIST sequence, and reading the BIST results are performed using the TAP. Access to the BIST architecture requires the ability to control the *BIST Start* to initialize the TPGs and ORAs and start the BIST sequence, as well as the ability to read the *BIST Done* and ORA *Pass/Fail* results from the BIST circuitry. This access must be provided within the confines of the typical FPGA/CPLD Boundary Scan circuitry architecture. Even in devices that do provide access to the core logic and routing resources of the FPGA/CPLD, the *Shift/Capture* and *Update* control signals are not made available to the core by the TAP circuitry in some devices. Instead, the only internal signals provided by the TAP for user-defined internal scan chains are *TCK*, *TDI*, a port to send data out on *TDO*, and usually some internal enable signal. Additional considerations in selecting a Boundary Scan access method for final implementation include total test time, PLB overhead, routability, and diagnostic resolution [100]. An overview of possible methods for access to the FPGA BIST architecture via the Boundary Scan interface is given in the following subsections. This includes the advantages and disadvantages of these approaches in terms of their impact on logic resources, and test time that can be achieved for the PLB and interconnect BIST techniques.

13.4.1 Using Existing Boundary Scan Cells

The Boundary Scan INTEST instruction and functionality provide a way to control the *BIST Start* input, but not all FPGAs and CPLDs may have an explicit INTEST capability in their Boundary Scan architecture and instruction set. Some FPGAs do provide an INTEST-like capability as a by-product of their EXTEST instruction. This capability is usually characterized by an I/O Boundary Scan cell which contains three complete double-latched registers (Shift/Capture flip-flop and Update flip-flop) for the input, the output, and the tri-state control signals associated with the bi-directional buffer for the programmable I/O cell. With this type of cell, the tri-state condition can be maintained at the FPGA I/O via the tri-state control register to prevent the bi-directional buffer from back driving other outputs on the PCB. Meanwhile, the input registers can be used to control the *BIST Start* signals while the output registers can be used to read the *BIST Done* and ORA *Pass/Fail* signal via the Boundary Scan chain.

This method requires no additional logic resources with little additional routing resources. This approach works well for both the logic BIST and interconnect BIST when the ORAs lie along the perimeter of the array of PLBs. When the ORAs lie in the middle of the array of PLBs, there can be excessive routing congestion when trying to bring the ORA *Pass/Fail* outputs to the Boundary Scan cells. Another disadvantage of this method is that the entire Boundary Scan chain must be read in order to retrieve the BIST results from even a small number of ORAs; this results in increased test time. To avoid reading the entire Boundary Scan chain, the ORA results can be routed to the Boundary Scan cells nearest the output of the scan chain (given that there are sufficient routing resources available) such that the ORA results shift out first. The shifting out of the remainder of the Boundary Scan chain can then be aborted.

13.4.2 Configuration Memory Readback

Most FPGAs/CPLDs that facilitate programming through the Boundary Scan interface (writing the configuration memory) also support reading the configuration memory. In some FPGAs/CPLDs, the contents of flip-flops are also read along with the configuration memory. In these devices, there exists a direct access to the ORA *Pass/Fail* and *BIST Done* flip-flops through the configuration memory readback instruction. This requires no additional logic or routing resources other than the ability to control the *BIST Start* signal for initialization and initiation of the BIST sequence. However, two problems exist that prevent this method from being as straightforward and attractive as it would appear at first glance.

First, reading the configuration memory via the Boundary Scan interface in most FPGAs requires a serial read of the entire memory. In a typical FPGA with array size of 20×20, this would amount to serially reading about 225,000 bits in order to obtain the BIST results.[1] In fact, it takes longer to read the configuration memory than to write it since there are more bits returned during the read than are configured during the write. This is due to the fact that the contents of the flip-flops are also returned during the configuration memory readback operation. Since the download time dominates the total testing time accounting for more than 95% of the test time, using configuration memory readback for access to BIST results doubles the testing time.

1. This accounts for the complete configuration memory including bits for over 100,000 CIPs, 1600 LUTs, over 200 I/O cells, miscellaneous logic in 400 PLBs including 1600 flip-flops

The second problem is determining where the BIST results are located in the stream of configuration memory bits being read back. Since the configuration memory map contains a great deal of architectural information about the device, FPGA and CPLD manufacturers are very reluctant to provide the required mapping information for fear of compromising their intellectual property. The memory map can be determined independently but, not without a considerable investment in time and effort to 'de-compile' or 'reverse engineer' the configuration memory map. Both of these problems could be overcome in the future if FPGA manufacturers provide the ability to selectively read small portions of the configuration memory (specifically the logic values in the flip-flops) and to give the address information for accessing the contents of the flip-flops to the customer. A nice alternative would be for the FPGA/CPLD manufacturers to provide a Boundary Scan instruction that simply reads the contents of all flip-flops (and maybe even writes the contents to facilitate control of *BIST Start*); this feature would be extremely valuable for BIST as well as for other types of system applications.

13.4.3 User-Defined Scan Chain

Some FPGAs/CPLDs provide instructions to access user-defined scan chains. By integrating an internal scan chain in the PLBs containing the ORAs, we obtain an architecture that allows us to observe the results of the comparisons done by every ORA, without additional logic resources and with only local routing resources for the scan chain. By using the data-select multiplexer that is part of the flip-flop circuitry in some PLBs, we can alternately use the flip-flop to latch mismatches in the ORA and to shift out the *Pass/Fail* results at the completion of the BIST sequence. As a result, we can create individual, independent ORAs with integrated scan registers in each ORA. This provides a significant enhancement to diagnostic resolution [8].

To illustrate the use of a minimal amount of control provided by the TAP to the core logic and interconnect, we will assume that there exists an internal enable signal, which we will refer to as *TEN*. This signal is active when the user-defined scan chain access instruction is decoded by the TAP instruction decoder and remains active until a different instruction is loaded into the TAP instruction register. Working within the confines of the typical FPGA boundary scan architecture, the *TEN* signal can be used for the *BIST Start* function. This *BIST Start* signal is used (when *TEN*=0) to reset the flip-flops in the TPGs, ORAs, and BUTs until the user-defined scan chain access instruction is loaded, at which time *TEN* goes high and the BIST sequence begins. The *TDI* input is used as the *Scan Mode* control to the internal scan chain containing the BIST results (*BIST Done* and the ORA *Pass/Fail* indications) as illustrated in Figure 13.14. By connecting the *BIST Done* output to the last register in the internal scan chain, this signal is immediately observable on *TDO* by holding *TDI* at the 'non-shift' logic value (logic 1 in this example implementation). As soon as *BIST Done* goes active, we begin shifting out the *Pass/Fail* indications by setting *TDI* to the 'shift' logic value (logic 0 in this case). $N_{ORA}+1$ clock cycles are then needed to retrieve all the BIST results. By supplying a logic 1 (via the active *TEN* signal) to the scan input of the first flip-flop in the scan chain, we are able to test the integrity of the internal scan chain as well. For example, a scan chain flip-flop stuck-at-0 will be detected by the absence of the logic 1 at the end of the BIST results-shifting sequence. Since a logic value of 1 represents an ORA failure, a flip-flop stuck-at-1 will be immediately detected.

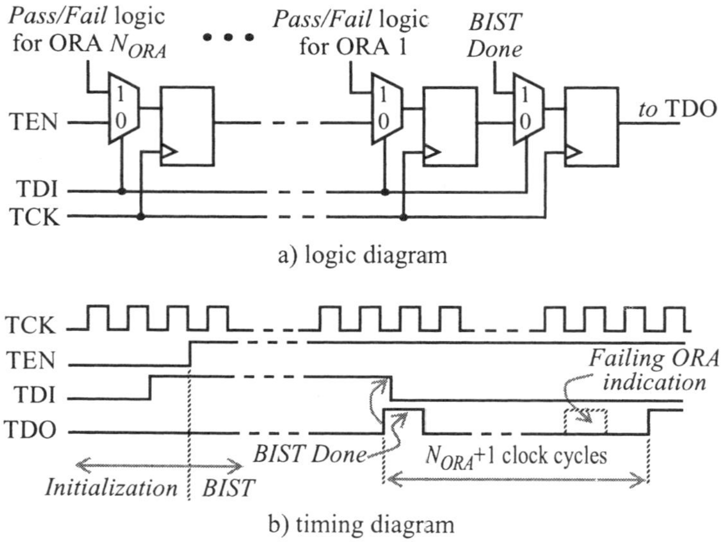

FIGURE 13.14 Integrated ORA and scan chain with relative timing diagram.

13.5 On-Line BIST

More recently, the re-programmability and the regular structure of FPGAs, in conjunction with run-time reconfiguration features supported by some FPGAs, have been exploited to develop on-line testing and diagnosis of the programmable logic and interconnect resources for fault tolerant operation of the FPGA [11]. Run-time reconfiguration allows a portion of the FPGA to be reconfigured while the remainder of the FPGA continues to operate without interruption. Therefore, portions of the FPGA can be reprogrammed to contain and execute BIST while the majority of the FPGA continues to execute the system function. In this on-line testing approach two rows and two columns of PLBs and their associated global and local routing resources are reserved for testing while the rest of the FPGA performs the system function. These two rows and two columns are referred to as self-testing areas (STARs). Figure 13.15a illustrates the floor plan of an FPGA containing an 8×8 array of PLBs with a working area for the system function (denoted by the white squares to represent the system function PLBs) along with a 2-column vertical STAR and a 2-row horizontal STAR. Vertical routing resources in the horizontal STAR are reserved for interconnecting the system function partitions located above and below the horizontal STAR. Similarly, horizontal routing resources in the vertical STAR are reserved for interconnecting the system function partitions to the left and right of the vertical STAR [281].

The STARs are used for testing both the PLBs and the programmable interconnect contained within the two columns or rows of the STAR. Once the logic and routing resources within the STAR have been completely tested, a portion of the system function logic and interconnect adjacent to the STAR is relocated to the current STAR position, creating a new STAR in the area vacated by that portion of the system function. Therefore, the vertical STAR sweeps back and forth across the FPGA testing the PLBs and the vertical routing resources while the horizontal STAR sweeps up and down the FPGA testing the horizontal routing resources. Only one STAR is needed to test the PLBs but both STARs are required to test the vertical and horizontal routing resources. The basic BIST structure is illustrated in Figure 13.15b and consists of multiple PLBs programmed to operate as a TPG which supplies pseudo-exhaustive test patterns to two identically programmed BUTs (for logic BIST) or two sets of WUTs (for interconnect BIST). The outputs of the BUTs or the destination end of the

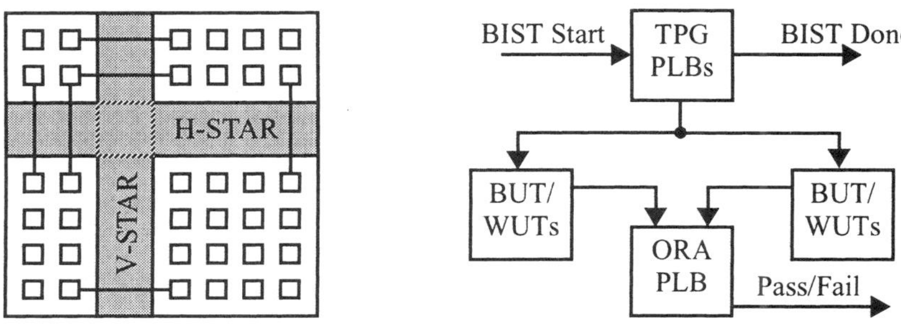

a) horizontal and vertical STARs b) basic BIST structure

FIGURE 13.15 FPGA self-test areas for on-line BIST of PLBs and interconnect.

two sets of WUTs are compared by a PLB programmed to function as a comparator-based ORA, like that illustrated in Figure 13.5. A STAR contains several of these disjoint BIST structures (or *tiles*) which operate in parallel. All access to the STAR and the BIST circuitry contained within the STAR is through the Boundary Scan interface in a similar manner as described in the previous section. This includes access to the configuration memory for moving the STARs and programming the STARs with the BIST structures as well as executing the BIST and retrieving the BIST results at the completion of the BIST sequence. The BIST circuitry is operated by TCK from the Boundary Scan interface. As a result, all testing and diagnosis and reconfiguration is performed transparent to the normal system function operation [10].

The on-line BIST for the PLBs is similar to that of the off-line BIST for LUT-based PLBs, described earlier in this chapter, in that the BUTs are repeatedly configured during each test phase, eventually to be tested in all modes of operation. The principle difference in the off-line and on-line BIST for PLBs is that more than two test sessions are required to completely test all the PLBs contained within the STAR. This is because there are only two BUTs in a BIST tile during any given test session. Only a single PLB is required for the comparator-based ORA in a BIST tile. Due to the number of BUT inputs that must be supplied with test vectors, multiple PLBs are required to implement the TPG. For the commercial FPGA used in the reported work, three PLBs were required for the TPG [11]. This required a minimum of six PLBs per BIST tile. To overcome routing congestion problems, two spare PLBs were incorporated into the BIST tile to provide additional routing resources. Figure 13.16 illustrates the floor plan of the 4×2 array of PLBs used for a single BIST tile in the vertical STAR, where T, B, O, and S denote a TPG PLB, a BUT, an ORA PLB, and a spare PLB, respectively. In order to test all PLBs in the 4×2 array, the BIST tile must be "rotated". The eight rotations shown in Figure 13.16 mean that, eventually, every PLB is completely tested twice, each time being compared by a different ORA and being compared with a different BUT.

This rotating strategy assures that any combination of multiple faulty PLBs is guaranteed to be detected. In addition, the BIST results from the eight rotations provide sufficient diagnostic information to identify any combination of up to two faulty PLBs and most combinations of more than two faulty PLBs [10]. Additional BIST-based diagnostic configurations can be applied to identify those combinations of multiple faulty PLBs that cannot be identified from the failing BIST results as well as to determine which part of the PLB is faulty (for example, which flip-flops or which bits in a LUT) for enhanced fault tolerance. While this BIST approach works well for on-line testing, it is not an efficient approach for off-line BIST (where the entire FPGA would be filled with STARs containing the BIST tiles with all tiles executing BIST in parallel). This is because the eight rotations are equivalent to eight test sessions compared to the two test sessions for the off-line BIST approach for PLBs described earlier in

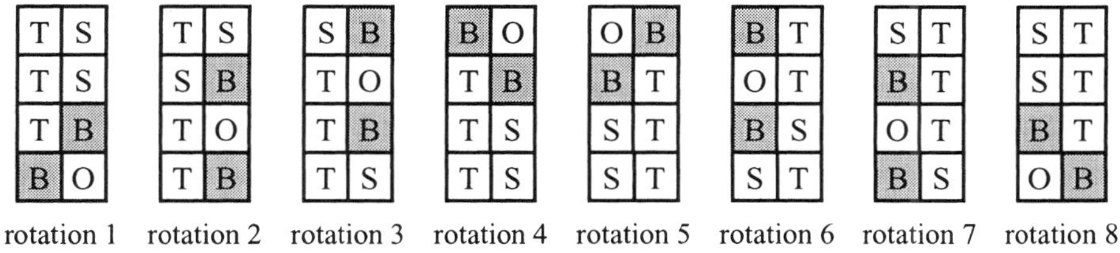

FIGURE 13.16 Rotations of on-line logic BIST tiles within STAR.

the chapter. The number of test phases per test session is the same for both BIST approaches, hence using the on-line approach for off-line testing increases the total testing time by a factor of four.

The on-line BIST for the FPGA programmable interconnect resources is similar to that of the off-line routing BIST, described earlier in this chapter, in that some of the PLBs are configured as TPGs and ORAs. The TPGs drive two sets of WUTs consisting of various wire segments and CIPs which are then compared by the ORA to detect mismatches caused by faults. Like the FPGA logic BIST, all WUTs are tested at least twice and each time compared to a different set of WUTs in order to minimize the possibility of equivalent faults escaping detection [281]. The principal difference between the on-line interconnect BIST and the comparison-based off-line interconnect BIST, described earlier in the chapter, is that testing is confined to the area of the STARs. The vertical STAR is limited to testing vertical global and local routing resources while the horizontal STAR is limited to testing horizontal global and local routing resources, as illustrated in Figure 13.17a. Testing the cross-point CIPs at the intersection of the horizontal and vertical global busses must involve both STARs, as illustrated in Figure 13.17b, since global horizontal routing resources in vertical STAR and global vertical routing resources in horizontal STAR are reserved for the system signals connecting the working areas separated by the STARs. Unlike logic BIST where the same number of test phases are required for each off-line and on-line test session, the on-line interconnect BIST requires about 50% more test phases than off-line BIST (41 test phases for on-line compared to 27 test phases for off-line for the particular FPGA architecture discussed) [281]. This is due to the restricted area of the STAR allowed for testing the interconnect. However, the diagnostic resolution obtained by this restricted area is much better than the earlier off-line interconnect BIST approaches. As a result, extending the on-line STAR-based interconnect BIST to off-line testing, by running multiple STARs in parallel as illustrated in Figure 13.17c, provides for a more effective BIST approach with better diagnostic resolution at the expense of a 50% increase in testing time.

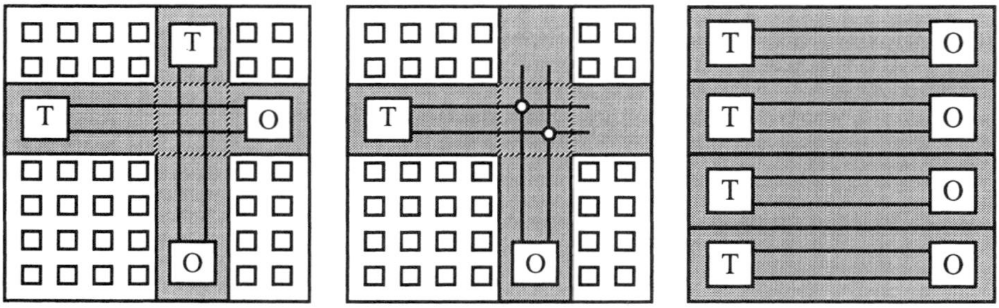

a) horizontal and vertical STARs for routing BIST

b) BIST structure for global-to-global cross-point CIPs

c) concurrent STARs for off-line routing BIST

FIGURE 13.17 FPGA self-test areas for on-line and off-line BIST of interconnect.

13.6 Benefits and Limitations

FPGAs and CPLDs are very regular in their structure, but testing is more complex than that of regular structures like RAMs, ROMs, and PLAs. This is primarily due to the programmability of the FPGA/CPLD along with the generalized architecture and features that support synthesis and implementation of a wide variety of digital system functions. ISR FPGAs and CPLDs facilitate downloading configurations specifically designed to provide BIST of the FPGA or CPLD at all levels of testing, including system-level testing. Once the device has been completely tested, it can be reprogrammed with the desired system function such that the system function does not incur any area overhead or performance penalty as a result of BIST. In fact, there is no need to consider BIST or DFT for any system function that will be programmed into an FPGA or CPLD. This type of BIST might be applied to some FPGAs or CPLDs that do not support ISR (such as electrically or UV erasable devices) but this would only be practical for wafer and/or device-level testing, and not for system-level testing since the device must be removed from the system for erasure and/or programming.

The idea of BIST for FPGAs and CPLDs can be extended to facilitate diagnosis of the programmable logic and interconnect resources. In some approaches, the diagnostic information is a by-product of the BIST results [8]. In other approaches, additional BIST configurations must be applied to accurately diagnose the fault [116]. Given a sufficient level of diagnostic resolution, the system function can be reconfigured to avoid the fault. As a result, FPGAs and CPLDs provide a practical approach to fault tolerant system operation when combined with BIST and BIST-based diagnostic configurations. This represents another unique benefit of BIST for FPGAs and CPLDs.

The main cost of BIST for FPGAs and CPLDs is the number of configurations required to completely test the device. These BIST configurations must be stored in the test machine or the system for downloading and execution in the programmable logic device. Since the time required to download and execute the BIST configuration is dominated by the programming time of the device, the test time can be long for many BIST configurations. Therefore, in order to minimize the testing time as well as the amount of memory required to store all of the BIST configurations, a primary goal during the development of the BIST is to minimize the number of test sessions and test phases.

For system-level testing, an alternative approach is to provide a single, high fault coverage test phase that detects the maximum number of faults that can be detected in the device with one execution of the BIST sequence [126]. This can significantly reduce the testing time as well as the amount of memory required to store the BIST configurations. This test phase would then be required once for each test session. Implementation of this type of test phase for a commercial FPGA obtained 82.9% fault coverage of the programmable logic resources when used once in each test session, as compared to 14 test phases per test session required to obtain 100% fault coverage of the programmable logic [283]. Extending this idea to two test phases per test session, the fault coverage was extended to 86.7%. While the high fault coverage BIST configuration does not ensure that the device is fault-free, it does provide a reasonably good indication of whether or not the device is working, which may be sufficient for some manufacturing (such as board-level test) and system-level (such as a fast system diagnosis) testing applications [126].

Testing the FPGA or CPLD at the system operating frequency may not be possible in all applications. This is due to the high fan out of test patterns from the TPG to the BUTs in the BIST approach for programmable logic and the many series CIPs configured in routing the test patterns to the BUTs. Similarly, the interconnect BIST approaches typically configure the WUTs with many wire segments and many series CIPs. This limits the maximum clock frequency that can be used to execute the BIST. As a result, some delay faults may not be detected. An alternative is to reduce the amount of logic or interconnect that is under test during a given BIST configuration to allow BIST execution at a higher clock frequency but this will, in turn, require more test phases and/or test sessions.

One additional limitation of BIST for FPGAs and CPLDs is that while the BIST architecture is generic, the specific test phases are not. In other words, the test phases must be developed for a specific PLB and/or interconnect architecture. Once these BIST configurations have been developed, they can be applied to all devices of the same size and type. While scaling the BIST configurations to fit different size devices in the same family or series is reasonably straightforward, each new family or series of devices requires development of new test phases specific to that architecture. However, like BIST for regular structures, BIST development for FPGAs and/or CPLDs is a good investment since the time and costs can be recovered by many products and projects.

CHAPTER 14

Applying Digital BIST to Mixed-Signal Systems

BIST for digital circuits has experienced considerably more research and development than BIST for analog and mixed-signal circuits. There have been a number of BIST approaches developed for specific types of analog circuits and references to some of these approaches are included in the Bibliography. In this chapter, we discuss some of the proposed BIST architectures that use digital BIST components for testing a wide variety of analog circuits in mixed-signal systems. The BIST approaches we have discussed thus far can be implemented in the digital portion of the mixed-signal system for testing the digital domain. However, testing the analog circuitry along with the digital-to-analog converters (DACs) and analog-to-digital converters (ADCs) introduces a number of new issues that are not considered in digital testing. In this chapter we take a look at some of these issues from the standpoint of applying digital BIST circuitry to testing the analog portion of mixed-signal systems.

Mixed-signal circuits provide a good environment for the development of BIST approaches for analog circuits by allowing some of the experience and expertise of BIST development in digital circuitry to be used as a platform for testing the analog portion of the mixed-signal system. The basic components of the BIST architecture can be incorporated into the digital portion of the design without adverse effects on the analog circuit performance. These digital components include the TPG and ORA functions as well as the necessary input isolation and test controller functions to initialize and control the BIST sequence for system-level access to the BIST. However, there are aspects of analog circuit testing which prevent the straightforward application of conventional digital TPG and ORA functions. For example, traditional signature analysis using LFSR or CA-based SARs and MISRs is unsuitable for application to analog BIST since the good circuit signature for these digital ORA functions is based on the assumption that an exact output response sequence is produced for every fault-free execution of the BIST sequence. In a mixed-signal circuit, the sampling noise in the DAC and ADC as well as processing variations which result in component parameter tolerances and environmental variations from temperature and voltage will prevent an exact repetition of the output response sequence. Therefore, reproducible BIST signatures are not possible from one execution of the BIST sequence to the next and new considerations must be given to ORAs.

The traditional digital pseudo-random patterns from TPGs based on LFSRs or CA registers will produce an analog waveform that is similar to noise after passing through a DAC. However, ramp input test signals have been used in analog testing techniques and have been found to provide good fault detection results and, in some cases, better results than sinusoidal test waveforms [55]. It has been observed that faults in analog circuits can cause detectable variations in output response delay, rise/fall times, and overshoot when stimulated by certain input test signals. It has also been observed that the detectability of faults with respect to the input test waveform can vary with the type of analog circuit under test [28].

14.1 Mixed-Signal BIST Architecture

A basic mixed-signal BIST architecture is shown in Figure 14.1 where the majority of the BIST circuitry has been added to the digital portion of the mixed-signal circuitry [279]. The normal mixed-signal system components include the digital and analog system functions along with the DACs and ADCs that are required to convert the dig-

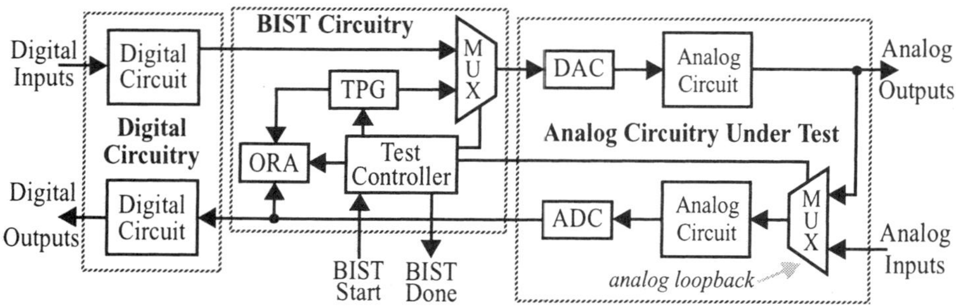

FIGURE 14.1 Basic BIST architecture for mixed-signal systems.

ital signals to analog waveforms and vice versa. The BIST circuitry additions include the digital TPG and ORA functions as well as input isolation multiplexers, a test controller, and analog loopback functions. The analog loopbacks (analog multiplexers) are the only circuits associated with the BIST architecture to be inserted in the analog domain. This minimizes the impact of the BIST circuitry on the operation and performance of the analog circuity. The purpose of the analog loopback is to facilitate a return path for the test waveforms from the TPG, through the analog CUT, and back to the ORA. An input isolation multiplexer is required for the insertion of the digital test patterns into the input data stream to the DAC. Since the target circuitry under test is the analog system circuits, including the DACs and ADCs, the digital TPG and its associated input isolation multiplexer are incorporated immediately prior to the digital inputs of the DAC. Similarly, the digital ORA is incorporated at the output of the ADC. This basic architecture has been proposed and implemented in a number of applications [199]. The architecture has been referred to as the ADC/DAC loopback BIST or simply as ADC/DAC BIST [49].

In order to make the BIST circuitry usable in the system for off-line testing and system diagnostics, the BIST circuitry must be capable of proper initialization of the analog circuitry under test, isolation of system data inputs, and reproducible results from one execution of the BIST sequence to the next. The length of the initialization sequence must be sufficient to clear the effects of previous system signals in the analog circuitry. Faults can be effectively isolated to a given section of analog circuitry within the diagnostic resolution of the analog loopback multiplexers [279]. Therefore, by adding more analog loopback multiplexers at strategic locations, the diagnostic resolution can be improved at the expense of area overhead and possible performance penalties in the analog portion of the mixed-signal system. For example, with analog loopback #1 in Figure 14.2 activated, any faults detected are isolated to that path from the TPG to the ORA. If the first BIST sequence indicates a good circuit, then analog loopback #1 can be deactivated while the analog loopback #2 is activated and the

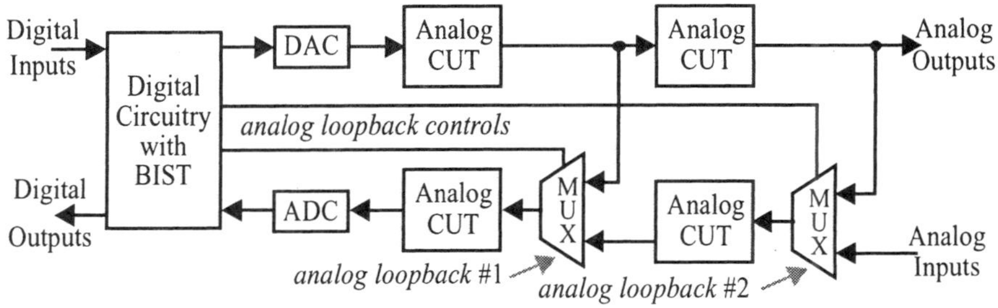

FIGURE 14.2 BIST architecture for system-level diagnosis.

BIST sequence re-executed. Faults detected during this second BIST sequence would be isolated to the analog circuitry in the right-hand portion of Figure 14.2.

14.1.1 TPGs for Mixed-Signal BIST

The first TPG implemented using this BIST architecture was a LFSR-based TPG as used in digital BIST [199]. The waveform produced by the pseudo-random sequence after it passes through the DAC has a flat spectrum which in turn implies an infinite number of tones. This 'white noise' type of test waveform has been considered by some to provide a universal stimulus [204]. While this type of test waveform has proven to be effective in detecting faults in the analog portion of mixed-signal system using this BIST architecture, other waveforms are known to be effective in detecting faults in analog circuits including ramps, impulse functions, step functions, sine waves, square waves, and DC signals [49].

The TPG illustrated in Figure 14.3 provides many other types of test waveforms in addition to the 'white noise' of the pseudo-random sequence produced by an LFSR or CA register [279]. The size of the TPG is based on the number of bits, *N*, to the DAC. The TPG circuitry includes a binary counter that also functions as an LFSR with a primitive characteristic polynomial. The counter is an up/down counter with active high synchronous parallel load and active high carry-out. Operating the Counter/

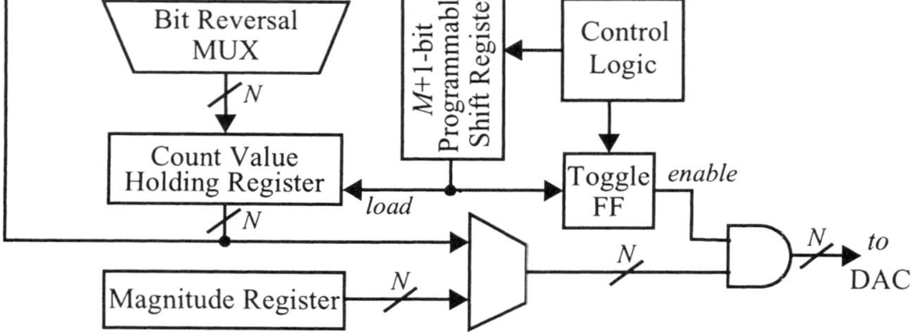

FIGURE 14.3 TPG functional block diagram.

LFSR in different modes provides a variety of analog test waveforms once the TPG output patterns pass through the DAC. For example, a single pass through the up-count range produces a ramp signal while multiple passes through the up-count range produce a saw-tooth analog test waveform. Combining a series of up/down counts generates triangular waveforms at the output of the DAC as illustrated in Figure 14.4a for a 4-bit up-down counter. The Bit Reversal Multiplexer reverses the order of bits to the DAC (MSB becomes LSB and vice versa) and has the effect for the counter modes of operation of producing high frequency test waveforms as illustrated in Figure 14.4a for bit reversal of the triangular waveform [279]. During any of these modes of operation, the outputs of the Programmable Shift Register is a logic 1 such that the Count Value Holding Register is continuously loaded and its output is passed to the DAC. In addition, the output of the toggle flip-flop is forced to a logic 1 such that the test patterns proceed through the AND gates to the DAC (via the input isolation multiplexer).

Since the frequency response of analog circuits is important in terms of fault detection capability, waveforms that sweep through a range of frequencies are produced by this TPG design. The frequency sweep mode of operation in the TPG provides a square wave test pattern which progressively increases in frequency. The magnitude of the square wave can either be constant and programmable by selecting the Magnitude Register, or the magnitude can progressively increase with the frequency by selecting the output of the Count Value Holding Register [157]. When the frequency sweep function is enabled, the AND gates are used to zero the magnitude of the generated square wave when the output of the toggle flip-flop is a logic 0, otherwise the magnitude is set by either the Counter Value Holding Register or the Magnitude Register.

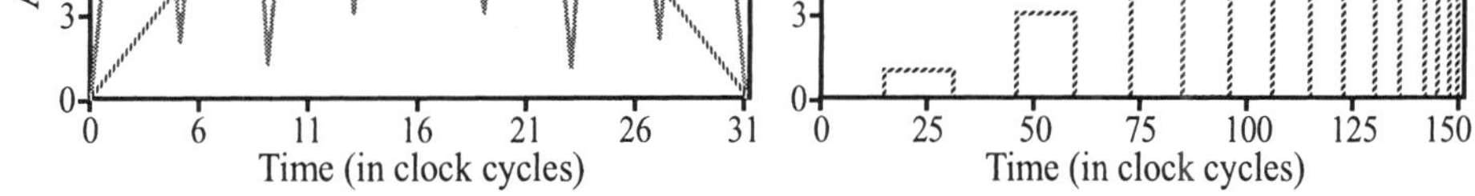

(a) triangular waveform and bit reversal

(b) varying amplitude frequency sweep waveform

FIGURE 14.4 Example test waveforms.

In the following example illustrated in Figure 14.4b, the operation of the varying amplitude frequency sweep will be described for a 4-bit TPG implementation. For this example, the M+1-bit Programmable Shift Register will be assumed to be a 2-bit shift register (M=1). The counter, toggle flip-flop, and Count Value Holding Register are initialized to all 0s at the beginning of the frequency sweep BIST sequence. The counter then counts through a complete N-bit count sequence and produces a carry-out. During this count cycle the output of the toggle-flip-flop is a logic 0 so the AND gate outputs produce all 0s to the DAC. The carry-out initiates a parallel load of the data in the Count Value Holding Register (which is all 0s at this point) into the counter and the counter begins to count from that point. Two clock cycles after the carry-out is active, the Count Value Holding Register is loaded with the contents of the counter as a result of the carry-out shifting through the Programmable Shift Register. At this point the counter has a count value of 0001 which is loaded into the Count Value Holding Register. At the same time the delayed carry-out emerging from the Programmable Shift Register causing the toggle flip-flop to generate a logic 1 which enables the AND gates to pass the data from the Count Value Holding Register to the DAC (a value of 0001 at this point). When the counter has completed the next count sequence and produced a carry-out, the data from the Count Value Holding Register is loaded into the counter so that the counter will begin its next count sequence from the value 0001, resulting in a shorter count sequence before the next carry-out is generated. When the delayed carry-out emerges from the shift register, the Count Value Holding Register is then loaded with the counter value 0010 and the toggle flip-flop disables the AND gates, sending all 0s to the DAC. At the next active carry-out the counter will be loaded with 0010, the Count Value Holding Register will subsequently be loaded with 0011, and the toggle flip-flop will send a logic 1 to the AND gates allowing the 0011 in the holding register to pass to the DAC. This process continues with the counter loading larger values from the Count Value Holding Register and counting through shorter count sequences before producing the next carry-out. This sequence of events produces the complete varying amplitude frequency sweep waveform shown in Figure 14.4b. When the data passed to the DAC is taken from the Magnitude Register, the frequency sweep is the same, but the magnitude is constant and determined by the value in the Magnitude Register whenever the output of the toggle flip-flop is a logic 1.

The Programmable Shift Register controls the number of clock cycles by which the carry-out is delayed before the Count Value Holding Register is loaded. The longer the delay, more increments of the counter take place and a larger count value is loaded into the Count Value Holding Register. Consequently, the count value is incremented M times by the counter prior to being loaded into the holding register, where it is held until the end of the current count cycle. This controls how quickly the square wave sweeps through the frequencies and the resultant length of the test sequence. Therefore, a larger value of M produces a faster sweep and a shorter BIST sequence.

Enabling the bit reversal during a frequency sweep mode will load non-sequential values into the Count Value Holding Register. When this reversed order bit value is loaded into the counter, the counter will begin counting from that value. As a result, the frequencies (and amplitudes in the case of the varying amplitude frequency sweep) will appear to be in a pseudo-random order.

Variations of the test waveforms can easily be incorporated into this TPG design. For example, by forcing the toggle flip-flop to remain at a logic 1 during the entire varying amplitude frequency sweep mode, the test waveform produced by the DAC looks like a parabolic ramp. Single clock cycle pulse waveforms and step functions are also easily produced by this TPG to test the impulse and step responses of the analog circuitry. The various test waveforms produced by the TPG are summarized in Table 14.1. Note that the magnitude of the constant amplitude frequency sweep, DC, pulse, and step waveforms are programmable by the value written into the Magnitude Register. The area overhead can be reduced by eliminating the Magnitude Register and forcing any desired fixed magnitude value at the multiplexer inputs.

TABLE 14.1 Summary of test waveforms produced by TPG.

Digital Test Pattern	Analog Waveform	Pictorial	Bit Reversal Waveform
count-up	saw-tooth	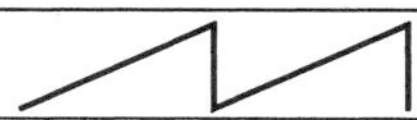	noise-like
count-down	saw-tooth	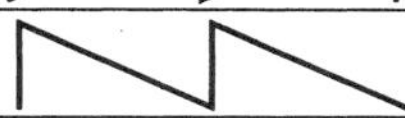	noise-like
count-up/down	triangular wave	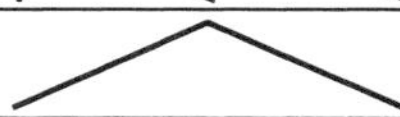	noise-like
LFSR or CAR	noise	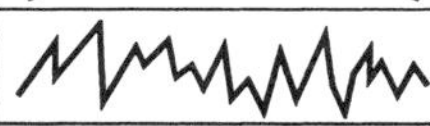	noise
constant magnitude	DC		DC
pulse	pulse	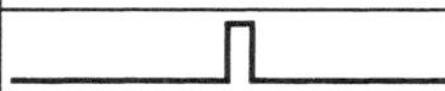	pulse
step function	step function	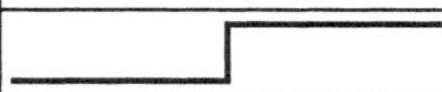	step function
Frequency Sweep (varying amplitude)	increasing amplitude with decreasing period	see Figure 14.4b	pseudo-random amplitudes & pseudo-random frequencies
Frequency Sweep (constant amplitude)	constant amplitude with decreasing period	see Figure 14.4b but with constant amplitude	constant amplitude & pseudo-random frequencies
parabolic ramp	increasing amplitude	see Figure 14.4b but with no return to zero	pseudo-random amplitudes with pseudo-random durations

14.1.2 ORAs for Mixed-Signal BIST

Traditional signature analysis using LFSR or CA-based SARs and MISRs is unsuitable for application to mixed-signal BIST. This is due to the sampling error in the DAC and ADC, component parameter variations, and environmental variations which prevent exact repetition of the output response sequence from one execution of the BIST to the next. As a result, mixed-signal BIST requires that we be able to look for a range of acceptable values for the fault-free circuit. An obvious choice for providing this capability is the accumulator. Summing the magnitudes produced by the output response at the ADC facilitates establishing upper and lower bounds on the final sum that will be considered to be a "good" circuit. Any BIST results falling outside of this range will be considered to be the result of a faulty CUT [199].[1] An analog checksum circuit has been previously proposed for BIST of analog circuits [55]. The advantage of a digital ORA is that the results can be read directly through digital system software interfaces during system-level testing without the need for an ADC to retrieve the BIST results.

Simply summing the magnitudes of the output responses of the analog CUT may not detect some faults [204]. For example, if the analog output response is a sine wave centered around 0 volts and the ADC produces the 2s complement value of negative magnitudes, the resultant sum in the accumulator would be zero. This would be indistinguishable from faulty circuit producing a DC output of 0 volts. One solution to this problem is to square the magnitude using a multiplier prior to summing the output response in the accumulator [204]. A similar problem exists with faults that result in the superposition of noise on the analog signal. The noise could average to zero such that there is no change in the resultant accumulator value from that of the fault-free circuit. In addition, faults that result in phase shifts may not be detected. Summing the magnitudes of the sampled analog signal may only detect the fault at the beginning and/or end of the BIST sequence, which may not be sufficient to cause a failing BIST signature. However, summing the absolute value of the difference between the input test patterns (from the TPG) and the output response of the analog circuit (from the ADC) facilitates detection of both of these fault cases (noise and phase shift) [157]. In addition, the circuitry to determine the absolute value of the difference between the input test pattern and the output response requires less area overhead than a multiplier to square the magnitude of the CUT output response.

1. The first mixed-signal BIST approach we implemented at Bell Labs in 1987 was based on the architecture of Figure 14.1 and used an LFSR-based TPG with a single precision accumulator for the ORA. While the approach was never analyzed with fault simulations, it was analyzed in the system using physical fault insertion and in the field during system operation and found to do a good job of detecting faulty analog lines and trunks in the switching system.

A final consideration is the type of accumulator used for the ORA. The size of the accumulator used for earlier applications was not explicitly stated [204]. However, a study of the fault detection capabilities of single precision, residue, and double precision accumulators indicated that single precision and residue accumulators were effective in discriminating faulty circuits only for short BIST sequences and small signal amplitudes [279]. The double precision accumulator, on the other hand, was found to be the best choice for use as an ORA in mixed-signal BIST. The ORA illustrated in Figure 14.5 consists of a double precision accumulator used to sum the magnitudes of the sampled output responses selected by the multiplexer. The absolute value subtractor produces the absolute value of the difference between the input test pattern and the output response of the CUT. The accumulator is designed for three modes of operation: summing, clear, and hold. The clear and hold modes of operation are used for BIST initialization and maintaining the final BIST results until read by the system, respectively. The size of the ORA is based on the number of bits, N, from the ADC.

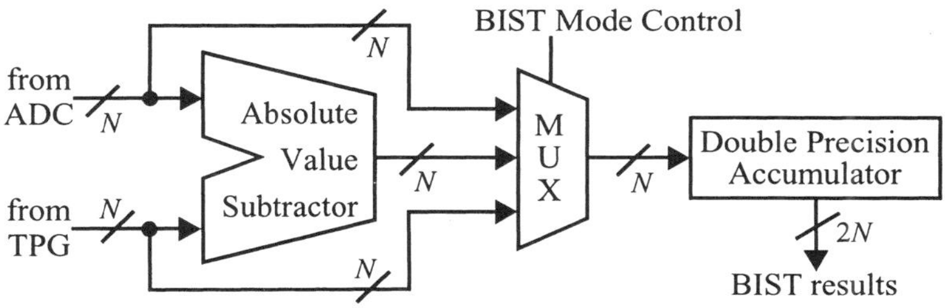

FIGURE 14.5 ORA block diagram.

14.2 Mixed-Signal BIST Operation

Unlike digital BIST approaches where the BIST circuitry is tested as part of the BIST sequence, the digital BIST circuitry in the mixed-signal BIST architecture must be tested independent of the analog circuitry. This is due to the variation in the analog output response sequence which requires a range of acceptable 'good' circuit signatures (final sums in the accumulator in this case). Faults in the digital BIST circuitry can go undetected within this range of acceptable BIST results. For example, a fault in the least significant bit of the TPG could create less variation in the input test waveform than is expected in the analog output response allowing the fault to go undetected. By incorporating the ability to directly sum the output of the TPG in the ORA accumulator as illustrated in Figure 14.5, all of the TPG and most of the ORA can be tested prior to testing the analog circuitry. The portion of the ORA that is not tested in this BIST mode is the absolute value subtractor and a portion of the multiplexer. During this digital BIST circuitry test, an exact BIST signature would be expected in the ORA for the fault-free case as opposed to a range of values.

The complete BIST consists of three separate test sessions. During each test session, multiple test phases are executed with each test phase consisting of generating one of the test waveforms produced by the TPG and applying it to the CUT. The accumulator sums the responses and the resultant ORA sum is read at the end of the test phase to determine the faulty/fault-free status of the circuitry being tested with a given test waveform for that test phase. If any test phase produces an incorrect accumulator value (a value outside the acceptable range of values for that test waveform), the circuit is considered to be faulty. The three test sessions include:

1. direct connection of the TPG output to the ORA input to sum the TPG patterns magnitudes in the ORA accumulator to test the digital BIST circuitry,
2. summing the magnitudes of the analog output response from the ADC, and
3. summing the absolute value of the difference between the input test pattern magnitude from the TPG and the analog output response from the ADC.

The determination of the range of accumulator values that reflect fault-free circuit behavior can be obtained by Monte-Carlo simulation of the fault-free analog circuit including the DAC and ADC circuits. During these simulations, the analog component parameter values must be varied within their specified tolerances along with voltage and temperature. In this way, the maximum and minimum values of accumulator results that reflect the fault-free circuit operation can be established. The range must be established for each test waveform applied to the CUT. As a result, there may be considerable simulation time required to establish the various ranges. An alternative approach is to determine these values during normal system operation with a large number of CUTs (to emulate component parameter variation) and with various voltages and temperatures.[1] While this approach works well for large CUTs that maybe difficult to simulate, it is difficult to ensure that the determined range of accumulator values covers all possible and acceptable variations.

Investigating the resultant signatures (values in the accumulator) at the end of the BIST sequences, it has been observed that the fault-free circuit signatures followed a

1. This in-system approach was used to establish the range of fault-free circuit results for the first mixed-signal BIST approach we implemented at Bell Labs. The method worked well enough and we were able to establish the range with no ill-effects during manufacturing and system-level testing.

normal distribution when normal probability distributions were used for component variations [157]. Since the final sum of magnitudes may be greater than 2^N, where N is the total number of bits in the accumulator, the 'wrap-around' effect of the sum can place the range of values such that the maximum acceptable value is greater than 0 but the minimum acceptable value is less than 2^N. As a result, the maximum 'good' circuit value is actually less than the minimum 'good' circuit value, with all other values representing a faulty CUT. Therefore, one must be careful not to simply consider the minimum and maximum values obtained from simulation or the actual system. Once the acceptable range of fault-free circuit signatures has been established for each test waveform and for each ORA mode of operation (magnitude summing as well as summing the absolute value of the difference in the input test waveform and the analog output response), a CUT producing a signature outside of the acceptable range for any given test waveform for either ORA mode would be considered as faulty.

14.3 Analysis of BIST Approach

In a similar manner as the fault-free circuit is simulated to establish the fault-free circuit signature range, the various test waveforms are applied to the faulty circuit in a simulation environment. Due to the possible component parameter and environmental variations of the fault-free components in the CUT, multiple simulations should be performed to determine the range of resultant signatures for the faulty circuit. If the resultant range of faulty circuit signatures lies outside the acceptable range for the fault-free circuit for any test waveform in either ORA summing mode, the fault or set of faults can be considered to always be detected with that waveform and summing mode during actual testing. The fault or set of faults must be considered as undetected if the resultant signatures of the faulty circuit always falls within the range of acceptable values for the fault-free circuit for all test waveforms and ORA summing modes. However, if some of the faulty circuit signatures fall outside and some fall inside the good circuit range for all test waveforms and ORA summing modes, the fault or set of faults can only be considered as potentially detected. In this case, the probability of fault detection is proportional to the percentage of faulty circuit signatures that lie outside the acceptable range of values for the fault-free circuit. Given a probability of detection for the ith fault, P_{D_i}, the fault coverage for an analog circuit under test with a given set of test waveforms and ORA techniques is given by:

$$FC = \frac{\sum\limits_{\text{all faults}} P_{D_i}}{\text{total number of faults}} \tag{14.1}$$

14.3.1 Benchmark Circuits

A set of analog and mixed-signal benchmark circuits was proposed at the 1997 IEEE International Test Conference (ITC'97) which consisted of eight circuits [128]. The intent of the benchmark circuits was to facilitate evaluation and comparison of various analog and mixed-signal testing approaches. One problem with the first release of the benchmark circuits was that there were no specified component parameter variations. The set of ITC'97 benchmark circuits was extended to include specified component parameter tolerances as well as other benchmark circuits [137]. To establish the range of acceptable component parameter variations, the output phase and gain frequency response was limited to a variation of ±10% of the output response using the nominal component values. These benchmark circuits are summarized in Table 14.2 in terms of their source, along with the numbers and types of components.[1] There are two types of fault models that may be considered for analog circuits: catastrophic and parametric faults [180]. Catastrophic faults (or hard faults as they are sometimes called) result from opens and shorts in the analog component terminals. Parametric faults (also known as soft faults) result from component parameter values being outside their specified tolerance. Table 14.2 also give the number and types of faults for each benchmark circuit.

TABLE 14.2 Summary of analog testing benchmark circuits [137].

Benchmark Circuit (Source)	Total Components	No. Rs	No. Cs	Other Components	Hard Faults	Soft Faults
Op Amp #1 (ITC'97 [128])	11	2	1	8 N&P MOSFETs	22	6
Continuous Time State Variable Filter (ITC'97 [128])	42	7	2	3 Op Amp #1s	84	36
Op Amp #2 (ITC'97 [128])	10	0	1	9 N&P MOSFETs	20	2
Leapfrog Filter (ITC'97 [128])	77	13	4	6 Op Amp #2s	154	46
Low Pass Filter	15	3	1	1 Op Amp #1	30	14
Elliptical Filter (SFA [81])	55	15	7	3 Op Amp #1s	110	62
Comparator (SFA [81])	13	3	0	1 Op Amp #2	26	8
Differential Pair (SFA [81])	9	5	0	4 BJTs	34	18
Single Stage Common-Emitter Amp (SFA [222])	6	5	0	1 BJT	16	12

1. These benchmark circuits along with detailed information including schematics, SPICE files, component variations, simulation results, and fault models are currently available at www.coe.uncc.edu/~cestroud/analogbc/mixtest.html.

14.3.2 Fault Simulation Results

Faults are easily detected by the mixed-signal BIST architecture when there are no component parameter variations in the analog CUT during fault simulations. In this case, there is no range of good circuit signatures such that any deviation of the analog output response is seen as a detection of the fault being emulated in the circuit. But for any practical implementation, component parameters will vary due to processing tolerances, voltage and temperature. To illustrate this, the results of fault simulations for the benchmark circuits using the hard fault models are summarized in Table 14.3 for the mixed-signal BIST approach that uses only an LFSR for the TPG and an accumulator for the ORA which only sums the output response from the ADC [199]. The information is given in terms of the number of faults simulated and detected along with the overall fault coverage with and without component parameter variation. The simulation results without component variation are denoted in the table as 'No Variation'. For the fault simulations with component parameter variations, the number of faults that were potentially detected has been included with the fault coverage calculated using Equation (14.1). As can be seen from the table, with the exception of one circuit, all faults were detected when component variation was not simulated. There is a significant drop in fault coverage (an average of about 15%) for most circuits when component variation is added, indicating that realistic component variation is necessary for accurate evaluation and comparison of mixed-signal BIST techniques.

TABLE 14.3 Fault simulations for hard faults with LFSR TPG.

Benchmark Circuit (Source)	Faults Simulated	No Variation		Parameter Variation		
		Faults Detected	FC	Faults Potential Detected	Faults Detected	FC
Op Amp 1 (ITC'97 [128])	22	22	100%	1	21	87.3%
Continuous Time State Variable Filter (ITC'97 [128])	18	16	88.9%	4	10	63.9%
Op Amp 2 (ITC'97 [128])	20	20	100%	0	20	100%
Leapfrog Filter (ITC'97 [128])	34	34	100%	3	29	93.4%
Low Pass Filter (Lucent Tech)	30	30	100%	6	22	93.3%
Elliptical Filter (SFA [81])	18	18	100%	3	8	52.2%
Comparator (SFA [81])	26	26	100%	5	19	85.0%
Differential Pair (SFA [81])	42	42	100%	15	27	74.0%
Single Stage Common-Emitter Amp (SFA [222])	16	16	100%	3	13	90.6%

The fault simulation data is given in Table 14.4 for the multi-waveform TPG from Figure 14.3 and the ORA from Figure 14.5 [157]. All faults were detected with no component variation. Even with component variation, all faults were either detected or potentially detected for these benchmark circuits; no faults were found to be undetectable. In addition, the mixed-signal BIST approach was able to obtain fault coverage in excess of 95% for all benchmark circuits. Therefore, it is apparent from comparing the data in these two tables that the fault detection capabilities of a given test waveform is circuit dependent and that a suite of test waveforms increases the fault detection capabilities of the BIST approach.

TABLE 14.4 Fault simulations for hard faults with multi-waveform TPG.

Benchmark Circuit (Source)	Faults Simulated	No Variation		Parameter Variation		
		Faults Detected	FC	Faults Potential Detected	Faults Detected	FC
Op Amp 1 (ITC'97 [128])	22	22	100%	1	21	98.6%
Continuous Time State Variable Filter (ITC'97 [128])	84	84	100%	20	64	97.4%
Op Amp 2 (ITC'97 [128])	20	20	100%	0	20	100%
Leapfrog Filter (ITC'97 [128])	34	34	100%	2	32	98.8%
Low Pass Filter	30	30	100%	0	30	100%
Elliptical Filter (SFA [81])	18	18	100%	0	18	100%
Comparator (SFA [81])	26	26	100%	2	24	95.4%
Differential Pair (SFA [81])	42	42	100%	9	33	95.0%
Single Stage Common-Emitter Amp (SFA [222])	16	16	100%	0	16	100%

The ORA mode of operation which sums the absolute value of the difference between the input test patterns (from the TPG) and analog CUT output response (from the ADC) is intended to detect faults that lead to phase shifts and/or noise in an otherwise good circuit response. This fault detection capability was investigated by injecting a noise signal source in the CUT and varying the amplitude of the noise source with respect to the amplitude of the test waveform. Multiple Monte-Carlo simulations were performed for each noise amplitude setting to facilitate component parameter variation. The fault detection probability is compared for the magnitude summing and the difference summing modes of operation of the ORA in Figure 14.6a. As can be seen, the difference summing mode detects any noise level greater than 5% of the signal amplitude, while the magnitude summing mode has only about a 0.2 probability of detecting the fault causing noise at 5% of the signal amplitude. Similar simulations for faults resulting in phase shifts in the analog output response indicates that the difference summing mode detects phase shifts greater than 29° while the magnitude summing mode has only about a 0.35 probability of detecting faults causing phase shifts of 29° as illustrated in Figure 14.6b.

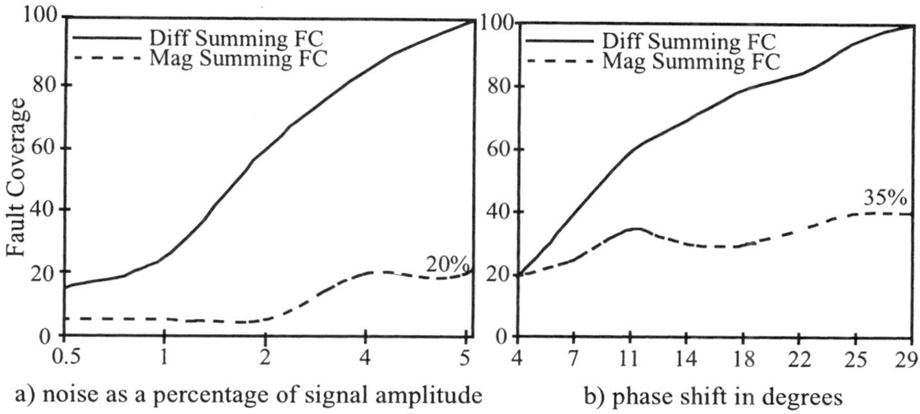

FIGURE 14.6 Noise and phase shift detection with difference summing mode.

14.4 Benefits and Limitations

The mixed-signal BIST approaches discussed in this chapter have the advantage of using the same BIST circuity as used for digital circuits. As a result, the TPG, ORA, and test controller can also be used to test the digital portion of the mixed-signal system as well as the analog portion. An additional advantage is that the majority of the BIST circuitry (excluding the analog loopback circuits) resides in the digital domain which minimizes the impact on the performance of the analog circuitry. This is similar to the non-intrusive digital BIST approaches and, like those BIST approaches, we can expect the fault coverage obtained with this BIST approach to be circuit dependent. However, the multiple test waveforms that can be generated by the TPG of Figure 14.3 facilitate the effective application of this BIST approach to a wide variety of analog CUTs as compared to the LFSR and magnitude summing accumulator approach (although the latter approach may prove to be effective and area efficient for some applications).

The digital logic area overhead in terms of the number of flip-flops and gates (including multiplexers, exclusive-OR gates and elementary logic gates) is summarized in Table 14.5 for the various BIST components (excluding the BIST controller) as a function of the number of bits to/from the DAC/ADC. It has been observed that the frequency sweep test waveforms provide better fault detection in filter-type circuits while the sawtooth, triangular, and LFSR waveforms provide better fault detection in amplifier circuits [137]. Therefore, when the analog CUT falls into one of these categories, digital logic area overhead can be reduced by only implementing TPG circuitry to generate those test waveforms that are most effective for detecting faults in that type of CUT. In the ORA, both magnitude summing and absolute value difference summing were found to be important to fault detection and, as a result, both functions should be implemented.

TABLE 14.5 Area overhead of mixed-signal BIST approaches.

BIST Circuitry		Area Overhead			
		Flip-Flops	**Exclusive-ORs**	**Multiplexers**	**Gates**
TPG	Counter/LFSR	N_{DAC}	$2N_{DAC}+3$	N_{DAC}	$2N_{DAC}$
	Frequency Sweep	$2N_{DAC}+1$	0	$3N_{DAC}$	N_{DAC}
	Bit Reversal	0	0	N_{DAC}	0
ORA	Accumulator (single precision)	N_{ADC}	$2N_{ADC}$ -2	N_{ADC}	$4N_{ADC}$ -4
	Accumulator (residue)	$N_{DAC}+1$	$2N_{ADC}$	N_{ADC}	$4N_{ADC}$
	Accumulator (double precision)	$2N_{ADC}$	$3N_{ADC}$	$2N_{DAC}$	$6N_{ADC}$
	Absolute Value Subtractor	0	$4N_{ADC}$	0	$4N_{ADC}$
	Mode Selector	0	0	$2N_{ADC}$	0
where		N_{DAC}= number of bit inputs to DAC N_{ADC}= number of bit outputs from ADC			

One limitation of the mixed-signal BIST approaches discussed in this chapter is that mixed-signal BIST implementations and analog fault simulation are relatively new compared to the wealth of experience in digital BIST. Analog testing has traditionally been primarily specification-oriented testing (SPOT) while digital testing moved into the defect-oriented testing (DOT) world many years ago [347]. Strict adherence to the SPOT methodology has prevented acceptance of fault models and benchmark circuits like those that propelled advances in the digital BIST realm. Since the analog test jury is reluctant to come to a verdict on this type of mixed-signal BIST approach, it is best to count on more traditional analog testing techniques during manufacturing testing used in parallel with a BIST approach in order to obtain actual testing results that can be used to better understand the fault and defect detection capabilities of the BIST approach. Regardless of the ultimate ability for the BIST approach to isolate faulty circuits, the BIST feature will always provide a cost effective technique for verifying the basic functionality of the chip or PCB before investing traditional testing time on expensive analog ATE. Once a chip or PCB is deemed to be a 'good' circuit by traditional manufacturing analog testing techniques, the BIST approach can again be used for system-level testing.

CHAPTER 15

Merging BIST and Concurrent Fault Detection

Concurrent fault detection circuits (CFDCs) are essential components for on-line testing in systems designed for high reliability, high availability, and designed to be diagnosable to a replaceable unit such as a PCB or chip [145][193]. CFDCs are also referred to as concurrent error detection (CED) circuits [183]. CFDCs are typically incorporated in VLSI and PCB designs to support system-level fault recovery and maintenance strategies. These circuits are typically applied to ASIC designs in an ad-hoc manner and usually only to the data path circuitry. Extensive use of CFDCs has been made in many digital systems such as electronic switching systems.

There are many types of error detecting codes (EDCs) and error correcting codes (ECCs) used in the design of CFDCs [122]. The different types of CFDCs require varying degrees of information redundancy (extra bits) in the system circuitry for the EDC/ECC code word. However, not all of these codes are useful in practical system applications because of the large area and performance penalties associated with their hardware implementations. Therefore, the choice of a CFDC has considerable impact on the overall area and cost of the final system.

Systems which incorporate CFDCs can incur area overhead penalties of as much as 30% to 40% due to these on-line testing functions, but these figures exclude any BIST or other DFT circuitry [274]. The same VLSI devices and PCBs that incorporate CFDCs often incorporate BIST for off-line testing. During the implementation of BIST, the CFDCs and maintenance registers are typically treated as any other system circuitry to be tested. As a result, the addition of BIST circuitry to test the CFDCs and maintenance registers (used to capture and report the errors detected by the CFDCs) further increase the area, performance, and cost penalties associated with overall testing. Therefore, if CFDCs can be used during off-line testing to test the circuitry they are designed to test during on-line testing, then the additional BIST circuitry applied to the CFDCs and their circuits under test can be reduced, thus reducing the total amount of BIST circuitry that would otherwise be needed.

15.1 System Use of CFDCs

Unlike fault tolerant hardware structures that use hardware redundancy such as *N*-tuple Modular Redundancy (NMR), CFDCs are based on information redundancy using EDCs or ECCs. While there are many types of EDCs and ECCs, not all of these are useful in practical system applications because area and performance penalties that result from the circuitry required to generate the code words. Code word generation is performed on the data at the data source before entering the CUT and the code words are checked (which requires regeneration of the code word) at the output of the CUT. Partitioning a system into subcircuits and inserting the code word check and regeneration circuits to detect faults at intermediate points facilitates effective fault isolation and diagnosis of replaceable units as illustrated in Figure 15.1. In this example, a code word generate circuit is present at the input to *System Circuit A*. A code word check/regenerate circuit is inserted between *System Circuit A* and *System Circuit B*, and a code word check circuit is present at the output of *System Circuit B*. Assuming that a fault exists in *System Circuit A*, regeneration of the code word at the output of *System Circuit A* prevents the propagation of an incorrect code word to *System Circuit B*. This, in turn, prevents an error from being produced at the output check circuit (assuming that *System Circuit B* is fault-free). The data that is passed to *System Circuit B* may be of no value since the data word bits and not the code word bits may have been changed by the fault. However, this technique provides fast and effective isolation of the fault at the time an error is detected, thereby reducing the time

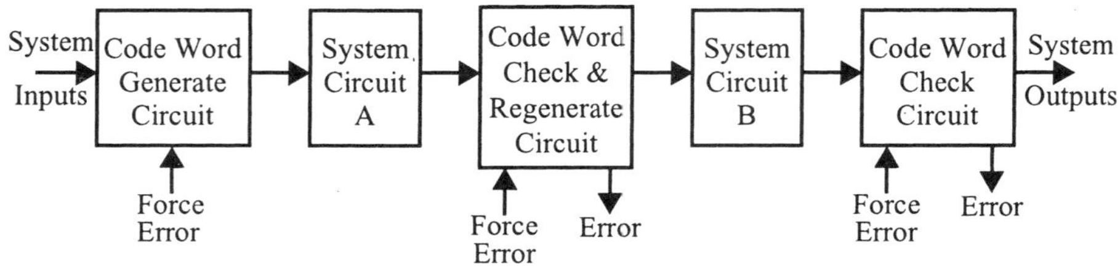

FIGURE 15.1 System partitioning and fault isolation with CFDCs.

required to execute the fault recovery strategy. In most CFDCs, the check circuit must be tested during off-line diagnostics to determine that it is fault-free and has not failed in a non-error indication mode. This requires the ability to force an error to be detected by the check circuit. Forcing the error is typically controlled via a diagnostic control input to either the generate or the check circuit as shown in Figure 15.1. The ability to force errors during off-line system diagnostics also tests the other maintenance register support functions including error source register (ESR), error threshold functions, and interrupt circuits that will be described shortly.

The general structure of a CFDC check/(re)generation circuit is illustrated in the block diagram of Figure 15.2a. The incoming/outgoing data may be serial data, parallel data, or may have both serial and parallel components; for example, an $N\times M$ data bus consists of a sequence of M words of N parallel bits each to construct the complete $N\times M$ bit word containing the data and code words. The latter format will be referred to as serial/parallel data in this chapter. The EDC/ECC bits are first regenerated for the incoming data. This regenerated code word is then inserted in the outgoing data and, at the same time, is compared with the incoming EDC/ECC bits in order to determine if an error has occurred. An error indication may be produced for each mismatching EDC/ECC bit as illustrated in Figure 15.2b for serial data. Alternatively, there may be one error indication for the entire code word as is the case with parallel

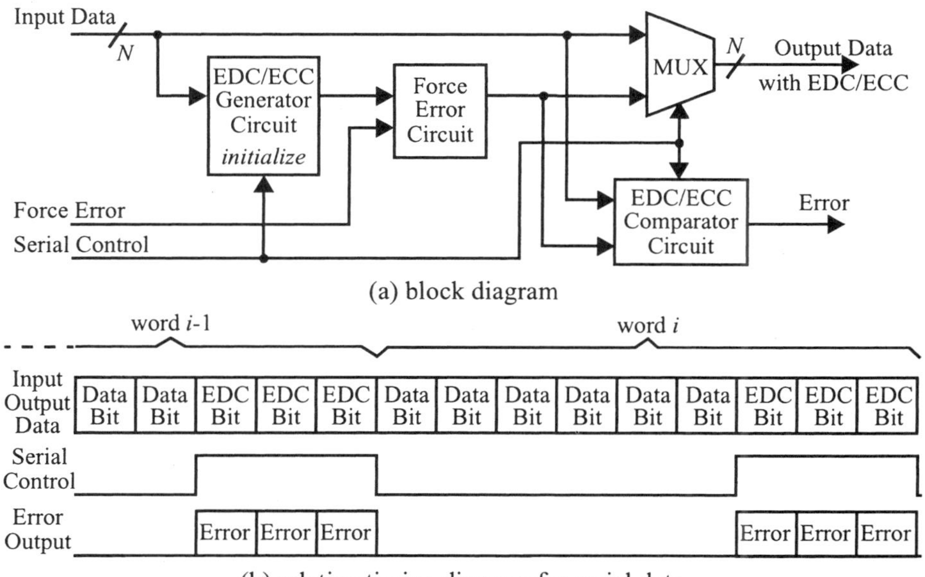

FIGURE 15.2 General structure and timing of CFDCs.

and serial/parallel data. Many CFDC designs include an input which can be used to force the generation of an incorrect EDC for system diagnosis of the error detection and reporting hierarchy during off-line testing of the system. By forcing an incorrect EDC/ECC word to be generated (this is typically accomplished by simply inverting one or more bits of the code word), a mismatch will occur in the CFDC error code word comparator and an incorrect code word will be transmitted with the outgoing data such that two errors will be observed (one in the forcing CFDC and the other at the destination CFDC). Alternatively, the force error circuit can be incorporated at the input to the code word insertion multiplexer to transmit an incorrect code word, or it can be placed at the input to the EDC/ECC comparator circuit. In either of these latter cases, only one error will be observed. In Figure 15.2, it is assumed that the *Serial Control* and *Force Error* inputs are active high along with the *Error* indication.

For CFDCs with a serial component to the data bus (as assumed in the timing diagram of Figure 15.2b), a serial control signal must be supplied to the CFDC which indicates the position(s) of the EDC/ECC bit(s) in the serial data stream. In most cases, the EDC/ECC bit(s) are located at the end of the data word since the regeneration of the EDC/ECC bit(s) in the check circuit is performed as the data bits pass through the check/regeneration circuity such that the newly regenerated EDC/ECC bit(s) are ready for comparison with the incoming EDC/ECC bit(s). Otherwise, the incoming EDC/ECC bits must be stored while the regeneration process is taking place. This serial control signal is used for several purposes: 1) initialization of the CFDC between data words, 2) insertion of the regenerated EDC/ECC bit(s) into the output data stream via the multiplexer, and 3) enabling errors, resulting from the comparison of the regenerated EDC/ECC bit(s) with the incoming EDC/ECC bit(s), to proceed to the ESR.

The errors produced by the CFDCs present the need for a systematic error reporting mechanism including an ESR, as well as a circuit to produce an interrupt to the controlling system software. This interrupt initiates fault recovery actions when an error is encountered. Priority encoded interrupts offer the system software the ability to determine whether to take fault recovery actions based solely on the information obtained from the interrupt without having to stop and read the ESR producing the interrupt. This is accomplished by reporting the binary value of the active error with the highest priority. Since some errors may not be of high priority in the fault recovery strategy (such as parity errors when a given bit error rate is expected), certain non-critical errors may need to be masked from the system interrupt; this requires a mask register and mask circuit. Another common method used to process low priority errors, or errors that are expected to occur periodically, is to provide an error threshold function. The error threshold circuit takes the form of a counter or rate counter such that an interrupt is generated only after a certain number of errors have been detected or the error detection rate has exceeded some predefined threshold. The

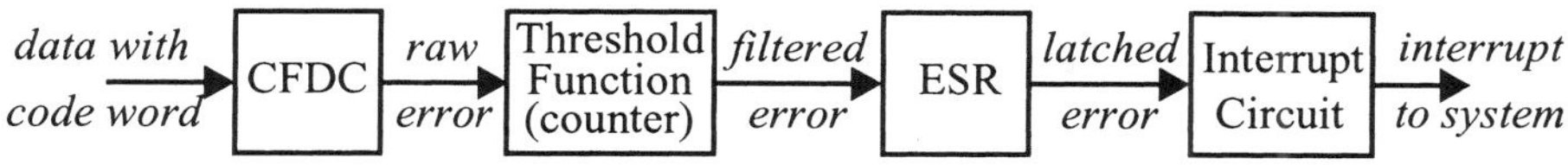

FIGURE 15.3 Threshold function used for error filtering.

insertion of a threshold function between the CFDC and the ESR filters the errors before entering the ESR as illustrated in Figure 15.3. Software routines are sometimes used to threshold these errors but these are generally inefficient in terms of system performance since, not only is the threshold function being performed by the software, but the interrupt must be serviced by reading and clearing the ESR to obtain the information for incrementing the error count.

15.2 Overview of CFDCs

Since a designer's primary objective is the design of the system function, selection of CFDCs and maintenance support functions for specific applications is often of secondary concern (as is often the case for BIST and DFT). The complexity of some EDCs/ECCs (such as Hamming) can be difficult to understand and can make the task of CFDC design more complex and error prone. In this section, a brief overview is given for each of the various CFDCs that are most frequently used in production VLSI and PCB designs.

15.2.1 Parity Circuits

Parity is determined by the number of logic 1s in the data word [145]. The parity code word is a single bit which produces either even or odd parity (an even or odd number of 1s, respectively) across the complete word, including the data bits and the parity bit. As a result, parity is the simplest and most commonly used CFDC in the majority of system applications. It also has the least overhead in terms of information redundancy (only one bit) and complexity of the circuitry for EDC generation and checking. Parity can only detect an odd number of bit errors and cannot provide error correction. Example parity check/regenerate circuits are illustrated in Figure 15.4 for 4-bit parallel data, serial data, and serial/parallel data with a 4-bit parallel component. In the serial and serial/parallel circuits, the parity "sense" (even or odd) is set by the value of the *Force Error* signal whenever the *Serial Control* signal is high, logic 0 for even parity and logic 1 for odd parity. Providing the opposite logic value on *Force Error* will force the opposite parity sense to be calculated and compared at the end of the data word. While *Serial Control* is low, the circuit calculates parity over the incoming data. When *Serial Control* goes high at the end of the data word, the incoming parity bit is blocked from the parity calculation in the serial/parallel circuit and is compared with the newly calculated parity bit to determine if an error exists in the data word. During the parity calculation, while *Serial Control* is low, the output of the comparison exclusive-OR gate is blocked from producing error indications. The newly calculated parity bit is inserted into the outgoing data stream when *Serial Control* is high. The operation of the serial parity circuit is similar to that of the serial/parallel.

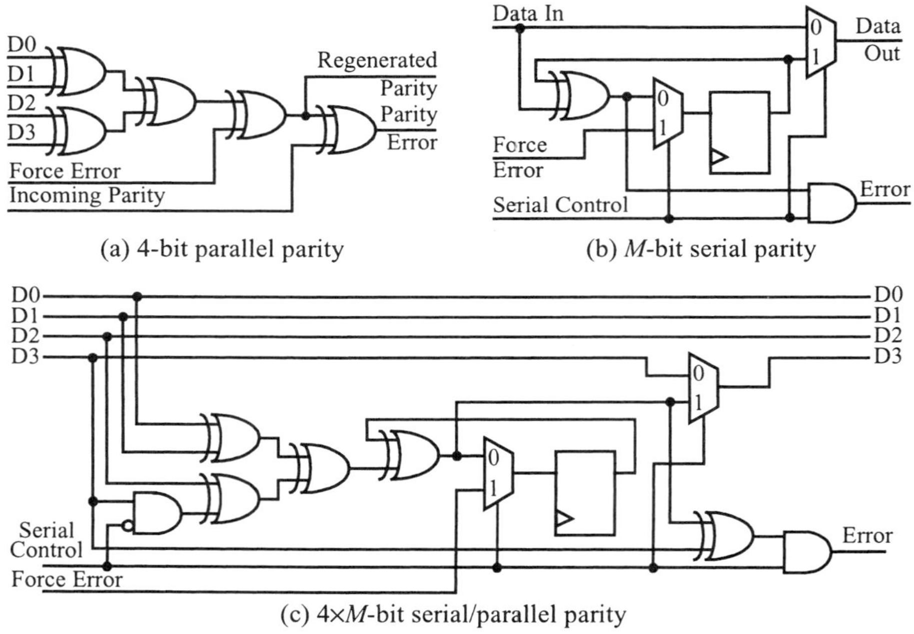

FIGURE 15.4 Parity check/regenerate circuits.

15.2.2 Cyclic Redundancy Check (CRC) Circuits

CRC is capable of multiple error detection including burst errors (consecutive erroneous bits) [122]. CRC requires more code word bits per data word than parity, as well as more complex circuitry to perform the code word regeneration function. CRC is also capable of error correction, but the error correction algorithms are complex and typically not practical for real-time applications [70]. Therefore, the most common applications of CRC are for error detection only. Like signature analysis discussed in

Chapter 5, CRC is based on polynomial division where the polynomial of the data word is divided by the CRC polynomial with the remainder used as the code word. The number of data bits to be serviced by an *N*-bit CRC word can be variable but should be less than 2^N-1 to ensure optimal error detection. The characteristic polynomial can be selected to provide detection of any number of bit errors (greater than the double bit error detection with a primitive polynomial of odd degree) [71]. The design of CRC check and regeneration is best suited for serial data, as illustrated in Figure 15.5. However, a serial/parallel check/regeneration circuit is easily implemented if the *N*×*M* data format coincides with an *N*-bit CRC word where the CRC register is constructed as an *N*-bit MISR. While the *Serial Control* input is low, the data word passes through the multiplexer to the output and, at the same time, goes into the LFSR (constructed as a SAR) for polynomial division by the characteristic polynomial. During this time, mismatches at the output of the lower exclusive-OR gate are blocked from propagating to the ESR. When the *Serial Control* input goes high, the CRC code word produced by the polynomial division in the LFSR is shifted out and inserted into the outgoing data stream. During this time, the regenerated CRC code word is compared bit by bit with the incoming CRC code word for any mismatches that would indicate an error. If the *Force Error* input is high while the *Serial Control* input is high, the regenerated CRC code word will be inverted, producing an error at the check circuit and inserting an incorrect CRC code word into the outgoing data stream. The *Serial Control* input is also used to generate a clear signal to re-initialize the LFSR before processing the next data word. Since CFDCs are often used system wide, as opposed to being self-contained and localized like a BIST approach, it is important to pay attention to the details of the design. This is particularly true in the case of CRC where the designer needs to ensure that the initialization value, characteristic polynomial, and implementation of the LFSR (internal or external feedback) are consistent with the design of other CRC circuits in the system. Otherwise, errors may be encountered in a fault-free system.[1]

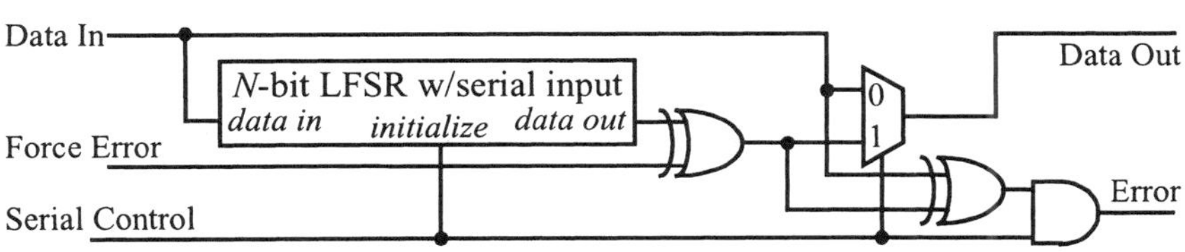

FIGURE 15.5 CRC serial check/regenerate circuit.

1. In one design, I implemented an external feedback LFSR while the other CRC circuits in the system used internal feedback; the problem arose from the ambiguous Type 1/Type 2 LFSR terminology. Fortunately, this design error was discovered early via system-level simulation corrected before the final system where the impact would have been significant.

15.2.3 Checksum Circuits

The basic idea of checksum circuits is to add the binary values of the set of data words in a block transfer of data and to use the sum as the EDC code word. As a result, checksum provides error detection of varying quality (depending on the type of checksum circuit used) but not error correction [122]. Four types of checksum check/(re)generation functions are commonly used: single precision, residue, double precision, and Honeywell. The core circuitry of the code word (re)generation for three of the four checksum circuits were illustrated in Figure 5.7. The circuitry required to implement a checksum CFDC is slightly more complex in terms of the number of gates than the CRC circuit since the checksum circuit requires some type of full adder.

In single precision checksum (Figure 5.7a), all data words are added together with any carry-out information from the addition ignored, while in residue checksum (Figure 5.7b), the carry-out of each addition is used as the carry-in for the next addition. It should be noted that there are two types of residue checksum circuits: 1) the final carry-out bit of the addition of the last data word is added as a carry-in to the final checksum word and 2) the final carry-out bit is ignored which reduces the hardware complexity and the delay through the checksum circuit without seriously effecting the fault detection capabilities. In double precision checksum (Figure 5.7c), the carry-out of each addition is accumulated such that the final checksum word occupies $2N$-bits. In Honeywell (represented by Figure 5.7a but with bus size of $2N$ instead of N), the data words are combined as $2N$-bit words and then added with the carry-outs ignored. In single precision and residue checksum circuits, the checksum word is a single N-bit word while in double precision and Honeywell checksum circuits, the code word is composed of two N-bit words. Therefore, checksum CFDCs are best suited for serial/parallel formats as illustrated by the block diagram in Figure 15.6.

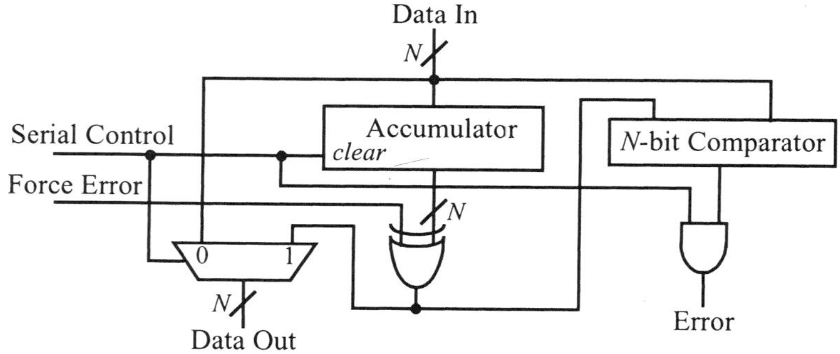

FIGURE 15.6 Checksum serial/parallel check/regenerate circuit.

15.2.4 Hamming Circuits

Hamming code provides not only error detection but also error correction capability based on an extension of the principles of parity. The Hamming code word is constructed from the parity bits of various combinations of the data bits determined by the parity check matrix, as illustrated in Figure 15.7a [145]. Note that the decimal value of each bit position in the parity check matrix corresponds to the binary value of the parity check matrix below (with MSB at the bottom). Also note that the Hamming code bits, H_i, occupy the 2^n positions in the parity check matrix and, as a result, have only a single 1 in any position in the column below the H_i. It is easy to extend this matrix to accommodate any desired size data word with a new Hamming code bit introduced each time a new 2^n value position is encountered. Each Hamming code bit is generated by the exclusive-OR of all the data bits, D_i, that have a 1 in the same row at the corresponding Hamming bit. The Hamming code word regeneration circuit is shown in Figure 15.7b. It should be noted that there are some duplicate gates in the circuit (if one rearranges the inputs to the exclusive-OR gates) that could be removed to reduce the total number of gates. To detect errors, the regenerated Hamming bits, $outH_i$, are compared to the incoming Hamming, inH_i, as shown in the figure.

Bit Position	**1**	**2**	**3**	**4**	**5**	**6**	**7**	**8**	**9**	**10**	**11**	**12**
Hamming/Data Bit	$\mathbf{H_1}$	$\mathbf{H_2}$	$\mathbf{D_1}$	$\mathbf{H_3}$	$\mathbf{D_2}$	$\mathbf{D_3}$	$\mathbf{D_4}$	$\mathbf{H_4}$	$\mathbf{D_5}$	$\mathbf{D_6}$	$\mathbf{D_7}$	$\mathbf{D_8}$
Parity Check Matrix	1	0	1	0	1	0	1	0	1	0	1	0
	0	1	1	0	0	1	1	0	0	1	1	0
	0	0	0	1	1	1	1	0	0	0	0	1
	0	0	0	0	0	0	0	1	1	1	1	1

(a) parity check matrix

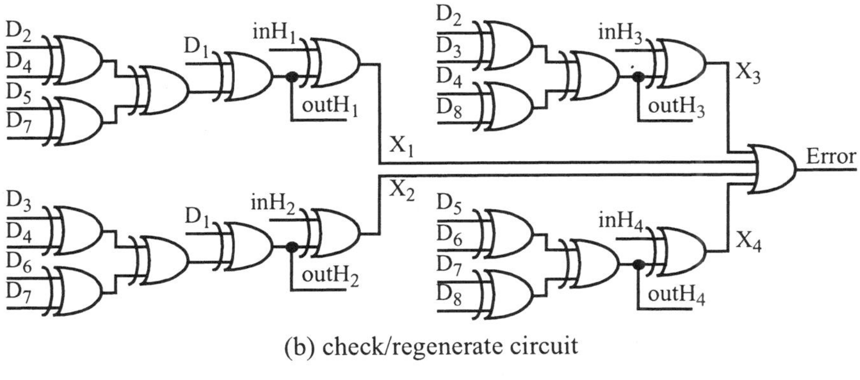

(b) check/regenerate circuit

FIGURE 15.7 8-bit Hamming check/regenerate circuit with parity check matrix.

Hamming circuits are more complex than parity circuits in terms of the number of gates and the number of additional bits required for the code word. However, Hamming code is quite efficient compared to other error correcting codes in terms of the area overhead and performance penalty required for the error correction process. The Hamming circuit models are based on single-bit error-correcting parity codes which use M Hamming bits to detect and correct single-bit errors in N data bits. Given N data bits, the required value of M can be calculated by the relationship $2^M > M+N$. If a single bit error is detected, the error can be corrected in the output data bus and the presence of the error is indicated by the active error signal.

To illustrate the use of Hamming codes for error correction, consider the examples in Figure 15.8 where the data bits are D_1 through D_8 and the Hamming bits are H_1 through H_4. Using the same starting 8-bit data word and its associated error-free Hamming code word obtained from the exclusive-OR network of Figure 15.7b, we begin with a single bit error in the position of D_2 (highlighted in gray) in the example of Figure 15.8a. Regenerating the Hamming bits based on the incoming, incorrect data, we obtain the new set of Hamming bits which can be exclusive-ORed with the

	D_8	D_7	D_6	D_5	D_4	D_3	D_2	D_1	H_4	H_3	H_2	H_1
D_i (correct) & H_i (correct) **sent**	1	1	0	0	1	1	1	0	0	0	1	1
D_i (incorrect) & H_i (correct) **received**	1	1	0	0	1	1	0	0	0	0	1	1
H_i **regenerated**									0	1	1	0
exclusive-OR of received H_i and regenerated $H_i \Rightarrow$ bit position 5 $\Rightarrow D_2$									0	1	0	1

(a) single data bit error

	D_8	D_7	D_6	D_5	D_4	D_3	D_2	D_1	H_4	H_3	H_2	H_1
D_i (correct) & H_i (correct) **sent**	1	1	0	0	1	1	1	0	0	0	1	1
D_i (correct) & H_i (incorrect) **received**	1	1	0	0	1	1	1	0	1	0	1	1
H_i **regenerated**									0	0	1	1
exclusive-OR of received H_i and regenerated $H_i \Rightarrow$ bit position 8 $\Rightarrow H_4$									1	0	0	0

(b) single Hamming bit error

	D_8	D_7	D_6	D_5	D_4	D_3	D_2	D_1	H_4	H_3	H_2	H_1
D_i (correct) & H_i (correct) **sent**	1	1	0	0	1	1	1	0	0	0	1	1
D_i (incorrect) & H_i (correct) **received**	1	1	0	1	1	1	0	0	0	0	1	1
H_i **regenerated**									1	1	1	1
exclusive-OR of received H_i and regenerated $H_i \Rightarrow$ bit position 12 $\Rightarrow D_8$									1	1	0	0

(c) double data bit error

FIGURE 15.8 Examples of Hamming error detection and correction.

incoming Hamming bits to indicate that an error exists. The value 0101, or 5, obtained from the exclusive-OR of the two sets of Hamming bits indicates that the fifth bit position with respect to the ordering of the bits in the parity check matrix in Figure 15.7a, or D_2, is in error such that inverting this bit would correct the error. A similar example is given in Figure 15.8b where the data remains error-free but an error is in one of the Hamming bits. The exclusive-OR of the regenerated Hamming bits with the incoming Hamming bits indicates an error in bit position 8 which, from the parity check matrix in Figure 15.7a, corresponds to Hamming bit H_4. Since the Hamming bits have been regenerated, there is no need for inverting the bit in error in this case. The regenerated Hamming bits can be substituted for the incoming Hamming bits. An example of the error correction circuit that would be used in conjunction with the Hamming code check/regeneration circuit of Figure 15.7b is illustrated in Figure 15.9. This circuit consists of a decoder driving exclusive-OR gates that will invert (correct) the bit indicated by the exclusive-OR of the Hamming bits generated in Figure 15.7b with the incoming Hamming bits.

Double bit errors can be detected but not corrected [145]. This is illustrated in the example in Figure 15.8c which has a double bit error (highlighted in gray). The double bit error is detected by the mismatch in the two sets of Hamming bits but cannot be corrected because the value obtained from the exclusive-OR of the Hamming bits incorrectly indicates that there is a single bit error in bit position 12, or D_8. The occurrence of non-correctable multiple bit errors can be detected by including an additional parity bit over the entire data and code word. Note that we have odd parity over the entire data and code word (in the error-free case) in our example. This will be true of any data word value and its corresponding Hamming bits for a data word of this size; the overall parity sense will change with different size data words but will remain constant for all error-free data words and their associated Hamming bits. By checking parity over the entire word, along with any mismatches in the incoming and regenerated Hamming bits, we can encounter possible cases given in Table 15.1. Under the

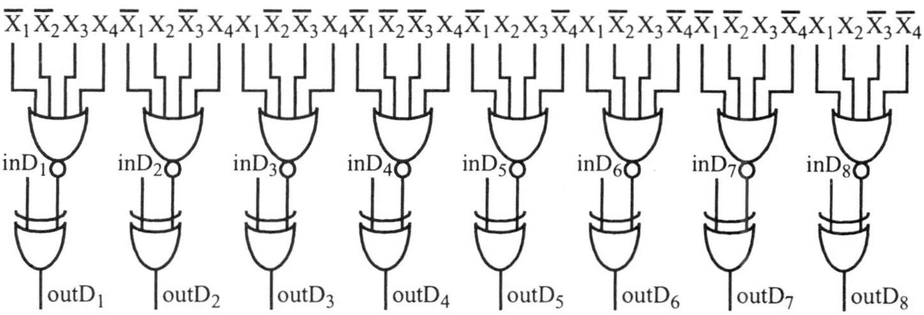

FIGURE 15.9 Error correction circuit for 8-bit Hamming check/regenerate circuit.

conditions that indicate multiple bit errors, the data must be considered to be invalid. It should also be noted that an odd number of multiple bit errors can alias as a correctable single bit error.

TABLE 15.1 Multiple bit error detection with Hamming code.

Comparison of Hamming Bits	Overall Parity	Errors
no mismatch	correct parity	no errors
mismatch	incorrect parity	correctable single bit error
mismatch	correct parity	non-correctable double bit error

15.2.5 Berger and Bose-Lin Circuits

Berger code can detects all multiple unidirectional errors (where the bits in error fail as logic 1s or logic 0s but not both in the same data word) but provides no error correction capability [145]. Berger code circuits are of about the same complexity (in terms of the number of gates) as CRC circuits but are conceptually simpler. The basic idea is to count the number of logic 1s in the data word and use the inverted binary count value as the code word. By inverting the count value for use as the code word, we are able to detect a stuck-at-0 fault on a serial data line since the Berger code bits would be all 1s to indicate an all 0s data word. The number of data bits to be serviced by the Berger code word can be variable but should be less than 2^N, where N is the number of Berger code bits, to ensure optimal error detection. Bose-Lin code is similar to Berger code with the exception that the number of data bits in the data word is greater than 2^N. Therefore, Bose-Lin code uses fewer information redundancy bits and fewer counter bits than Berger code [42]. As opposed to detecting all unidirectional errors in the case of Berger codes, an N-bit Bose-Lin code can detect up to and including all N-bit unidirectional errors.

Berger and Bose-Lin codes are best suited for serial data formats since the area overhead is excessive for data with a parallel component. An example check/regeneration circuit for Berger and Bose-Lin codes is illustrated in Figure 15.10 where an N-bit binary counter is used for the regeneration of the code word. The counter includes an active high *count enable* for counting the number of logic 1s in the data word. In addition, the counter has a shift mode, controlled by the active high *shift* input (driven by the active high *Serial Control* input), which allows the count value to be shifted out for inversion by the exclusive-NOR gate and insertion into the outgoing data stream while it is being compared to the incoming code word for detection of errors. To initialize the counter for the next data word, the shift data input is tied to a logic 0 (or the counter logic can be design to produce logic 0s during the shift mode of operation to avoid undetectable faults caused by tying the *shift data in* to logic 0).

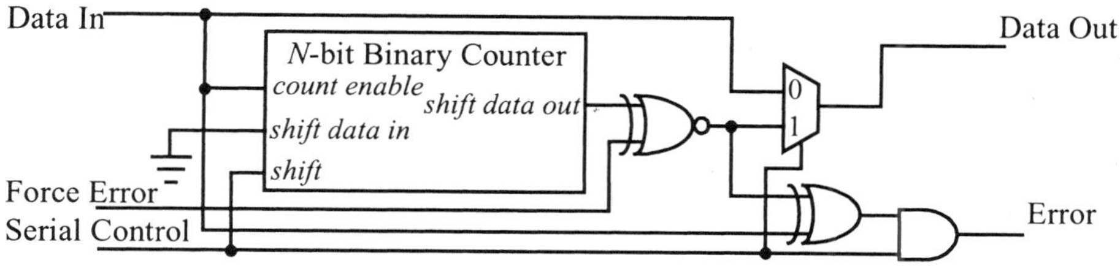

FIGURE 15.10 Berger and Bose-Lin serial check/regenerate circuit.

15.2.6 Comparing CFDCs

The various CFDCs discussed above are compared in Table 15.2 in terms of the number of code word bits, logic gates and flip-flops that would be required to implement the check/regenerate CFDCs. The logic gates are broken into the number of exclusive-OR gates, multiplexers (MUXs), and elementary logic gates (Gates). The parallel data format considered is an N-bit parallel data word (for Hamming code circuits this includes D data bits and H Hamming bits). The serial data format assumes a serial data word less than 2^N-bits. Finally the serial/parallel data format assumes an $N\times(M-1)$-bit or $N\times(M-2)$-bit data word with an N-bit or $2N$-bit code word added, respectively, to obtain a complete $N\times M$-bit data format.

TABLE 15.2 Comparison of area overhead for CFDCs.

CFDC Approach		Data Format	Bits	Flip-Flops	Exclusive-OR gates	MUXs	Gates
Parity		parallel	1	0	N+1	0	0
		serial	1	1	1	2	1
		serial/parallel	1	1	N	2	2
CRC		serial	N	N+1	6	1	2
		serial/parallel	N	N+1	N+5	N	2
Berger		serial	N	N	N+2	N+1	N
Checksum	single precision	serial/parallel	N	N	3N	N	4N-3
	residue	serial/parallel	N	N+1	3N+1	N	4N+3
	double precision	serial/parallel	2N	2N	4N	2N	6N-4
	Honeywell	serial/parallel	2N	3N	6N	2N	8N-6
Hamming	check only	parallel	H	0	H+DH/2	0	1
	check & correct	parallel	H	0	D+H+DH/2	0	D+1
	2-bit check & 1-bit correct	parallel	H+1	0	2D+H+DH/2	0	D+2

Note that some of CFDCs are not suited for certain formats. For example, Hamming is most effective in the parallel mode and the checksum circuits are only suited for serial/parallel mode. The Berger/Bose-Lin code circuits are only suited for serial mode, while the CRC works well for serial or serial/parallel data formats. When selecting a CFDC for a given application, the error detection and/or correction capabilities must be considered for the particular application. While Hamming is not efficient in terms of area overhead and number of code word bits, it is capable of error correction. While CRC is capable of error correction, the error correction algorithms are complex and not practical for hardware implementation. One additional point that should be noted is that the underlying EDC/ECC theory for CFDCs is based on the number of bit errors in the data word (which result from faults) and not on the number of faults (where single faults may produce multiple bit errors).

15.3 Using CFDCs for Off-Line BIST

The maintenance registers provide a hardware interface between the errors detected by the CFDCs and the system software. The functions supported by maintenance registers typically include reading, setting, clearing, masking, and forcing errors as well as providing system interrupts when unmasked errors are active. These functions are essential to system operation and can account for considerable design time and logic area in a VLSI device or on a PCB. The maintenance registers include an ESR to latch errors produced by the CFDCs, a mask register to mask errors in the ESR from the interrupt output, a diagnostic control register to force errors to be produced by the CFDCs, and an interrupt circuit (with or without priority encoding). The ESR, mask register, and diagnostic control register typically have a bit for each CFDC in the circuit or system with a traditional system processor interface or with a Boundary Scan interface for access to the registers.

During normal on-line system operation, the diagnostic control register is not used such that no errors are forced in the CFDCs. During off-line system-level testing, however, the diagnostic control register is used to force errors in the CFDCs in order to test the CFDCs and the interface between the CFDCs and the ESR. Investigations of the fault coverage obtained with CFDCs during normal on-line system operation and off-line system-level testing indicate that improvements can be obtained for both of these testing situations [275]. More importantly, these improvements give valuable insight into the most effective way to merge CFDCs with BIST. To understand these improvements, we consider an example circuit to which CFDCs are commonly applied in switching and transmission system applications. The elastic store (or FIFO) has a RAM structure at its core but also has a considerable amount of additional logic for the write and read pointer circuitry as illustrated in Figure 15.11. The Write and Read Pointers consist of binary counters that are used to access the next RAM address to be written (during a Push operation) or read (during a Pop operation), respectively. The CFDCs are typically applied at the data inputs and outputs of the elastic store to ensure the integrity of the data through the elastic store [275].

The fault coverage obtained for the elastic store circuit alone as well as the elastic store combined with various CFDCs is given in the upper half of the entries in Table 15.3. The *off-line maximum* entries correspond to the maximum fault coverage that one could expect during off-line testing with high data activity applied to the data inputs and random forcing of errors at the code word generate circuit while monitoring the data outputs of the CUT as well as the error output of the CFDC. These same conditions apply to the *off-line* fault coverage entries with the exception that the data outputs of the elastic store are not monitored during the fault simulation (only the CFDC error output is monitored). This corresponds to the fault coverage one would obtain at the system-level where observability is limited to only the CFDC error output. Finally, the *on-line* fault coverage entries imply monitoring only the CFDC error output, without forcing errors, and corresponds to the fault coverage that would be obtained during on-line testing via observability of the interrupt circuit and subsequent reads of the ESR. An inspection of the undetected faults reveals that they are all in the pointer logic and the RAM core of the elastic store is well tested.

When investigating parity, it was observed that by periodically changing the sense of parity in the elastic store (such that no error signal is produced by the check circuit for the fault-free circuit) the on-line fault coverage increased significantly [275]. This was accomplished by including an additional counter bit in the pointer logic (both Write and Read pointers) to control the sense of the parity being written into and read from the RAM core of the elastic store as illustrated by the dashed lines in Figure 15.11. These two additional bits are then used to drive the parity control input

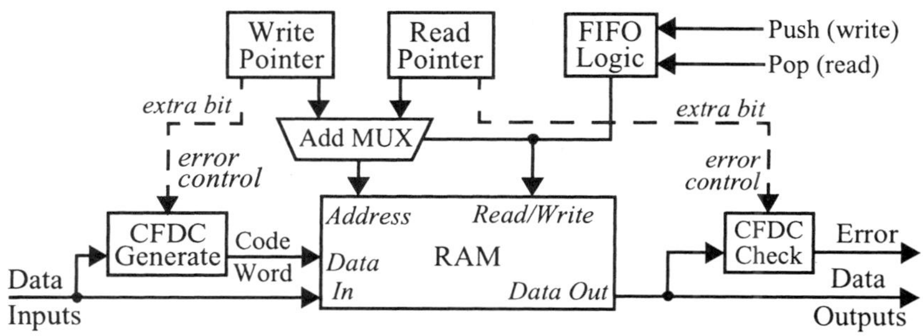

FIGURE 15.11 Elastic store with CFDCs.

of their respective generate/check circuits. By toggling the force error input at both the CFDC generate and check circuits (via the additional pointer bits), errors are not generated by the check circuit as a result of the manner in which the CFDCs in Figure 15.2 are constructed. This small increase in the area overhead to the elastic store circuit leads to a significant increase in both on-line and off-line testing fault coverage (as indicated in Table 15.3) over that of the original application of CFDCs to the elastic store circuit described above. While this technique improves fault coverage for both off-line BIST and on-line testing, there are many cases where it cannot be applied to on-line testing. For example, disturbance of the Hamming code bits could cause the wrong data bit to be inverted such that instead of correcting a single bit error, a double bit error would be introduced. But it is important to note that this technique can always be applied to off-line testing without any problems.

TABLE 15.3 Elastic store fault coverage for CFDCs.

CFDC	Elastic Store			Elastic Store and CFDCs		
	off-line max	off-line	on-line	off-line max	off-line	on-line
normal fault coverage						
Parity	95.6%	95.5%	86.9%	95.8%	95.6%	87.6%
Berger	95.6%	95.4%	87.3%	97.1%	95.0%	88.2%
Checksum	91.1%	91.0%	86.5%	92.3%	90.4%	82.3%
CRC	90.9%	90.8%	88.0%	89.8%	83.9%	79.5%
Hamming	95.4%	95.1%	87.0%	95.4%	95.2%	87.7%
improved fault coverage by periodically forcing errors						
Parity	98.9%	96.8%	96.8%	98.6%	96.7%	96.7%
Berger	99.0%	97.0%	96.8%	97.4%	95.9%	95.8%
Checksum	98.9%	98.8%	95.8%	96.8%	93.7%	91.8%
CRC	98.9%	98.7%	94.5%	96.3%	96.0%	92.3%
Hamming	98.9%	96.8%	96.7%	98.1%	96.3%	96.2%

Because of the high fault coverage obtained with CFDCs, it has been shown that fault coverage is increased by monitoring the error outputs of the CFDCs during off-line BIST [97]. However, as can be seen from the high off-line fault coverage in Table 15.3, systems which incorporate CFDCs may have no need to add BIST circuitry to the circuit under test or the CFDC circuity as long as random data (including control, data, and force error inputs) is applied by other BIST circuitry in the device and the CFDC error output is continuously monitored by other BIST compaction circuitry during off-line testing. By monitoring the CFDC error output only (and not the data outputs of the circuit under test), very little fault coverage is given up as shown in the table and only the CFDC error output needs to be observed or compacted during off-line testing.

During the architectural design process, the system circuitry would be partitioned for on-line testing with no additional BIST circuitry added to the circuitry tested by the CFDCs or the CFDCs themselves. The CFDCs can then be used for off-line testing as long as there is a TPG function to provide high data activity test patterns to all inputs of the circuit covered by the CFDCs, including the force error inputs to the CFDCs. Most BIST TPG techniques produce good pseudo-random patterns that can be applied to the circuit partitions with CFDCs. However, by forcing errors pseudo-randomly during off-line testing, the error outputs from the CFDCs must be monitored continuously since these errors will be actively occurring during off-line testing. The latching action of the typical ESR will not facilitate adequate output response analysis since once the error is latched all observability to the error output of the CFDC is blocked. One option is to use off-line BIST data compaction circuits to compact the error outputs from the CFDCs.

Alternatively, this issue can be resolved by modifying the ESR to operate as a MISR during the BIST mode of operation to facilitate output response compaction of the error outputs from the various CFDCs in the chip or system [275]. This may be a better approach than trying to feed the error outputs to the typical ORA since the ESR already has system access by the maintenance processor, either through a traditional processor interface bus or through a Boundary Scan interface. An example bit-sliced design of this modified ESR is illustrated in Figure 15.12. Note that the exclusive-OR circuitry for implementing the characteristic polynomial of the MISR is not shown in the figure. Since random activation of the force error input produces a significant increase in the fault coverage obtained by the CFDCs during off-line testing, the diagnostic control register can be modified to act as an LFSR during BIST to supply pseudo-random patterns to the force error inputs of the CFDCs. Therefore, the designer need only be concerned with ensuring high data activity during BIST at the

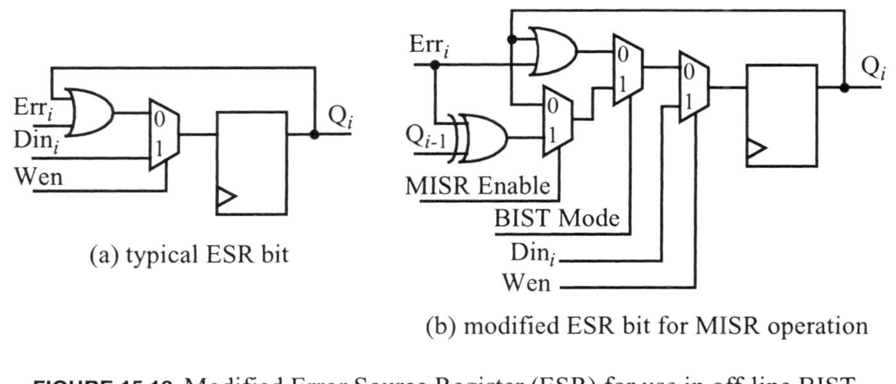

(b) modified ESR bit for MISR operation

FIGURE 15.12 Modified Error Source Register (ESR) for use in off-line BIST.

data inputs to the circuitry covered by the CFDCs. This can be done with traditional BIST TPG techniques.

15.4 On-Line BIST Approaches

There have been a number of approaches to bring BIST into the realm of on-line testing. The basis for the idea is illustrated by the BIST approach in Figure 15.13 which is referred to as concurrent BIST [275]. A TPG and ORA are added to the circuit with an input isolation multiplexer in the same manner as the basic, non-intrusive BIST architecture. The main differences are in the additional comparator, the design of the BIST controller, and the assumption that the CUT is combinational logic. The basic idea for on-line testing is to compare the incoming system data with the test pattern being produced by the TPG. When these patterns match, the ORA is enabled to compact the output response of the CUT and the TPG is enabled to advance to the next test pattern to be applied to the CUT. When there is no match between the incoming system data and the TPG test pattern, the ORA is disabled and the TPG remains in its current state. The on-line testing sequence is complete when the TPG has cycled through all the patterns it is designed to produce, at which time the resultant BIST signature can be read from the ORA to determine the faulty/fault-free status of the CUT. The ORA is re-initialized and the on-line BIST sequence can begin again.

The concurrent BIST approach can also be used for normal off-line testing where the comparison logic is disabled and the test patterns are continuously applied to the CUT via the input isolation multiplexer. It should be noted that the concurrent BIST approach can also be applied to a pipe-lined sequential logic structure that does not contain feedback loops such that the output response is not dependent on previous data passing through the system. In addition, multiplexers can be added to the com-

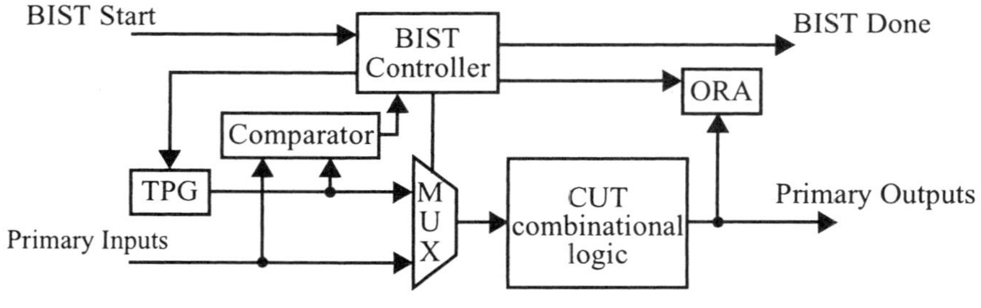

FIGURE 15.13 Concurrent BIST architecture.

parator and ORA to facilitate testing different CUTs in turn for a distributed BIST architecture like that shown in Figure 11.1b.

The potential problem with the concurrent BIST approach as described above is the long time that may be required to cycle through the test patterns of the TPG before the final signature can be obtained to determine if a fault exists [275]. An alternative suggestion has been to incorporate multiple TPGs and ORAs that are at different phases of the test pattern generation cycle, but this incurs significant area overhead. An alternative to waiting for the system data to match the next test pattern from the TPG is to 'steal' unused cycles during normal operation to apply test patterns to the CUT [130]. The idea is that in many systems some form of idle code is inserted into data paths when essential system data is not being processed. However, many systems require continuous data such that there are no idle clock cycles to steal for on-line testing.

To overcome the limitation of combinational logic or feedback-free, pipe-lined sequential CUTs as well as the potentially long test times, another approach uses the opposite edge of the system clock along with multiplexers to test general sequential logic on-line. The basic idea of the approach is illustrated in Figure 15.14. During the active edge of the system clock (rising edge in this example), the normal system data is processed. On the opposite edge of the system clock, the test patterns are applied to the CUT via the multiplexer and compacted in the test logic. The test logic alternately functions as a TPG and an ORA in different test sessions, in much the same manner as in the case of the BILBO. As a result, this BIST approach inherits many of the same problems with test session scheduling and register self-adjacency as the BILBO approach. In addition, since the data has to propagate through the combinational logic in a half clock cycle in order to time-share the combinational logic, the effective data rate of the entire design is doubled. This further limits the practical application of this on-line BIST approach.

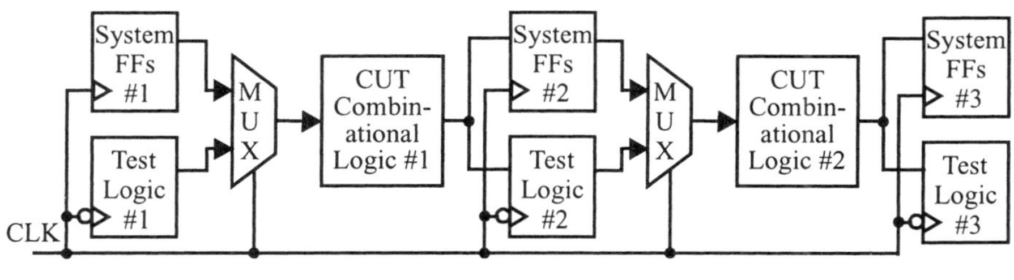

FIGURE 15.14 Another concurrent BIST architecture.

15.5 Benefits and Limitations

The CFDCs that are commonly applied in high reliability, high availability, high maintainability systems are analogous to the ORA structures discussed in Chapter 5. This analogy is summarized in Table 15.3. For those systems that incorporate CFDCs, these circuits can also be used for ORA functions during off-line BIST. High fault coverage can be obtained by continuously monitoring the error outputs of the CFDCs during the BIST mode such that no additional BIST logic is required in the portion of the CUT that is covered by the CFDCs. Of course, one does need to apply test patterns to the data inputs of the portion of the CUT covered by the CFDCs but this can be done by incorporating non-intrusive TPG techniques at the data inputs. In addition, it has been observed that fault coverage is improved by randomly forcing errors in the CFDCs during the testing sequence. This can be accomplished by operating the diagnostic control register as an LFSR during the BIST mode. Continuously monitoring the error outputs of the CFDCs during off-line BIST is best accomplished by designing the ESR to operate as a MISR during BIST. If the CFDCs are put to use during off-line BIST, there is no need to include any BIST circuitry in the CFDCs or the CUTs covered by CFDCs. This can facilitate a significant reduction in the area overhead and performance penalties associated with BIST.

TABLE 15.4 CFDC correlation to ORA structures.

CFDC	ORA Counterpart
Parity	Concentrator Parity Check
CRC	Signature Analysis SAR or MISR
Checksum	Accumulators
Berger Bose-Lin	Counting Techniques
Hamming	Concentrator Parity Check

Conversely, it is worth considering putting off-line BIST structures to use during on-line operation by making them operate as CFDCs during normal on-line operation. This helps to maximize the value and effectiveness of BIST circuitry added to a chip or PCB. The only problem here is that a single CFDC by itself is worthless without a counterpart to generate the code word. Therefore, the commitment to incorporating CFDCs implies selecting and sticking with a specific type throughout the circuit.

Acronyms

ADC	Analog-to-Digital Converter
ASIC	Application Specific Integrated Circuit
ATE	Automatic Test Equipment
ATPG	Automatic Test Pattern Generation
AOI	And-Or Invert
BILBO	Built-In Logic Block Observer
BIST	Built-In Self-Test
BUT	Block Under Test
CA	Cellular Automata
CAD	Computer-Aided Design
CAR	Cellular Automata Register
CBILBO	Concurrent Built-In Logic Block Observer
CED	Concurrent Error Detection
CFDC	Concurrent Fault Detection Circuit
CFSR	Complete Feedback Shift Register

CIP	Configuration Interconnect Point
CRC	Cyclic Redundancy Check
CUT	Circuit Under Test
CPLD	Complex Programmable Logic Device
CSTP	Circular Self-Test Path
DAC	Digital-to-Analog Converter
DFT	Design For Testability
DOT	Defect-Oriented Test
ECC	Error Correction Code
EDC	Error Detection Code
EJTAG	European Joint Test Action Group
EEPLA	Electrically Erasable Programmable Logic Array
ESR	Error Source Register
FB	Feedback
FC	Fault Coverage
FF	Flip-Flop
FIFO	First-In First-Out
FILO	First-In Last-Out
FMA	Failure Mode Analysis
FPGA	Field Programmable Gate Array
FSM	Finite State Machine
HDL	Hardware Description Language
ILA	Iterative Logic Array
I/O	Inputs/Outputs
ISCAS	International Symposium on Circuits And Systems
ISDN	Integrated Services Data Network
ISR	In-System Re-programmable
ITC	International Test Conference

JTAG	Joint Test Action Group
LCAR	Linear Cellular Automata Register
LFSR	Linear Feedback Shift Register
LHCA	Linear Hybrid Cellular Automata
LOCST	LSSD On-Chip Self-Test
LSB	Least Significant Bit
LSSD	Level Sensitive Scan Design
LUT	Look-Up Table
MATS	Modified Algorithmic Test Sequence
MCM	Multi-Chip Module
MICA	Multiple Input Cellular Automata
MISR	Multiple Input Signature Register
MSB	Most Significant Bit
MTTR	Mean Time To Repair
MUX	Multiplexer
NFET	N-channel MOS Field Effect Transistor
NMR	*N*-tuple Modular Redundancy
OAI	Or-And Invert
ORA	Output Response Analyzer
ODC	Output Data Compaction
PCB	Printed Circuit Board
PEST	Pseudo-Exhaustive Self-Test
PFET	P-channel MOS Field Effect Transistor
PLA	Programmable Logic Array
PLB	Programmable Logic Block
PRPG	Pseudo-Random Pattern Generator
RAM	Random Access Memory
ROM	Read Only Memory

RS	Reset/Set
RTL	Register Transfer Level
RTS	Random Test Socket
sa0	stuck-at-0
sa1	stuck-at-1
SAR	Signature Analysis Register
SPOT	Specification-Oriented Test
SRSG	Shift Register Sequence Generator
SST	Simultaneous Self-Test
STAR	Self-Test Area
STUMPS	Self-Test Using MISR/Parallel SRSG
TAP	Test Access Port
TCK	Test Clock
TDI	Test Data In
TDO	Test Data Out
TMR	Triple Modular Redundancy
TMS	Test Mode Select
TPG	Test Pattern Generator
VHDL	VHSIC Hardware Description Language
VHSIC	Very High Speed Integrated Circuit
VLSI	Very Large Scale Integration
WLFSR	Weighted Linear Feedback Shift Register
WUT	Wire Under Test

Bibliography

[1] M. Abadir, "TIGER: Testability Insertion Guidance Expert System," *Proc. IEEE International Conf. on Computer-Aided Design*, 1989, pp. 562-565.

[2] M. Abadir and M. Breuer, "Test Schedules for VLSI Circuits," *IEEE Trans. on Computers*, Vol. 35, No. 4, pp. 361-367, April 1986.

[3] M. Abadir and M. Breuer, "A Knowledge-Based System for Designing Testable VLSI Chips," *IEEE Design & Test of Computers*, Vol. 2, No. 4, pp. 56-68, Aug. 1985.

[4] M. Abramovici, M. Breuer and A. Friedman, *Digital Systems Testing and Testable Design*, Piscataway, New Jersey: IEEE Press, 1994.

[5] M. Abramovici, J. Emmert and C. Stroud, "Roving STARs: An Integrated Approach to On-Line Testing, Diagnosis, and Fault Tolerance for FPGAs in Adaptive Computing Systems," *Proc. NASA/DoD Workshop on Evolvable Hardware*, 2001, pp. 73-92.

[6] M. Abramovici, J. Kulikowski and R. Roy, "The Best Flip-Flops to Scan," *Proc. IEEE International Test Conf.*, 1991, pp. 166-173.

[7] M. Abramovici and P. Menon, "A Practical Approach to Fault Simulation and Test Generation for Bridging Faults," *IEEE Trans. on Computers*, Vol. 34, No. 7, pp. 658-662, July 1985.

[8] M. Abramovici and C. Stroud, "BIST-Based Test and Diagnosis of FPGA Logic Blocks," *IEEE Trans. on VLSI Systems*, Vol. 9, No. 1, pp. 159-172, Feb. 2001.

[9] M. Abramovici and C. Stroud, "BIST-Based Detection and Diagnosis of Multiple Faults in FPGAs," *Proc. IEEE International Test Conf.*, 2000, pp. 785-794.

[10] M. Abramovici, C. Stroud, B. Skaggs and J. Emmert, "Improving BIST-Based Diagnosis for Roving STARs," *Proc. IEEE International On-Line Testing Workshop*, 2000, pp. 31-39.

[11] M. Abramovici, C. Stroud, S. Wijesuriya, C. Hamilton and V. Verma, "Using Roving STARs for On-Line Testing and Diagnosis of FPGAs in Fault-Tolerant Applications," *Proc. IEEE International Test Conf.*, 1999, pp. 973-982.

[12] V. Agarwal, "Multiple Fault Detection in Programmable Logic Arrays," *IEEE Trans. on Computers*, Vol. 29, No. 6, pp. 518-522, June 1980.

[13] V. Agrawal, K. Cheng, D. Johnson and T. Lin, "Designing Circuits with Partial Scan," *IEEE Design & Test of Computers*, Vol. 10, No. 2, pp. 8-15, April 1988.

[14] V. Agrawal, S. Jain and D. Singer, "Automation in Design for Testability," *Proc. Custom Integrated Circuits Conf.*, 1984, pp. 159-163.

[15] V. Agrawal, C. Kime and K Saluja, "A Tutorial on Built-In Self-Test: Principles," *IEEE Design & Test of Computers*, Vol. 10, No. 1, pp. 73-82, March 1993.

[16] V. Agrawal, C. Kime and K Saluja, "A Tutorial on Built-In Self-Test: Applications," *IEEE Design & Test of Computers*, Vol. 10, No. 2, pp. 69-77, June 1993.

[17] V. Agrawal and S. Seth, *Tutorial: Test Generation for VLSI Chips*, Los Alamitos, California, IEEE Computer Society Press, 1988.

[18] P. Agrawal, "Test Generation at Switch Level," *Proc. IEEE International Conf. on Computer-Aided Design*, 1984, pp. 128-130.

[19] S. Akers, "A Parity Bit Signature for Exhaustive Testing," *Proc. IEEE International Test Conf.*, 1986, pp. 48-53.

[20] M. AlShaibi and C. Kime, "Fixed-Biased Pseudorandom Built-In Self-Test for Random Pattern Resistant Circuits," *Proc. IEEE International Test Conf.*, 1994, pp. 929-938.

[21] A. Albicki and A. Krasniewski, "Speed vs. Testability in NMOS VLSI Systems," *Proc. IEEE International Conf. on Computer-Aided Design*, 1984, pp. 105-107.

[22] K. Arabi and B. Kaminska, "Oscillation-Test Strategy for Analog and Mixed-Signal Circuits," *Proc. IEEE VLSI Test Symp.*, 1996, pp. 476-482.

[23] K. Arabi and B. Kaminska, "Oscillation Built-In Self-Test (OBIST) Scheme for Functional and Structural Testing of Analog and Mixed-Signal Integrated Circuits," *Proc. IEEE International Test Conf.*, 1997, pp. 786-795.

[24] L. Avra, "Allocation and Assignment in High-Level Synthesis for Self-Testable Data Paths," *Proc. IEEE International Test Conf.*, 1991, pp. 463-472.

[25] L. Avra and E. McCluskey, "Synthesizing for Scan Dependence in Built-In Self-Testable Designs," *Proc. IEEE International Test Conf.*, 1993, pp. 734-743.

[26] F. Azais, S. Bernard, Y. Bertrand, X. Michel and M. Renovell, "A Low-Cost Adaptive Ramp Generator for Analog BIST Applications," *Proc. IEEE VLSI Test Symp.*, 2001, pp. 266-271.

[27] J. Bailey, C. Stroud, K. Chhor, N. Lau, and W. Orso, "A Method for Testing Partially Programmable Logic Arrays," *Proc. IEEE Automatic Test Conf.*, 2000, pp. 175-180.

[28] A. Balivada, J. Chen and J. Abraham, "Analog Testing with Time Response Parameters," *IEEE Design and Test of Computers*, Vol. 13, No. 2, pp. 18-25, Apr.-June1996.

[29] P. Bardell, "Calculating the Effects of Linear Dependencies on *m*-Sequences Used as Test Stimuli," *Proc. IEEE International Test Conf.*, 1989, pp. 252-256.

[30] P. Bardell, "Analysis of Cellular Automata Used in Pseudorandom Pattern Generation," *Proc. IEEE International Test Conf.*, 1990, pp. 762-768.

[31] P. Bardell and W. McAnney, "Self-Testing of Multi-chip Logic Modules," *Proc. IEEE International Test Conf.*, 1982, pp. 200-204.

[32] P. Bardell, W. McAnney and J. Savir, *Built-In Self-Test for VLSI: Pseudorandom Sequences*, Somerset, New Jersey: John Wiley & Sons, 1987.

[33] Z. Barzilai, J. Savir, G. Markowsky and M. Smith, "VLSI Self-Testing Based on Syndrome Techniques," *Proc. IEEE International Test Conf.*, 1981, pp. 102-109.

[34] J. Bateson, *In-Circuit Testing*, New York: Van Nostrand Reinhold Co., 1985.

[35] E. Behnke, "3B21D BIST/Boundary-Scan System Diagnostic Test Story," *Proc. IEEE International Test Conf.*, 1994, pp. 120-126.

[36] N. Benowitz, D. Calhoun, G. Alderson, J. Bauer and C. Joeckel, "An Advanced Fault Isolation System for Digital Logic," *IEEE Trans. on Computers*, Vol. 24, No. 5, pp. 489-487, May 1975.

[37] J. Bergeron, *Writing Test Benches: Functional Verification of HDL Models*, Boston: Kluwer Academic Publishers, 2000.

[38] M. Bershteyn, "Calculation of Multiple Sets of Weights for Weighted Random Testing," *Proc. IEEE International Test Conf.*, 1993, pp. 1031-1040.

[39] F. Beucler and M/ Manner, "HILDO: the Highly Integrated Logic Device Observer," *VLSI Systems Design*, Vol. 5, No. 6, pp. 88-96, June 1984.

[40] D. Bhavsar and R. Heckelman, "Self-Testing by Polynomial Division," *Proc. IEEE International Test Conf.*, 1981, pp. 208-216.

[41] P. Bose and J. Abraham, "Test Generation for Programmable Logic Arrays," *Proc. ACM/IEEE Design Automation Conf.*, 1982, pp. 574-580.

[42] B. Bose and D. Lin, "Systematic Unidirectional Error Detecting Codes," *IEEE Trans. on Computers*, Vol. 34, No. 11, pp. 1026-1032, Nov. 1985.

[43] M. Breuer, R. Gupta and J. Lien, "Concurrent Control of Multiple BIT Structures," *Proc. IEEE International Test Conf.*, 1988, pp. 431-442.

[44] A. Briers and K. Totton, "Random Pattern Testability by Fast Fault Simulation," *Proc. IEEE International Test Conf.*, 1986, pp. 274-281.

[45] F. Brglez, D. Bryan, and K. Kominski, "Combinational Profiles of Sequential Benchmark Circuits," *Proc. IEEE International Symp. on Circuits and Systems*, 1989, pp. 1929-1934.

[46] A. Brosa and J. Figueras, "Digital Signature Proposal for Mixed-Signal Circuits," *Proc. IEEE International Test Conf.*, 2000, pp. 1041-1050.

[47] O. Brynestad, E. Aas, and A. Vallestad, "State Transition Graph Analysis as a Key to BIST Fault Coverage," *Proc. IEEE International Test Conf.*, 1990, pp. 537-543.

[48] C. Buck, "The Economic Benefits of Test During Burn-In: Real-World Experiences," *Proc. IEEE International Test Conf.*, 1987, pp. 1086-1093.

[49] M. Burns and G. Roberts, *Introduction to Mixed-Signal IC Test and Measurement*, New York: Oxford University Press, 2000.

[50] M. Bushnell and V. Agrawal, *Essentials of Electronic Testing for Digital, Memory and Mixed-Signal VLSI Circuits*, Boston: Kluwer Academic Publishers, 2000.

[51] K. Butler, "Estimating the Economic Benefits of DFT," *IEEE Design & Test of Computers*, Vol. 16, No. 1, pp. 71-79, Jan.-Mar. 1999.

[52] J. Carletta and C. Papachristou, "Structural Constraints of Circular Self-Test Paths," *Proc. IEEE VLSI Test Symp.*, 1994, pp. 87-92.

[53] C. Cha, "A Testing Strategy for PLAs," *Proc. ACM/IEEE Design Automation Conf.*, 1978, pp. 326-331.

[54] J. Chang, C. Tseng, C. Li, M. Purtell and E. McCluskey, "Analysis of Pattern-Dependent and Timing Dependent Failures in an Experimental Test Chip," *Proc. IEEE International Test Conf.*, 1998, pp. 184-193.

[55] A. Chatterjee, B. Kim and N. Nagi, "DC Built-In Self-Test for Linear Analog Circuits," *IEEE Design & Test of Computers*, Vol. 13, No. 2, pp. 26-33, Apr.-June 1996.

[56] P. Chaudhuri, D. Chowdhury, S. Nandi and S. Chattopadhyay, *Additive Cellular Automata Theory and Applications*, Los Alamitos, California: IEEE Computer Society Press, 1997.

[57] C. Chen, "BISTSYN - A Built-In Self-Test Synthesizer," *Proc. IEEE International Conf. on Computer-Aided Design*, 1991, pp. 240-243.

[58] K. Cheng and V. Agrawal, *Unified Methods for VLSI Simulation and Test Generation*, Boston: Kluwer Academic Publishers, 1989.

[59] K. Cheng and V. Agrawal, "A Partial Scan Method for Sequential Circuits with Feedback," *IEEE Trans. on Computers*, Vol. 39, No. 4, pp. 544-548, April 1990.

[60] K. Cheng and C. Lin, "Timing-Driven Test Point Insertion for Full-Scan and Partial-Scan BIST," *Proc. IEEE International Test Conf.*, 1995, pp. 506-514.

[61] B. Chess and T. Larrabee, "Bridge Fault Simulation Strategies for CMOS Integrated Circuits," *Proc. ACM/IEEE Design Automation Conf.*, 1993, pp. 458-462.

[62] C. Chin and E. McCluskey, "Test Length for Pseudorandom Testing," *IEEE Trans. on Computers*, Vol. 36, No. 2, pp. 252-256, Feb. 1987.

[63] D. Clarke and I. Lee, "Testing-Based Analysis of Real-Time System Models," *Proc. IEEE International Test Conf.*, 1996, pp. 894-903.

[64] F. Corno, N. Gaudenzi, P. Prinetto, and M. Reorda, "On the Identification of Optimal Cellular Automata for Built-In Self-Test of Sequential Circuits," *Proc. IEEE VLSI Test Symp.*, 1998, pp. 424-429.

[65] F. Corno, P. Prinetto and M. Reorda, "Making the Circular Self-Test Path Technique Effective for Real Circuits," *Proc. IEEE International Test Conf.*, 1994, pp. 949-957.

[66] F. Corno, P. Prinetto, and M. Reorda, "Circular Self-Test Path for FSMs," *IEEE Design & Test of Computers*, Vol. 13, No. 4, pp. 50-60, Oct.-Dec. 1996.

[67] E. Crane, S. Davidson, J. Kane, C. Stroud and S. Wu, "Trends in Digital Device Test Methodologies," *AT&T Technical J.*, Vol. 73, No. 2, pp. 10-18, Feb. 1994.

[68] C. Craig, C. Kime and K. Saluja, "Test Scheduling and Control for VLSI Built-In Self-Test," *IEEE Trans. on Computers*, Vol. 37, No. 9, pp. 1099-1109, Sept. 1988.

[69] A. Crouch, *Design for Test for Digital ICs and Embedded Core Systems*, Upper Saddle River, New Jersey: Prentice Hall PTR, 1999.

[70] T. Damarla and C. Stroud, "Design of Signature Registers for Double Bit Error Identification," *Proc. IEEE International Symp. on Circuits and Systems*, Vol.4, 1996, pp. 65-68.

[71] T. Damarla, C. Stroud and A. Sathaye, "Multiple Error Detection and Identification via Signature Analysis," *J. of Electronic Testing: Theory and Applications*, Vol. 6, No. 6, pp. 193-207, June 1995.

[72] T. Damarla, C. Stroud, S. Tungate, K. Keyes, W. Su and G. Michael, "Improving the Effectiveness of Circular Built-In Self-Test," *Tech. Dig. Government Microcircuit Applications Conf.*, 1996, pp. 347-350.

[73] D. Das and N. Touba, "Synthesis of Circuits with Low-Cost Concurrent Error Detection Based on Bose-Lin Codes", *Proc. IEEE VLSI Test Symp.*, 1998, pp. 309-315.

[74] B. Davis, *The Economics of Automatic Testing*, London: McGraw Hill, 1982.

[75] I. Dear, C. Dislis, A. Ambler and J. Dick, "Economic Effects in Design and Test," *IEEE Design & Test of Computers*, Vol. 8, No. 4, pp. 64-77, Dec. 1991.

[76] S, Devadas and K. Keutzer, "Design of Integrated Circuits Fully Testable for Delay Faults and Multifaults," *Proc. IEEE International Test Conf.*, 1990, pp. 221-227.

[77] S. Duncan, "A BIST Design Methodology Experiment," *Proc. IEEE International Test Conf.*, 1989, pp. 755-762.

[78] T. Eberle, B. McVay, C. Meyers and J. Moore, "ASIC BIST Synthesis: a VHDL Approach," *Proc. IEEE International Test Conf.*, 1996, pp. 741-750.

[79] E. Eichelberger and E. Lindbloom, "Random-Pattern Coverage Enhancement and Diagnosis for LSSD Logic Self-Test", *IBM J. of Research and Development*, Vol. 27, No. 3, pp. 265-274, May 1983.

[80] E. Eichelberger and T. Williams, "A Logic Design Structure for LSI Testability," *Proc. ACM/IEEE Design Automation Conf.*, 1977, pp. 462-468.

[81] B. Epstein, M. Czigler and S. Miller, "Fault Detection and Classification in Linear Integrated Circuits: An Application of Discrimination Analysis and Hypothesis Testing," *IEEE Trans. on Computer-Aided Design*, Vol. 12, No. 1, pp. 102-113, Jan. 1993.

[82] D. Farren and T. Ambler, "The Economics of System-Level Testing," *IEEE Design & Test of Computers*, Vol. 14, No. 3, pp. 51-58, July-Sept. 1997.

[83] J. Ferguson, "A CMOS Fault Extractor for Inductive Fault Analysis," *IEEE Trans. on Computer-Aided Design*, Vol. 7, No. 11, pp. 1181-1194, Nov. 1988.

[84] F. Ferguson and T. Larrabee, "Test Pattern Generation for Realistic Bridge Faults in CMOS ICs," *Proc. IEEE International Test Conf.*, 1991, pp. 492-499.

[85] F. Frayman, M. Tegethoff and B. White, "Issues In Optimizing the Test Process - A Telecom Case Study," *Proc. IEEE International Test Conf.*, 1996, pp. 800-808.

[86] A. Frisch and T. Almy, "HABIST: Histogram-Based Analog Built-In Self-Test," *Proc. IEEE International Test Conf.*, 1997, pp. 760-767.

[87] H. Fujiwara, "A New PLA Design for Universal Testability," IEEE Trans. on Computers, Vol. 33, No. 8, pp. 745-750, Aug. 1984.

[88] T. Gabara, G. Cyr and C. Stroud, "Metastability of CMOS Master/Slave Flip-Flops," *IEEE Trans on Circuits and Systems*, Vol. 39, No. 10, pp. 734-740, Oct. 1992.

[89] M. Gericota, G Alves, M. Silva and J. Ferreira, "DRAFT: An On-Line Fault Detection Method for Dynamic and Partially Reconfigurable FPGAs," Proc. IEEE International On-Line Testing Workshop, 2001, pp. 34-36.

[90] S. Gerstendorfer and H. Wunderlich, "Minimized Power Consumption for Scan-Based BIST," *Proc. IEEE International Test Conf.*, 1999, pp. 77-84.

[91] P. Girard, L. Guiller, C. Landrault, S. Pravossoudovitch and H. Wunderlich, "A Modified Clock Scheme for a Low Power BIST Test Pattern Generator," *Proc. IEEE VLSI Test Symp.*, 2001, pp. 306-311.

[92] C. Gloster and S. Subramanian, "A Fault Coverage-Driven Partial Scan Chain Selection Technique", *Proc. IEEE International ASIC Conf.*, 1994, pp. 366-369.

[93] C. Gloster and F. Brglez, "Boundary Scan with Cellular-Based Built-In Self-Test," *Proc. IEEE International Test Conf.*, 1988, pp. 138-145.

[94] S. Golomb, *Shift Register Sequences*, Launa Hills, California: Aegean Park Press, 1982.

[95] R. Gulati and C. Hawkins, *I_{DDQ} Testing of VLSI Circuits*, Boston: Kluwer Academic Publishers, 1993.

[96] R. Gupta, R. Gupta and M. Breuer, "A BALLAST Methodology for Structured Partial Scan," *IEEE Trans. on Computers*, Vol. 39, No. 4, pp. 538-544, April 1990.

[97] S. Gupta and D. Pradhan, "Can Concurrent Checkers Help BIST?," *Proc. IEEE International Test Conf.*, 1992, pp. 140-150.

[98] S. Gupta, J. Rajski and J. Tyszer, "Test Pattern Generation Based on Arithmetic Operations", *Proc. IEEE International Conf. on Computer-Aided Design*, 1994, pp. 117-124.

[99] D. Ha and S. Reddy, "On BIST PLAs," *Proc. IEEE International Test Conf.*, 1987, pp. 342-351.

[100] C. Hamilton, G. Gibson, S. Wijesuriya and C. Stroud, "Enhanced BIST-Based Diagnosis of FPGAs via Boundary Scan Access," *Proc. IEEE VLSI Test Symp.*, 1999, pp. 413-418.

[101] I. Harris and R. Tessier, "Interconnect Testing in Cluster-Based FPGA Architectures," *Proc. ACM/IEEE Design Automation Conf.*, 2000, pp. 49-54.

[102] I. Harris and R. Tessier, "Diagnosis of Interconnect Faults in Cluster-Based FPGA Architectures," *Proc. IEEE International Conf. on Computer-Aided Design*, 2000, pp. 472-476.

[103] S. Hassan and E. McCluskey, "Testing PLAs Using Multiple Parallel Signature Analyzers," *Proc. International Fault Tolerant Computing Symp.*, 1983, pp. 422-425.

[104] S. Hassan and E. McCluskey, "Increased Fault Coverage Through Multiple Signatures," *Proc. International Fault Tolerant Computing Symp.*, 1984, pp. 354-359.

[105] S. Hellebrand, S. Tarnick, J. Rajski and B. Courtois, "Generation of Vector Patterns Through Reseeding of Multiple Polynomial Linear Feedback Shift Registers," *Proc. IEEE International Test Conf.*, 1992, pp. 120-129.

[106] S. Hellebrand and H. Wunderlich, "An Efficient Procedure for the Synthesis of Fast Self-Testable Controller Structures," *Proc. IEEE International Conf. on Computer-Aided Design*, 1994, pp. 110-116.

[107] G. Hetherington, T. Fryars, N. Tamarapali, M. Kassab, A. Hassan and J. Rajski, "Logic BIST for Large Industrial Designs: Real Issues and Case Studies," *Proc. IEEE International Test Conf.*, 1999, pp. 358-367.

[108] P. Hortensius, R. McCleod, W. Pries, D. Miller and H. Card, "Cellular Automata-Based Pseudorandom Number Generators for Built-In Self-Test," *IEEE Trans. on Computer-Aided Design*, Vol. 8, No. 8., pp. 842-859, Aug. 1989.

[109] W. Huang and F. Lombardi, "An Approach to Testing Programmable/Configurable Field Programmable Gate Arrays," *Proc. IEEE VLSI Test Symp.*, 1996, pp. 450-455.

[110] W. Huang, F. Meyer, X. Chen and F. Lombardi, "Testing Configurable LUT-Based FPGAs," *IEEE Trans. on VLSI Systems*, Vol. 6, No. 2, pp. 276-283, April 1998.

[111] C. Hudson and G. Peterson, "Parallel Self-Test with Pseudorandom Test Patterns," *Proc. IEEE International Test Conf.*, 1987, pp. 954-963.

[112] G. Huertas, D. Vazquez, A. Rueda and J. Huertas, "Effective Oscillation-Based Test for Application to a DTMF Filter Bank," *Proc. IEEE International Test Conf.*, 1999, pp. 541-555.

[113] J. Hughes and E. McCluskey, "An Analysis of the Multiple Fault Detection Capabilities of Single Stuck-At Fault Test Sets", *Proc. IEEE International Test Conf.*, 1984, pp. 52-58.

[114] G. Illes, "ATE Cost Effectiveness: How Much Performance Can You Afford?," *Proc. IEEE International Test Conf.*, 1987, pp. 1005-1013.

[115] R. Illman, T. Bird, G. Catlow, S. Clarke, L. Theobald and G. Willets, "Built-In Self-Test of the VLSI Content Addressable Filestore," *Proc. IEEE International Test Conf.*, 1991, pp. 37-46.

[116] T. Inoue, S. Miyazaki and H. Fujiwara, "Universal Fault Diagnosis for Lookup Table FPGAs," *IEEE Design & Test of Computers*, Vol. 15, No. 1, pp. 39-44, Jan. 1998.

[117] S. Jain and V. Agrawal, "Modeling and Test Generation Algorithms for MOS Circuits," *IEEE Trans. on Computers*, Vol. 34, No. 5, pp. 426-433, May 1985.

[118] S. Jain and C. Stroud, "Self-Testing of Embedded Memories," *IEEE Design & Test of Computers*, Vol. 3, No. 5, pp. 27-37, Oct. 1986.

[119] A. Jas, C. Krishna and N. Touba, "Hybrid BIST Based on Weighted Pseudo-Random Testing: A New Test Resource," *Proc. IEEE VLSI Test Symp.*, 2001, pp. 2-8.

[120] N. Jarwala and C. Yau, "Achieving Board-Level BIST Using the Boundary Scan Master," *Proc. IEEE International Test Conf.*, 1991, pp. 649-658.

[121] F. Jensen and N. Petersen, *Burn-In*, Chichester, United Kingdom: John Wiley & Sons, 1982.

[122] B. Johnson, *Design and Analysis of Fault Tolerant Digital Systems*, Reading, Massachusetts: Addison-Wesley, 1988.

[123] J. Johnson, "Is DFT Right for You?," *Proc. IEEE International Test Conf.*, 1999, pp. 1090-1097.

[124] W. Jone and C. Papachristou, "A Coordinated Approach to Partitioning and Test Pattern Generation for Pseudoexhaustive Testing," *Proc. ACM/IEEE Design Automation Conf.*, 1989, pp. 525-530.

[125] W. Jone, C. Papachristou and M. Pereina, "A Scheme for Overlaying Concurrent Testing of VLSI Circuits," *Proc. ACM/IEEE Design Automation Conf.*, 1989, pp. 531-536.

[126] C. Jordan and W. Marnane, "Incoming Inspection of FPGAs," *Proc. European Test Conf.*, 1993, pp. 371-377.

[127] J. Kalinowski, A. Albicki and J. Beausang, "Test Control Line Distribution in Self-Testable VLSI Circuits," *Proc. International Conf. on Computer-Aided Design*, 1986, pp. 60-63.

[128] B. Kaminska, K. Arabi, I. Bell, J. Huertas, B. Kim, A. Rueda, M. Soma and P. Goteti, "Analog and Mixed-Signal Benchmark Circuits - First Release," *Proc. IEEE International Test Conf.*, 1997, pp. 183-190.

[129] R. Kapur, S. Patil, T. Snethen and T. Williams, "Design of an Efficient Weighted Random Pattern Generation System," *Proc. IEEE International Test Conf.*, 1994, pp. 491-500.

[130] R. Karri and N. Mukherjee, "Versatile BIST: An Integrated Approach to On-Line/Off-Line BIST," *Proc. IEEE International Test Conf.*, 1998, pp. 910-917.

[131] G. Kiefer and H. Wunderlich, "Using BIST Control for Pattern Generation," *Proc. IEEE International Test Conf.*, 1997, pp. 347-355.

[132] G. Kiefer and H. Wunderlich, "Deterministic BIST with Multiple Scan Chains," *Proc. IEEE International Test Conf.*, 1998, pp. 1057-1064.

[133] K. Kim, D. Ha and J. Tront, "On Using Signature Registers as Pseudorandom Pattern Generators in Built-In Self-Test," *IEEE Trans. on Computer-Aided Design*, Vol. 7, No. 8, pp. 919-928, August 1988.

[134] K. Kim, J. Tront and D. Ha, "Automatic Insertion of BIST Hardware Using VHDL," *Proc. ACM/IEEE Design Automation Conf.*, 1988, pp. 9-15.

[135] K. Kim, J. Tront and D. Ha, "BIDES: A BIST Design Expert System," *J. Electronic Testing: Theory and Applications*, Vol. 2, No. 2, pp. 165-179, June 1991.

[136] H. Klug, "Microprocessor Testing by Instruction Sequences Derived from Random Patterns," *Proc. IEEE International Test Conf.*, 1987, pp. 73-80.

[137] R. Kondagunturi, E. Bradley, K. Maggard, and C. Stroud, "Benchmark Circuits for Analog and Mixed-Signal Testing," *Proc. IEEE Southeast Regional Conf.*, 1999, pp. 217-220.

[138] B. Koenemann, J. Mucha and G. Zwiehoff, "Built-In Logic Block Observation Technique," *Proc. IEEE International Test Conf.*, 1979, pp. 37-41.

[139] B. Koenemann, J. Mucha and G. Zwiehoff, "Built-In Test for Complex Digital Integrated Circuits," *IEEE J. Solid State Circuits*, Vol. 15, No. 3, pp. 315-318, June 1980.

[140] A. Krasniewski and A. Albicki, "Automatic Design of Exhaustively Self-Testing Chips with BILBO Modules," *Proc. IEEE International Test Conf.*, 1985, pp. 363-371.

[141] A. Krasniewski and S. Pilarski, "Circular Self-Test Path: A Low Cost BIST Technique," *Proc. ACM/EEE Design Automation Conf.*, 1987, pp. 407-415.

[142] A. Krasniewski and S. Pilarski, "Circular Self-Test Path: A Low Cost BIST Technique for VLSI Circuits," *IEEE Trans. on Computer-Aided Design*, Vol. 8, No. 1, pp. 46-55, Jan. 1989.

[143] A. Krstic and K. Cheng, *Delay Fault Testing for VLSI Circuits*, Boston: Kluwer Academic Publishers, 1998.

[144] A. Kunzmann and H. Wunderlich, "An Analytical Approach to the Partial Scan Problem," *J. of Electronic Testing: Theory and Applications*, Vol. 1, No. 2, pp. 163-174, May 1990.

[145] P. Lala, *Fault Tolerant and Fault Testable Hardware Design*, London: Prentice Hall International, 1985.

[146] D. Lavo, B. Chess, T. Larrabee, and F. Ferguson, "Diagnosing Realistic Bridging Faults with Single Stuck-At Information," *IEEE Trans. on Computer-Aided Design*, Vol. 17, No. 3, pp. 255-267, March 1998.

[147] D. Lavo, T. Larrabee and B. Chess, "Beyond the Byzantine Generals: Unexpected Behavior and Bridging-Fault Diagnosis," *Proc. IEEE International Test Conf.*, 1994, pp. 611-619.

[148] J. LeBanc, "LOCST: A Built-In Self-Test Technique," *IEEE Design & Test of Computers*, Vol. 1, No. 4, pp. 42-52, Aug. 1984.

[149] S. Lejmi, B. Kaminska and B. Ayari, "Retiming, Resynthesis, and Partitioning for Pseudo-Exhaustive Testing of Sequential Circuits," *Proc. IEEE VLSI Test Symp.*, 1995, pp. 434-439.

[150] B. Lewis, S. Lim, R. Puckett and C. Stroud, "A Prototype Unit for Built-In Self-Test of Analog Circuits," *Proc. IEEE Southeast Regional Conf.*, 1999, pp. 221-224.

[151] J. Li, X. Sun and K. Soon, "Tree-Structured Linear Cellular Automata and Their Applications as PRPGs," *Proc. IEEE International Test Conf.*, 1997, pp. 858-867.

[152] C. Lin, Y. Zorian and S. Bhawmik, "PSBIST: A Partial Scan Based Built-In Self-Test Scheme," *Proc. IEEE International Test Conf.*, 1993, pp. 507-516.

[153] C. Lin, Y. Zorian and S. Bhawmik, "Integration of Partial Scan with Built-In Self-Test," *J. Electronic Testing: Theory and Applications*, Vol. 7, No. 2, pp. 125-137, August 1995.

[154] J. Liu, R. Makki and A. Kayssi, "Dynamic Power Supply Current Testing of CMOS SRAMs," *J. Electronic Testing: Theory and Applications*, Vol. 16, No. 5, pp. 499-511, June 2000.

[155] M. Lubaszewski, S. Mir and L. Pulz, "ABILBO: Analog Built-in Block Observer," *Proc. IEEE International Conf. on Computer-Aided Design*, 1996, pp. 600-603.

[156] S. Ma, I. Shaik, and R. Fetherston, "A Comparison of Bridging Fault Simulation Methods," *Proc. IEEE International Test Conf.*, 1999, pp. 587-595.

[157] K. Maggard and C. Stroud, "Built-In Self-Test for Analog Circuits in Mixed-Signal Systems," *Proc. IEEE Southeast Regional Conf.*, 1999, pp. 225-228.

[158] S. Makar and E. McCluskey, "On the Testing of Multiplexers," *Proc. IEEE International Test Conf.*, 1988, pp. 669-679.

[159] R. Makki, S. Su and T. Nagle, "Transient Power Supply Current Testing of Digital CMOS Circuits," *Proc. IEEE International Test Conf.*, 1995, pp. 892-901.

[160] R. Makki and K. Palaniswami, "Practical Partitioning for Testability with Time-Shared Boundary Scan," *Proc. IEEE International Test Conf.*, 1990, pp. 970-977.

[161] C. Maunder and F. Beenker, "Boundary Scan: A Framework for Structured Design-for-Test," *Proc. IEEE International Test Conf.*, 1987, pp. 714-723.

[162] C. Maunder and R. Tulloss, *The Test Access Port and Boundary Scan Architecture*, Los Alamitos, California: IEEE Computer Society Press, 1990.

[163] P. Maxwell and R. Aitken, "Biased Voting: A Method for Simulating CMOS Bridging Faults in the Presence of Variable Gate Logic Thresholds," *Proc. IEEE International Test Conf.*, 1993, pp. 63-72.

[164] P. Maxwell, R. Aitken, and L. Huismann, "The Effect on Quality of Non-Uniform Fault Coverage and Fault Probability," *Proc. IEEE International Test Conf.*, 1994, pp. 739-746.

[165] P. Maxwell, I. Hartanto and L. Bentz, "Comparing Functional and Structural Tests," *Proc. IEEE International Test Conf.*, 2000, pp. 400-406.

[166] P. Mazumder and J. Patel, "An Efficient Built-In Self Testing for Random Access Memories," *Proc. IEEE International Test Conf.*, 1987, pp. 1072-1077.

[167] P. Mazumder and J. Yih, "A Novel Built-In Self-Repair Approach to VLSI Memory Yield Enhancement," *Proc. IEEE International Test Conf.*, 1990, pp. 833-841.

[168] W. McAnney and J. Savir, "Built-In Checking of the Correct Self-Test Signature," *Proc. IEEE International Test Conf.*, 1986, pp. 54-58.

[169] W. McAnney and J. Savir, "There is Information in Faulty Signatures," *Proc. IEEE International Test Conf.*, 1987, pp. 630-636.

[170] E. McCluskey, *Logic Design Principles with Emphasis on Testable Semicustom Circuits*, Upper Saddle River, New Jersey: Prentice Hall, 1986.

[171] E. McCluskey, "Built-In Self-Test Techniques," *IEEE Design & Test of Computers*, Vol. 2, No. 2, pp. 21-28, March 1985.

[172] E. McCluskey, "Verification Testing - A Pseudoexhaustive Test Technique," *IEEE Trans. on Computers*, Vol. 33, No. 6, pp. 541-546, June 1984.

[173] E. McCluskey, "Quality and Single-Stuck Faults," *Proc. IEEE International Test Conf.*, 1993, pp. 597.

[174] E. McCluskey and S. Bozorgui-Nesbat, "Design for Autonomous Test," *Proc. IEEE International Test Conf.*, 1980, pp. 15-21.

[175] E. McCluskey and S. Bozorgui-Nesbat, "Design for Autonomous Test," *IEEE Trans. on Computers*, Vol. 30, No. 11, pp. 860-875, Nov. 1981.

[176] E. McCluskey and C. Tseng, "Stuck-Fault Tests vs. Actual Defects," *Proc. IEEE International Test Conf.*, 2000, pp. 336-343.

[177] G. McLeod, "BIST Techniques for ASIC Design," *Proc. IEEE International Test Conf.*, 1992, pp. 496-505.

[178] G. McLeod, "Built-In System Test and Fault Location," *Proc. IEEE International Test Conf.*, 1994, pp. 291-299.

[179] K. Mei, "Bridging and Stuck-At Faults," *IEEE Trans. on Computers*, Vol. 23, No. 7, pp. 720-727, July 1974.

[180] A. Meixner and W. Maly, "Fault Modeling for the Testing of Mixed Integrated Circuits," *Proc. IEEE International Test Conf.*, 1991, pp. 564-572.

[181] S. Millman and E. McCluskey, "Detecting Bridging Faults with Stuck-At Test Sets," *Proc. IEEE International Test Conf.*, 1988, pp. 773-783.

[182] J. Miranda, "A BIST and Boundary Scan Economics Framework," *IEEE Design & Test of Computers*, Vol. 14, No. 3, pp. 17-23, July-Sept.r 1997.

[183] S. Mitra and E. McCluskey, "Which Concurrent Error Detection Scheme to Choose?" *Proc. IEEE International Test Conf.*, 2000, pp. 985-994.

[184] S. Mourad and Y. Zorian, *Principles of Testing Electronic Systems*, New York: John Wiley & Sons, 2000.

[185] S. Mourad and E. McCluskey, "Testability of Parity Checkers," *IEEE Trans. on Industrial Electronics*, Vol. 36, No. 2, pp. 254-260, Feb. 1989.

[186] G. Mrugalski, J. Tyszer and J. Rajski, "Synthesis of Pattern Generators Based on Cellular Automata with Phase Shifters," *Proc. IEEE International Test Conf.*, 1999, pp. 368-377.

[187] D. Mukherjee, C. Njinda and M. Breuer, "Synthesis of Optimal 1-Hot Coded On-chip Controllers for BIST Hardware," *Proc. IEEE International Conf. on Computer-Aided Design*, 1991, pp. 236-239.

[188] A Munshi, F. Meyer and F. Lombardi, "A New Method for Testing EEPLAs," *Proc. IEEE International Symp. on Defect and Fault Tolerance in VLSI Systems*, 1998, pp. 146 -154.

[189] F. Muradali, V. Agarwal and B. Nadeau-Dostie, "A New Procedure for Weighted Random Built-In Self-Test," *Proc. IEEE International Test Conf.*, 1990, pp. 660-669.

[190] B. Nadeau-Dostie, *Design for At-Speed Test, Diagnosis and Measurement*, Boston: Kluwer Academic Publishers, 2000.

[191] B. Nadeau-Dostie, D. Burek and A. Hassan, "ScanBIST: A Multi-frequency Scan-based BIST Method," *Proc. IEEE International Test Conf.*, 1992, pp. 506-513.

[192] D. Neebel and C. Kime, "Multiple Weighted Cellular Automata", *Proc. IEEE VLSI Test Symp.*, 1994, pp. 81-86.

[193] M. Nicolaidis, Y. Zorian and D. Pradhan, *On-Line Testing for VLSI*, Boston: Kluwer Academic Publishers, 1998.

[194] P. Nigh and A. Gattiker, "Test Method Evaluation Experiments and Data," *Proc. IEEE International Test Conf.*, 2000, pp. 454-463.

[195] D. Niggemeyer and E. Rudnick, "Automatic Generation of Diagnostic March Tests," *Proc. IEEE VLSI Test Symp.*, 2001, pp. 299-304.

[196] J. Oldfield and R. Dorf, *Field Programmable Gate Arrays*, New York: Wiley Interscience, 1995.

[197] W. Olson and W. Chu, "Implementing Built-In Self-Test in a RISC Microprocessor," *Proc. IEEE International Test Conf.*, 1987, pp. 810-817.

[198] M. Ohletz, T. Williams and J. Mucha, "Overhead in Scan and Self-Testing Designs," *Proc. IEEE International Test Conf.*, 1987, pp. 460-470.

[199] M. Ohletz, "Hybrid Built-In Self-Test for Mixed Analog/Digital Integrated Circuits," *Proc. European Test Conf.*, 1991, pp. 307-316.

[200] A. Osseiran, *Analog and Mixed-Signal Boundary Scan*, Boston: Kluwer Academic Publishers, 1999.

[201] D. Ostapko and S. Hong, "Fault Analysis and Test Generation for Programmable Logic Arrays," *IEEE Trans. on Computers*, Vol. 28, No. 9, pp. 617-627, Sept. 1979.

[202] C. Ouyang and W. Maly, "Efficient Extraction of Critical Area in Large VLSI ICs," *Proc. International Symp. on Semiconductor Manufacturing*, 1996, pp. 301-304.

[203] J. Pabst, "Elements of VLSI Production Test Economics," *Proc. IEEE International Test Conf.*, 1987, pp. 982-986.

[204] C. Pan and K. Cheng, "Pseudo-Random Testing and Signature Analysis for Mixed-Signal Systems," *Proc. IEEE International Conf. on Computer-Aided Design*, 1995, pp. 102-107.

[205] C. Pan and K. Cheng, "Implicit Functional Testing for Analog Circuits," *Proc. IEEE VLSI Test Symp.*, 1996, pp. 489-494.

[206] K. Parker, *The Boundary Scan Handbook,* Boston: Kluwer Academic Publishers, 1998.

[207] W. Peterson and E. Weldon, *Error Correcting Codes*, New York: John Wiley & Sons, 1972.

[208] S. Pilarski, A. Krasniewski and T. Kameda, "Estimating Testing Effectiveness of the Circular Self-Test Path Technique," *IEEE Trans. on Computer-Aided Design*, Vol. 11, No. 10, pp. 1301-1316, Oct. 1992.

[209] D. Pok, C. Chen, J. Schamus, C. Montgomery and J. Tsui, "Chip Design for Monobit Receiver," *IEEE Trans. on Microwave Theory and Techniques*, Vol. 45, No. 12, pp. 2283-8895, Dec. 1997.

[210] I. Polian and B. Becker, "Multiple Scan Chain Design for Two-Pattern Testing," *Proc. IEEE VLSI Test Symp.*, 2001, pp. 88-93.

[211] I. Pomeranz and S Reddy, "The Multiple Observation Time Test Strategy," *IEEE Trans. on Computers*, Vol. 41, No. 5, pp. 627-637, May 1992.

[212] T. Powell, F. Hii and D. Cline, "A 256Meg SDRAM BIST for Disturb Test Application," *Proc. IEEE International Test Conf.*, 1997, pp. 200-208.

[213] D. Pradhan and K. Son, "The Effect of Undetectable Faults in PLAs and a Design for Testability," *Proc. IEEE International Test Conf.*, 1980, pp. 359-367.

[214] M. Pradhan, E. Obrien, S. Lam and J. Beausang, "Circular BIST with Partial Scan," *Proc. IEEE International Test Conf.*, 1988, pp. 719-729.

[215] C. Pynn, "Analyzing Manufacturing Test Costs," *IEEE Design & Test of Computers*, Vol. 14, No. 3, pp. 36-40, July-Sept. 1997.

[216] J. Rajski, G. Mrugalski and J. Tyszer, "Comparative Study of CA-based PRPGs and LFSRs with Phase Shifters," *Proc. IEEE VLSI Test Symp.*, 1999, pp. 236-245.

[217] J. Rajski, N. Tamarapalli and J. Tyszer, "Automated Synthesis of Large Phase Shifters for Built-In Self-Test," *Proc. IEEE International Test Conf.*, 1998, pp. 1047-1056.

[218] J. Rajski and J. Tyszer, "Design of Phase Shifters for BIST Applications," *Proc. IEEE VLSI Test Symp.*, 1998, pp. 218-224.

[219] J. Rajski and J. Tyszer, *Arithmetic Built-In Self-Test for Embedded Systems*, Englewood Cliffs, New Jersey: Prentice-Hall PTR, 1998.

[220] R. Rajsuman, "A New Testing Method for EEPLA," *IEEE Trans. on Computer-Aided Design*, Vol. 13, No. 7, pp. 935-939, July 1994.

[221] M. Raposa, "Dual Port Static RAM Testing," *Proc. IEEE International Test Conf.*, 1988, pp. 362-368.

[222] E. Ratazzi, "Fault Coverage Measurement for Analog Circuits," *Proc. IEEE Custom Integrated Circuits Conf.*, 1992, pp. 17.2.1-4.

[223] I. Rayane, J. Velasco-Medina and M. Nicolaidis, "A Digital BIST for Operational Amplifiers Embedded in Mixed-Signal Systems," *Proc. IEEE VLSI Test Symp.*, 1999, pp. 304-310.

[224] D. Raymond, "In-Circuit and Functional ATE: What's the Real Difference?," *Proc. IEEE International Test Conf.*, 1980, pp. 315-320.

[225] J. Rearick and J. Patel, "Fast and Accurate Bridging Fault Simulation," *Proc. IEEE International Test Conf.*, 1993, pp. 54-62.

[226] M. Renovell, J. Portal, J. Figueras and Y. Zorian, "SRAM-based FPGA: Testing the LUT/RAM Modules," *Proc. IEEE International. Test Conf.*, 1998, pp. 1102-1111.

[227] M. Renovell, J. Portal, J. Figueras and Y. Zorian, "Testing the Interconnect of RAM-Based FPGAs," *IEEE Design & Test of Computers*, Vol. 15, No. 1, pp. 45-50, Jan.-March 1998.

[228] M. Renovell, J. Portal, J. Figueras and Y. Zorian, "SRAM-based FPGA: Testing the Embedded RAM Modules," *J. of Electronic Testing: Theory and Applications*, Vol. 14, No. 1, pp. 159-167, Jan. 1999.

[229] D. Resnick, "Testability and Maintainability with a New 6K Gate Array," *VLSI Design*, Vol. 4, No. 2, pp. 34-38, March 1983.

[230] H. Ritter and B. Muller, "Built-In Test Processor for Self-Testing Repairable Random Access Memories," *Proc. IEEE International Test Conf.*, 1987, pp. 1078-1084.

[231] G. Roberts and A. Lu, *Analog Signal Generation for Built-In Self-Test of Mixed-Signal Integrated Circuits*, Boston: Kluwer Academic Publishers, 1995.

[232] G. Robinson, "DFT, Test Lifecycles and the Product Lifecycle," *Proc. IEEE International Test Conf.*, 1999, pp. 705-713.

[233] E. Rudnick, J. Patel and I. Pomeranz, "On Potential Fault Detection in Sequential Circuits," *Proc. IEEE International Test Conf.*, 1996, pp. 142-149.

[234] A. Russ and C. Stroud, "Non-Intrusive Built-In Self-Test for FPGA and MCM Applications," *Proc. IEEE Automatic Test Conf.*, 1995, pp. 480-485.

[235] K. Saluja, R. Sharma and C. Kime, "Concurrent Comparative Testing Using BIST Resources," *Proc. IEEE International Conf. on Computer-Aided Design*, 1987, pp. 336-339.

[236] R. Sankaralingam, B. Pouya and N. Touba, "Reducing Power Dissipation During Test Using Scan Chain Disable," *Proc. IEEE VLSI Test Symp.*, 2001, pp. 319-324.

[237] V. Sar-Dessai and D. Walker, "Resistive Bridge Fault Modeling, Simulation, and Test Generation," *Proc. IEEE International Test Conf.*, 1999, pp. 596-605.

[238] J. Savir, "Distributed Generation of Weighted Random Patterns," *Proc. IEEE VLSI Test Symp.*, 1998, pp. 225-232.

[239] J. Savir and W. McAnney, "Identification of Failing Tests with Cycling Registers," *Proc. IEEE International Test Conf.*, 1988, pp 322-327.

[240] J. Savir and W. McAnney, "A Multiple Seed Linear Feedback Shift Register," *Proc. IEEE International Test Conf.*, 1990, pp. 657-659.

[241] J. Sayah and C. Kime, "Test Scheduling for High Performance VLSI System Implementations," *Proc. IEEE International Test Conf.*, 1988, pp. 421-430.

[242] N. Saxena and E. McCluskey, "Extended Precision Checksums," *Proc. International Symp. on Fault-Tolerant Computing*, 1987, pp. 142-147.

[243] T. Schuele and A. Stroele, "Test Scheduling for Minimal Energy Consumption under Power Constraints," *Proc. IEEE VLSI Test Symp.*, 2001, pp. 312-318.

[244] R. Sedmak, "Design for Self-Verification: An Approach for Dealing with Testability Problems in VLSI-Based Designs," *Proc. IEEE International Test Conf.*, 1979, pp. 112-124.

[245] R. Sedmak, "Implementation Techniques for Self-Verification," *Proc. IEEE International Test Conf.*, 1980, pp. 267-278.

[246] R. Sedmak, "Boundary Scan: Beyond Production Test," *Proc. IEEE VLSI Test Symp.*, 1994, pp. 415-420.

[247] M. Segers, "A Self-Test Method for Digital Circuits," *Proc. IEEE International Test Conf.*, 1981, pp. 79-85.

[248] Semiconductor Industry Association, *International Technology Roadmap for Semiconductors: 1999 Edition*, Austin, TX: SEMATECH, 1999.

[249] M. Serra, T. Slater, J. Muzio and D. Miller, "Analysis of One Dimensional Linear Cellular Automata and Their Aliasing Properties," *IEEE Trans. on Computer-Aided Design*, Vol. 9, No. 7, pp. 767-778, July 1990.

[250] I. Shperling and E. McCluskey, "Circuit Segmentation for Pseudo-Exhaustive Testing via Simulated Annealing," *Proc. IEEE International Test Conf.*, 1987, pp. 58-66.

[251] D. Siewiorek and R. Swarz, *The Theory and Practice of Reliable System Design*, Bedford, Massachusetts: Digital Press, Digital Equipment Corp., 1982.

[252] M. Sivaraman and A. Strojwas, *A Unified Approach for Timing Verification and Delay Fault Testing*, Boston: Kluwer Academic Publishers, 1998.

[253] F. Siavoshi, "WTPGA: A Novel Weighted Test Pattern Generation Approach for VLSI Built-In Self-Test," *Proc. IEEE International Test Conf.*, 1988, pp. 256-262.

[254] L. Sigal and C. Kime, "Concurrent Off-Phase Built-In Self-Test of Dormant Logic," *Proc. IEEE International Test Conf.*, 1988, pp. 934-941.

[255] K. Skahill, *VHDL for Programmable Logic*, Reading, Massachusetts: Addison-Wesley, 1996.

[256] J. Smith, "Detection of Faults in Programmable Logic Arrays," *IEEE Trans. on Computers*, Vol. 28, No. 11, pp. 848-853, Nov. 1979.

[257] M. Soma, "A Design For Test Methodology for Active Analog Filter," *Proc. IEEE International Test Conf.*, 1993, pp. 183-192.

[258] "Standard Test Access Port and Boundary-Scan Architecture," *IEEE Standard P1149.1*, 1990.

[259] A. Steininger and C. Scherrer, "On the Necessity of On-Line BIST in Safety Critical Applications," *Proc. Fault-Tolerant Computing Symp.*, 1999, pp. 208-215.

[260] A. Stroele, "Self-Test Scheduling with Bounded Test Execution Time," *Proc. IEEE International Test Conf.*, 1992, pp. 130-139.

[261] A. Stroele, "Test Response Compaction Using Arithmetic Functions," *Proc. IEEE VLSI Test Symp.*, 1996, pp. 380-386.

[262] A. Stroele, "Bit Serial Pattern Generation and Response Compaction Using Arithmetic Functions," *Proc. IEEE VLSI Test Symp.*, 1998, pp. 78-84.

[263] A. Stroele and H. Wunderlich, "Configuring Flip-Flops to BIST Registers," *Proc. IEEE International Test Conf.*, 1994, pp. 939-938.

[264] C. Stroud, "An Automated Built-In Self-Test Approach for General Sequential Logic Synthesis," *Proc. of ACM/IEEE Design Automation Conf.*, 1988, pp. 3-8.

[265] C. Stroud, "Automated Built-In Self-Test for Sequential Logic Synthesis," *IEEE Design & Test of Computers*, Vol. 5, No. 6, pp. 22-32, Dec. 1988.

[266] C. Stroud, "Built-In Self-Test for High-Speed Data-Path Circuitry," *Proc. IEEE International Test Conf.*, 1991, pp. 47-56.

[267] C. Stroud, J. Bailey, and J. Emmert, "A New Method for Testing Re-Programmable PLAs," *J. Electronic Testing: Theory and Applications*, Vol. 11, No. 6, pp. 635-640, Dec. 2000.

[268] C. Stroud and A. Barbour, "Testability and Test Generation for Majority Voting Fault-Tolerant Circuits," *J. Electronic Testing: Theory and Applications*, Vol. 4, No. 3, pp. 201-214, March 1993.

[269] C. Stroud and A. Barbour, "Design for Testability and Test Generation for Static Redundancy System Level Fault Tolerant Networks," *Proc. of IEEE International Test Conf.*, 1989, pp. 812-818.

[270] C. Stroud, P. Chen, S. Konala and M. Abramovici, "Evaluation of FPGA Resources for Built-In Self-Test of Programmable Logic Blocks," *Proc. ACM International Symp. on FPGAs*, 1996, pp. 107-113.

[271] C. Stroud and G. Cyr, "Built-In Self-Test for Embedded RAMs, ROMs, and PLAs in Custom VLSI," *Proc. National Communications Forum*, 1986, pp. 1176-1179.

[272] C. Stroud and T. Damarla, "Improving the Efficiency of Error Identification via Signature Analysis," *Proc. IEEE VLSI Test Symp.*, 1995, pp 244-249.

[273] C. Stroud and T. Damarla, "Efficient Error Bit Identification from Failing Signatures," *Proc. IEEE International ASIC Conf.*, 1996, pp. 259-262.

[274] C. Stroud, M. Ding, S. Seshadri, I. Kim, S. Roy, S. Wu and R. Karri, "A Parameterized VHDL Library for On-Line Testing," *Proc. IEEE International Test Conf.*, 1997, pp. 479-488.

[275] C. Stroud, M. Ding, W. Long, Y. Yang, R. Karri and S. Wu, "Maximizing the Effectiveness of On-Line Testing Functions," *Proc. IEEE International On-Line Testing Workshop*, 1998, pp. 47-51.

[276] C. Stroud, J. Emmert and J. Bailey, "A New Bridging Fault Model for More Accurate Fault Behavior," *Proc. IEEE Automatic Test Conf.*, 2000, pp. 481-485.[1]

[277] C. Stroud, J. Emmert, J. Bailey, K. Chhor and D. Nickolic, "Bridging Fault Extraction from Physical Design Data for Manufacturing Test Development," *Proc. IEEE International Test Conf.*, 2000, pp. 760-769.

[278] C. Stroud, P. He and T. Damarla, "Register Size vs. Fault Coverage in Modified Circular Built-In Self-Test," *Proc. IEEE Automatic Test Conf.*, 1996, pp. 23-28.

[279] C. Stroud, P. Karunaratna and E. Bradley, "Digital Components for Built-In Self-Test of Analog Circuits," *Proc. IEEE International ASIC Conf.*, 1997, pp. 47-51.

[280] C. Stroud, S. Konala, P. Chen and M. Abramovici, "Built-In Self-Test for Programmable Logic Blocks in FPGAs," *Proc. IEEE VLSI Test Symp.*, 1996, pp. 387-392.

[281] C. Stroud, M. Lashinsky, J. Nall, J. Emmert and M. Abramovici, "On-Line BIST and Diagnosis of FPGA Interconnect Using Roving STARs," *Proc. IEEE International On-Line Testing Workshop*, 2001, pp. 31-39.

[282] C. Stroud, E. Lee and M. Abramovici, "BIST-Based Diagnostics of FPGA Logic Blocks," *Proc. IEEE International Test Conf.*, 1997, pp. 539-547.

[283] C. Stroud, E. Lee, S. Konala and M. Abramovici, "Selecting Built-In Self-Test Configurations for Field Programmable Gate Arrays," *Proc. IEEE Automatic Test Conf.*, 1996, pp. 29-35.

[284] C. Stroud, E. Lee, S. Konala and M. Abramovici, "Using ILA Testing for BIST in FPGAs," *Proc. IEEE International Test Conf.*, 1996, pp. 68-75.

1. IEEE Automatic Test Conf. may be more commonly known to some as AUTOTESTCON.

[285] C. Stroud and G. Liang, "Design Verification Techniques for System Level Testing Using ASIC Level BIST Implementations," *Proc. IEEE International ASIC Conf.*, 1993, pp. 140-143.

[286] C. Stroud, R. Munoz and D. Pierce, "Behavioral Model Synthesis with CONES," *IEEE Design & Test of Computers*, Vol. 5, No. 3, pp. 22-30, June 1988.

[287] C. Stroud, R. Munoz and D. Pierce, "CONES: A System for Automated Synthesis of VLSI and Programmable Logic from Behavioral Models," *Proc. IEEE International Conf. on Computer-Aided Design*, 1986, pp. 428-431.

[288] C. Stroud and C. Ryan, "Multiple Fault Simulation with Random and Cluster Fault Injection Capabilities," *Proc. IEEE International ASIC Conf.*, 1995, pp. 218-221.

[289] C. Stroud and R. Shaw, "An ASIC Level Built-In Self-Test Implementation for System Level Testing," *Proc. IEEE International ASIC Conf.*, 1991, pp. 641-644.

[290] C. Stroud and J. Tannehill, "Applying Built-In Self-Test to Majority Voting Fault Tolerant Circuits," *Proc. IEEE VLSI Test Symp.*, 1998, pp. 303-308.

[291] C. Stroud, S. Wijesuriya, C. Hamilton and M. Abramovici, "Built-In Self-Test of FPGA Interconnect," *Proc. IEEE International Test Conf.*, 1998, pp. 404-411.

[292] C. Su and C. Kime, "Computer-Aided Design of Pseudoexhaustive BIST for Semiregular Circuits," *Proc. IEEE International Test Conf.*, 1990, pp. 680-689.

[293] S. Su and R. Makki, "Testing of Random Access Memories by Monitoring Dynamic Power Supply Current," *J. Electronic Testing: Theory and Applications*, Vol. 3, No. 3, pp. 265-278, March 1992.

[294] S. Su, R. Makki and T. Nagle, "Transient Power Supply Current Monitoring: A New Test Method for CMOS VLSI Circuits," *J. Electronic Testing: Theory and Applications*, Vol. 6, No. 1, pp. 23-43, Jan. 1995.

[295] D. Suk and S. Reddy, "A March Test for Functional Faults in Semiconductor Random-Access Memories," *IEEE Trans. on Computers*, Vol. 30, No. 12, pp. 982-985, Dec. 1981.

[296] M. Sullivan and C. Stroud, "Reducing the Cost of Circular Built-In Self-Test by Selective Flip-Flop Replacement," *Proc. IEEE Automatic Test Conf.*, 1995, pp. 486-491.

[297] X. Sun and M. Serra, "Merging Concurrent Checking and Off-Line BIST," *Proc. IEEE International Test Conf.*, 1992, pp. 958-967.

[298] X. Sun, J. Xu, B. Chan and P. Trouborst, "Novel Technique for BIST of FPGA Interconnects," *Proc. IEEE International Test Conf.*, 2000, pp. 795-803

[299] S. Sunter and N. Nagi, "Test Metrics for Analog Parametric Faults," *Proc. IEEE VLSI Test Symp.*, 1999, pp. 226-234.

[300] S. Sunter and A. Roy, "BIST for Phase-Locked Loops in Digital Applications," *Proc. IEEE International Test Conf.*, 1999, pp. 532-540.

[301] N. Tamarapalli and J. Rajski, "Constructive Multi-Phase Test Point Insertion for Scan-Based BIST," *Proc. IEEE International Test Conf.*, 1996, pp. 649-658.

[302] A. Thaker, H. Sucar and P. Gelsinger, "INTEL's Test Quality System," *Proc. IEEE International Test Conf.*, 1987, pp. 1094-1097.

[303] D. Thompson, T. Gabara and C. Stroud, "A 180 MHz ASIC for High Speed Interfaces," *Proc. IEEE International Solid-State Circuits Conf.*, 1991, pp. 140-141.

[304] N. Touba and E. McCluskey, "Transformed Pseudo-Random Patterns for BIST," *Proc. IEEE VLSI Test Symp.*, 1995, pp. 410-416.

[305] N. Touba and E. McCluskey, "Synthesis of Mapping Logic for Generating Transformed Pseudo-Random Patterns for BIST," *Proc. IEEE International Test Conf.*, 1995, pp. 674-682.

[306] N. Touba and E. McCluskey, "Altering a Pseudo-Random Bit Sequence for Scan Based BIST," *Proc. IEEE International Test Conf.*, 1996, pp. 167-175.

[307] N. Touba and E. McCluskey, "Test Point Insertion Based on Path Tracing," *Proc. IEEE VLSI Test Symp.*, 1996, pp. 2-8.

[308] M. Toner and G. Roberts, "A BIST Scheme for an SNR, Gain Tracking, and Frequency Response Test of a Sigma-Delta ADC," *IEEE Trans. on Circuits and Systems II*, Vol. 41, No. 12, pp. 1-15, Dec. 1995.

[309] R. Truer, H. Fijiwara and V. Agarwal, "Implementing a Built-In Self-Test PLA Design," *IEEE Design & Test of Computers*, Vol. 2, No. 2, pp. 37-48, April 1985.

[310] H. Tsai, K. Cheng, C. Lin and S. Bhawmik, "A Hybrid Algorithm for Test Point Selection for Scan-Based BIST," *Proc. ACM/IEEE Design Automation Conf.*, 1997, pp. 478-483.

[311] H. Tsai, S. Bhawmik and K. Cheng, "An Almost Full-Scan BIST Solution - Higher Fault Coverage and Shorter Test Application Time," *Proc. IEEE International Test Conf.*, 1998, pp. 1065-1073.

[312] K. Tsai, S. Hellebrand, J. Rajski and M. Sadowsak, "STARBIST: Scan Autocorrelated Random Pattern Generation," *Proc. ACM/IEEE Design Automation Conf.*, 1997, pp. 472-477.

[313] C. Tseng, S. Mitra, S. Davidson and E. McCluskey, "An Evaluation of Pseudo-Random Testing for Detecting Real Defects," *Proc. IEEE VLSI Test Symp.*, 2001, pp. 404-409.

[314] S. Tungate, P. He, S. Seshadri, C. Stroud, M. Sullivan and T. Damarla, "Design Automation Tools for Built-In Self-Test Implementations," *Proc. IEEE Automatic Test Conf.*, 1996, pp. 113-119.

[315] J. Turino, "Test Economics in the 21st Century," *IEEE Design & Test of Computers*, Vol. 14, No. 3, pp. 41-44, July-Sept. 1997.

[316] J. Turino, "Design for Test and Time-to-Market - Friends or Foes?," *Proc. IEEE International Test Conf.*, 1999, pp. 1098-1101.

[317] J. Udell, "Test Set Generation for Pseudoexhaustive BIST," *Proc. IEEE International Conf. on Computer-Aided Design*, 1986, pp. 52-55.

[318] J. Udell and E. McCluskey, "Efficient Circuit Segmentation for Pseudo-Exhaustive Test," *Proc. IEEE International Conf. on Computer-Aided Design*, 1987, pp. 148-151.

[319] A. van de Goor, *Testing Semiconductor Memories: Theory and Practice*, Chichester, United Kingdom: Wiley-Interscience, 1991.

[320] A. van de Goor, "Using March Tests to Test SRAMs," *IEEE Design & Test of Computers*, Vol. 10, No. 1, pp. 8-14, March 1993.

[321] A. van de Goor and S. Hamdioui, "Fault Models and Tests for Two-Port Memories," *Proc. IEEE VLSI Test Symp.*, 1998, pp. 401-410.

[322] A. van de Goor and B. Smit, "Generating March Tests Automatically," *Proc. IEEE International Test Conf.*, 1994, pp. 870-878.

[323] A. van de Goor and I. Tlili, "March Tests for Word-Oriented Memories," *Proc. Design Automation and Test in Europe*, 1998, pp. 501-508.

[324] D. van de Lagemaat and H. Bleeker, "Testing a Board with Boundary Scan," *Proc. IEEE International Test Conf.*, 1987, pp. 724-729.

[325] J. van Sas, F. Catthoor and H. de Man, "Cellular Automata Based Self-Test for Programmable Data Paths," *Proc. IEEE International Test Conf.*, 1990, pp. 769-778.

[326] D. Vazquez, J. Huertas and A. Reuda, "Reducing the Impact of DFT on the Performance of Analog Integrated Circuits," *Proc. IEEE VLSI Test Symp.*, 1996, pp. 42-47.

[327] P. Veenstra, F. Beenker and J. Koomen, "Testing Random Access Memories: Theory and Practice," *IEE Proc. G*, Vol. 135, No. 1, pp. 24-28, Feb. 1988.

[328] M. Vilas and C. Gloster, "Automatic Built-In Self-Test Insertion for High Level Circuit Descriptions," *Proc. IEEE International ASIC Conf.*, 1995, pp. 222-226.

[329] B. Vinnakota, editor, *Analog and Mixed-Signal Test*, Upper Saddle River, New Jersey, Prentice-Hall PTR, 1998.

[330] R. Wadsack, "Fault Modeling and Logic Simulation of CMOS and NMOS Integrated Circuits," *Bell System Tech. J.*, Vol. 57, No. 5, pp. 1449-1474, May-June 1978.

[331] L. Wang and E. McCluskey, "Complete Feedback Shift Register Design for Built-In Self-Test," *Proc. IEEE International Conf. on Computer-Aided Design*, 1986, pp. 56-59.

[332] L. Wang and E. McCluskey, "Circuits for Pseudoexhaustive Test Pattern Generation," *Proc. IEEE International Test Conf.*, 1986, pp. 25-37.

[333] L. Wang and E. McCluskey, "Built-In Self-Test for Sequential Machines," *Proc. IEEE International Test Conf.*, 1987, pp. 334-341.

[334] S. Wang and T. Tsai, "Test and Diagnosis of Faulty Logic Blocks in FPGAs," *Proc. IEEE International Conf. on Computer-Aided Design,* 1997, pp. 722-727.

[335] D. Wheater, P. Nigh, J. Mechler and L. Lacroix, "ASIC Test Cost/Strategy Trade-offs," *Proc. IEEE International Test Conf.*, 1994, pp. 93-102.

[336] T. Williams, "Test Length in a Self-Testing Environment," *IEEE Design & Test of Computers*, Vol. 2, No. 2, March 1985.

[337] T. Williams and K. Parker, "Design for Testability - A Survey," *Proc. of IEEE*, Vol. 71, No. 1, pp. 98-112, Jan. 1983.

[338] T. Williams, W. Deahn, M. Gruetzner and C. Starke, "Comparison of Aliasing Errors for Primitive and Non-Primitive Polynomials," *Proc. IEEE International Test Conf.*, 1986, pp. 282-288.

[339] T. Williams, W. Deahn, M. Gruetzner and C. Starke, "Bounds and Analysis of Aliasing Errors in Linear Feedback Shift Registers," *IEEE Trans. on Computer-Aided Design*, Vol. 7, No. 1, pp. 75-83, January 1988.

[340] S. Wei, P. Nag, R. Blanton, A. Gattiker and W. Maly, "To DFT or Not to DFT?", *Proc. IEEE International Test Conf.*, 1997, pp. 557-566.

[341] N. Weste and K. Eshraghian, *Principles of CMOS VLSI Design: A System Perspective*, Reading, Massachusetts: Addison-Wesley, 1985.

[342] G. Westerman, J. Heath and C. Stroud, "Delay Fault Testability Modeling with Temporal Logic," *Proc. IEEE Automatic Test Conf.*, 1998, pp. 22-27.

[343] G. Westerman, R. Kumar, C. Stroud and J. Heath, "Discrete Event System Approach for Delay Fault Analysis in Digital Circuits," *Proc. American Control Conf.*, 1997, pp. 239-243.

[344] W. Wolf, *Modern VLSI Design: Systems on Silicon*, Upper Saddle River, New Jersey: Prentice Hall PTR, 1998.

[345] E. Wu, "PEST: A Tool for Implementing Pseudo-Exhaustive Self-Test," *AT&T Technical J.*, Vol. 70, No. 1, pp. 87-100, Jan.-Feb. 1991.

[346] H. Wunderlich and G. Kiefer, "Bit Flipping BIST," *Proc. IEEE International Conf. on Computer-Aided Design*, 1996, pp. 337-343.

[347] Y. Xing, "Defect-Oriented Testing of Mixed-Signal ICs: Some Industrial Experience," *Proc. IEEE International Test Conf.*, 1998, pp. 678-687.

[348] Y. Zorian and A. Ivanov, "EEODM: An Effective BIST Scheme for ROMs," *Proc. IEEE International Test Conf.*, 1990, pp. 871-879.

Index

O

P

R